THE TWENTY THIRD CHESAPEAKE SAILING YACHT SYMPOSIUM

C SYS

March 15 –16, 2019
Annapolis, Maryland, USA

Presented by:

The Society of Naval Architects and Marine Engineers

Sailing Yacht Research Foundation

TABLE OF CONTENTS

Papers Presented on Friday March 15th, 2019

Maneuver Simulation and Optimization for AC50 Class
Dr. Heikki Hansen, Oracle Team USA / DNV GL SE, Potsdam, Germany
Dr. Karsten Hochkirch, Oracle Team USA / DNV GL SE, Potsdam, Germany
Ian Burns, Oracle Team USA, San Francisco, USA
Scott Ferguson, Oracle Team USA, San Francisco, USA

6DOF behavior of an offshore racing trimaran in an unsteady environment.
Paul Kerdraon, VPLP Design, Vannes, France, and Ecole Centrale Nantes, France
Boris Horel, Ecole Centrale Nantes, LHEEA res. dept. (ECN and CNRS), Nantes, France
Patrick Bot, Naval Academy Research Institute, Brest, France
Adrien Letourneur, VPLP Design, Vannes, France
David Le Touzé, Ecole Centrale Nantes, LHEEA res. dept. (ECN and CNRS), Nantes, France

THE 23rd CHESAPEAKE SAILING YACHT SYMPOSIUM
ANNAPOLIS, MARYLAND, MARCH 15 – 16th, 2019

TABLE OF CONTENTS

Papers Presented on Saturday March 16th, 2019

THE 23rd CHESAPEAKE SAILING YACHT SYMPOSIUM
ANNAPOLIS, MARYLAND, MARCH 15 – 16th, 2019

Steering Committee:

Event Chair:
Jaye Falls BSc, MSc, PhD
United States Naval Academy

Papers Committee Chair:
Britton Ward, BSc, MSc
Farr Yacht Design, Ltd.

Proceedings Editor
Jaye Falls BSc, MSc, PhD
United States Naval Academy

Promotion/Arrangements:
Dobbs Davis, BSc, MSc, PhD
Sailing Yacht Research Foundation

Event Coordination:
Brenda Zelada, Kathy Hartness
Society of Naval Architects and Marine
Engineers

Technical Review Committee:

Britton Ward BS, MS
Farr Yacht Design, Ltd.
britton@farrdesign.com

William Lasher BSE, MSE, PhD
Penn State Behrend, Erie, PA
lasher@psu.edu

Frank W. DeBord Jr. PE, BE, ME
Chesapeake Marine Technology, LLC.
fdebord@atlanticbb net

David Kring, BS, PhD
Navatek, Ltd.
dkring@navatekltd.com

David Le Pelley, ME
Doyle Sails
david@vspars.com

Jaye Falls, BSc, MS, PhD
United States Naval Academy
falls@usna.edu

Jonathan Binns, BE (hons), MS, PhD
Australian Maritime College, Uni of Tasmania
jrbinns@amc.edu.au

Wm. Bryan Baker, SM, MS
Baker Performance Design
wbbaker15@gmail.com

Richard Royce BSIM, BSE, MS, MSE, PhD
Webb Institute
rroyce@webb.edu

Robert Ranzenbach, BS, MS, PhD
ORC International Technical Committee
rranzenbach@gmail.com

Rick Harris, BS, MS, PhD.
Oceaneering

Hannes Renzsch, PhD
Fluid Engineering Solutions
hannes@fluidengineeringsolutions.com

Ignazio Maria Viola, PhD
University of Edinburgh
i m.viola@ed.ac.uk

Chris Mairs, BA, MS, PhD.
Cardinal Engineering, LLC.
cmairs@cardinalengineeringllc.com

Jim Schmicker, BSc, MSc
Farr Yacht Design, Ltd.
jim@farrdesign.com

Sponsors

THE INTERNATIONAL COMMUNITY FOR MARITIME AND OCEAN PROFESSIONALS

The Society of Naval Architects and Marine Engineers
99 Canal Center Plaza, Suite 310
Alexandria, VA 22314
www.sname.org

Sailing Yacht Research Foundation
1643 Warwick Avenue, Box 300
Warwick, Rhode Island 02889
www.sailyachtresearch.org

The Twenty-Second CSYS was held on March 15-16, 2019
The papers were presented in Chauvenet Hall
Located on the campus of the United States Naval Academy
Annapolis, Maryland, USA.

ABSTRACTS IN ORDER OF PRESENTATION

Hydrodynamics of Three Slender Models Resembling Polynesian Canoe Hulls

Richard G.J. Flay, University of Auckland, Auckland, New Zealand
Ignazio Maria Viola, Institute for Energy Systems, University of Edinburgh, United Kingdom
Geoffrey J. Irwin, University of Auckland, Auckland, New Zealand

Towing tank tests were carried out on three slender models in order to obtain more information on the hydrodynamics of such shapes, and in particular, how the shapes could generate side force when operating at a leeway angle. Slender hulls are of much interest for multi-hull vessels. There is increased interest in such vessels at present due to the use of wing-sail multi-hulls in the recent America's Cup races in San Francisco and Bermuda, although the AC yachts use foils to generate a lot of the side force, and only at low speeds are the hulls in the water. On the other hand, ancient Polynesian multi-hull vessels did not appear to have keels, and so the side-force had to be generated by the hulls. The authors speculates that the earliest vessels had rounded hulls (from trees) and were probably used mainly for sailing downwind. However, with time, it would have been realized that sail powered vessels could also maneuver across the wind, and then the importance of side force as well as drag would have begun to be appreciated. Modern Polynesian multi-hull vessels often have one or both hulls which are Vee-shaped in cross-section. Hence, it appeared that evolution may have caused a change in shape from circular to Vee, presumably because such shapes are better able to generate side force. A CFD study with ANSYS-CFX using three different hulls was carried out as suggested by the first author and it showed that sharper Vee sections were better at generating side-force than a rounded hull. The purpose of the present tests is to investigate whether such behavior could also be observed in physical testing. Three models were manufactured and were tested in the Towing Tank at Newcastle University in July and August 2013. One set of tests were carried out with fixed sink and trim, and the August tests had free sink and trim, with a thrust moment applied which approximated the sail force at part height up the mast. It was found that there was good agreement between the CFD and tank test results, and indeed that the hypothesis that narrower Vee shaped hulls would generate more side-force when at leeway than a rounded hull was proved.

A Comparative Study of Program FloSim Results against SYRF Wide Light Project Data

Brian Maskew, Computational Flow Simulations LLC, Winthrop, WA
Frank DeBord, Chesapeake Marine Technology LLC, Easton, MD

In November, 2015, the Sailing Yacht Research Foundation (SYRF) published the tank test data from their "Wide Light Yacht Project" for the hydrodynamics of a modern, high performance, semi-planing yacht. This comprehensive data set, comprising canoe body with and without appendages in upright, heeled and yawed conditions, provides an important validation base for CFD codes; previously, such data were not readily accessible mainly due to proprietary issues. The SYRF report includes a number of comparative results from commercial CFD codes, the RANS program Star-CCM+ results in general showing the best correlations with the measured data, albeit with some significant departures.

In this paper, we present results computed using an advanced Boundary Element Code, FloSim, these calculations being compared against the test matrix of "Wide Light" measured data and also Star-CCM+ calculated results. Times for computer model preparation and case execution are discussed together with computer requirements.

An outline of the FloSim method is presented; it is an unsteady code with a coupled integral boundary layer analysis for viscous effects such as skin friction resistance and boundary layer displacement effect. Its time-stepping procedure has free convection and rollup of vortex wake elements providing non-linear lift properties and includes a number of modeling techniques for treating "real world" effects, such as flow separation and wave breaking. The free surface wave development uses a non-linear mixed Eulerian-Lagrangian treatment at each time step. For the higher speed cases in the SYRF data set that have a

breaking bow wave crest, FloSim's Wave-Breaker treatment is applied to convert excessive energy in the bow wave to a "dead-weight" pressure applied on the free surface; this effectively attenuates downstream wave amplitudes consistent with the loss of energy at the breaking crest.

FloSim, already used in America's Cup and Volvo racing yacht analyses, was developed specifically to bridge the gap between basic potential flow panel methods and RANS codes, with the objective of providing accurate, practical solutions on a laptop computer within a reasonable turnaround time and cost. In essence, the results presented so far in this paper demonstrate these objectives have been achieved. The comparisons of FloSim's essentially "low-order" results against test data and Star-CCM+ RANS calculations, provide a measure of tradeoff between calculation accuracy versus cost and turnaround time for a case. Throughout the discussions presented below, the reader should keep in mind that this was not a "blind" comparison as it was for the original Wide Light Project participants who published their numerical results.

The Un-restrained Sailing Yacht Model Tests – A New Approach and Technology Appropriate to Modern Sailing Yacht Seakeeping

Etienne Gauvain, Wolfson Unit MTIA, Southampton, UK

Over the recent years the Wolfson Unit has seen a greater impetus from yacht designers and their clients to quantify and compare the seakeeping qualities of their sailing yacht design choices. Modern high performance yachts, fitted with a wide range of appendages generating lift and creating large moments, provide a number of complex challenges for designers.

Assessing seakeeping behavior and performance in a seaway is, indeed, important during the design process since the motions cause unsteady effects on the yacht hydrodynamic characteristics, for instance on the lift generating capabilities of the appendages. Hence it may not be justifiable to assume during the optimization process that the yacht outperforming other design candidates in calm water would also perform well in waves.

Therefore, the Wolfson Unit developed an innovative experimental model testing approach that would be an improvement over existing methods, simulating the 6 degrees of freedom motions and accelerations.

The unique un-restrained sailing test approach uses a mast mounted air screw device to simulate the aerodynamic propulsion from the sails allowing a scaled model of the yacht to be tested at a range of conditions, sea states and wave directions (from head seas to following waves). These tests can be used as a comparative tool to assess controllability and seakeeping characteristics of multiple configurations (e.g. hull shape, appendages, inertias) by quantifying induced motions and providing an estimate of added resistance in waves. Non-linearity attitudes such as surfing can be investigated.

Free-running model testing is a technique frequently used in the development of power vessels, but little adoption is made in the sailing yacht world. Furthermore dynamic and seakeeping studies are at present challenging for computational fluid dynamics based tools, encouraging an experimental based approach.

This paper introduces an un-restrained model testing method on sailing yachts. Discussion will also be made on how this new method can be implemented in design decisions and add value during the design and performance evaluation process for sailing yachts.

Experimental and numerical prediction of foiling monohull dinghy performance

A. H. Day, W. Shi, T. King, Naval Architecture Ocean & Marine Engineering, University of Strathclyde
F.M. Fresneda, S.R.Turnock, Performance Sports Engineering Lab, University of Southampton

The rise in interest in large foiling yachts, such as those in the America's Cup, has spurred a corresponding interest in foiling applications in monohull sailing dinghies, both for boats designed specifically for foiling, and through retro-fit foil kits. The present study considers the assessment of foil systems for such boats, and specifically the prediction of the necessary foil performance of t-foils with and without flaps. Towing tank and wind tunnel experiments are used alongside numerical simulations. The

accuracy of different simulation approaches is examined and the corresponding sensitivity of the predicted boat speed to the numerical method adopted is identified. Results are presented via two case studies, predicting performance for a foiling International Moth, and for a Europe dinghy retro-fitted with hydrofoils.

Benchmark data for the numerical simulation is obtained using two approaches. The main lifting foil for a moth dinghy with flap was tested at full-scale in a towing tank at a range of speeds, trim angles and flap angles. Results are compared with published experiment data for a similar moth foil. In a complementary set of tests, the rudder of a Europe dinghy was tested in a wind-tunnel, initially in the original design condition, and subsequently with a T-foil rudder designed for a retro-fit foiling solution.

A range of numerical techniques of varying levels of complexity, fidelity, and numerical intensity were then used to predict the performance of the foils. The techniques adopted include 2D predictions using XFOIL allied to a simple correction for 3D effects, a numerical lifting line theory, a fully 3D panel code and RANS CFD. The data from these simulations is used in two ways. The data is first used to examine the accuracy of the different simulation techniques by comparison of predictions with the measured data. The simulated data is then deployed in a 5-DOF Velocity Prediction Program (VPP) in order to predict the performance of a foiling moth and a concept design for a foiling Europe over a windward-leeward course, in order to illustrate the sensitivity of the predicted boat performance to the choice of foil simulation technique.

Conclusions are drawn regarding the accuracy of the simulation techniques, and the pros and cons of wind tunnel and towing tank testing of T-foils are discussed. The VPP study demonstrates the trade-offs which may be exploited in the design process when computational resource is limited: between a small number of high fidelity simulations with high numerical intensity and a large number of moderate-fidelity simulations which are computationally less intensive.

Impact of Composite Layup on the Hydrodynamic Performance of a Surface Piercing Hydrofoil

V. Temtching, Ecole Navale/SEAAIR Foil Resource Center
B. Augier, IFREMER, Laboratoire du comportement de structures en Mer
B. Paillard, Alternative Current Energy
T. Dalmas, Ecole Navale/ IFREMER, Laboratoire du comportement de structures en Mer
N. Dumerge, IFREMER, Laboratoire du comportement de structures en Mer

Composite materials are good candidates for hydrofoils manufacturing, ensuring a good balance between strength and weight. In the high performances sailing yacht domain, hydrofoils are thin structures, highly loaded that experience significant displacements. This study investigates experimentally and numerically the influence of the laminate layup on the hydrodynamic performances of a surface piercing hydrofoil. Four hydrofoils with a constant chord, geometrically identical with different composite layups are mechanically characterized and tested in a hydrodynamic flume. The foils are designed to have a significant tip displacement of 5 to 10% of the span. Experimental results highlight a bending-twisting effect that leads to significant change in the hydrodynamic performances of the structures. Two different FSI numerical approaches: from a potential code coupled with beam theory to the full coupling of a shell structural code and a VOF hydro model with free surface are compared to the experiments with great results. The two approaches are two complementary bricks in the design process to compute the effect of passive deformation on hydrodynamic performances of the foils and therefore the yacht stability.

Application of System-based Modelling and Simplified-FSI to a Foiling Open 60 Monohull

Boris Horel, Ecole Centrale Nantes, LHEEA res. dept. (ECN and CNRS), France
Mathieu Durand, SIRLI Innovations, LUNA ROSSA Challenge, France

The increasing number of foiling yachts in offshore and inshore races has driven engineers and researchers to significantly improve the current modelling methods to face new design challenges such as flight analysis and control (Heppel, 2015). Following the publication of the AC75 Class Rules for the 36th America's Cup (RNZYS, 2018) and since the brand new Open 60 Class yachts are all equipped with hydrofoils, the presented study will propose a system-based modelling coupled with a simplified FSI (fluid-structure interaction) method that leads to better understand the dynamic behavior of monohulls with deformable hydrofoils.

The aim of the presented paper is to establish an innovative approach to assess appendage behavior in a dynamic VPP (velocity prediction program). For that purpose, dynamic computations are based on a 6DOF mathematical model derived from the general non-linear maneuvering equations (Horel, 2016). The force model is expressed as the superposition of 7 major force components expressed at the center of gravity of the yacht: gravity, hydrostatic, maneuvering, damping, propulsion (wind), control (rudders, daggerboard, foils …) and wave (Froude-Krylov and diffraction phenomenon).

As test cases, course keeping simulations are performed on an Open 60 yacht with control loops to simulate the wing trimmer, helmsman and foil trimmer when finding the optimal foil settings is needed. In first hand, IMOCA's polar diagrams are used as reference.

In calm water and in waves, the influence of foil's shapes (foil with shaft pointing downward and tip pointing upward, foil with shaft pointing upward and tip pointing downward) and stiffness (non-deformable, realistic, flexible) on the global behavior of the yacht is presented.

Maneuver Simulation and Optimization for AC50 Class

Dr. Heikki Hansen, Oracle Team USA / DNV GL SE, Potsdam, Germany
Dr. Karsten Hochkirch, Oracle Team USA / DNV GL SE, Potsdam, Germany
Ian Burns, Oracle Team USA, San Francisco, USA
Scott Ferguson, Oracle Team USA, San Francisco, USA

The stability and the dynamic behavior is an integral part of designing hydrofoil supported sailing vessels, such as the America's Cup (AC) 50 class. The foil design and the control systems have an important influence on the performance and stability of the vessel. Both foil and control system design also drive the maneuverability of the vessel and determine maneuvering procedures. The AC50 class requirements lead to complex foil control systems and the maneuvering procedures become sophisticated and multifaceted.

Sailing and maintaining AC50 class yachts is a complex, expensive and time-consuming task. A dynamic velocity prediction program (DVPP) for the AC50 is therefore developed to assess the dynamic stability of different foil configurations and to simulate and optimize maneuvers. The goal is to evaluate certain design ideas and maneuvering procedures with this simulator so that sailing time on the water can be saved.

The paper describes the principal concepts of developing a AC50 model in the DVPP FS-Equilibrium. The force components acting on the yacht are defined based on physical principles, computational fluid dynamics (CFD) simulations and experimental investigations. The control systems for adjusting the aero- and hydrodynamic surfaces are modelled. Controllers are utilized to simulate the human behavior of performing sailing tasks. Maneuvers are then defined as sequences of crew actions and crew behaviors.

In the paper examples of utilizing the DVPP in preparation for the 35th America's Cup in Bermuda are described. The DVPP is for example used to investigate the effect of different boat set-ups on stability and handling during maneuvers. With the sailing team, maneuver procedures are developed and tested. Procedures such as dagger board and rudder elevator movement and crew position are investigated and

evaluated to minimize the distance lost during tacking and gybing. The DVPP is also employed for trajectory optimization during maneuvers.

6DOF behavior of an offshore racing trimaran in an unsteady environment.

Paul Kerdraon, VPLP Design, Vannes, France, and Ecole Centrale Nantes, France
Boris Horel, Ecole Centrale Nantes, LHEEA res. dept. (ECN and CNRS), Nantes, France
Patrick Bot, Naval Academy Research Institute, Brest, France
Adrien Letourneur, VPLP Design, Vannes, France
David Le Touzé, Ecole Centrale Nantes, LHEEA res. dept. (ECN and CNRS), Nantes, France

While in recent years the use of hydrofoils has experienced a substantial growth, traditional design tools such as Velocity Prediction Programs (VPP) have proven inadequate to help architects and engineers with performance trade offs which now include specific stability issues related to these foils. The quest for performance also demands a better account of the unsteadiness of the environment in which the offshore yachts evolve. Time-domain analysis and system-based modeling allow for an improved understanding of the controllability and dynamic stability of given geometries, enabling to adapt and refine the design. This paper presents such a dynamical unsteady model, based on the superposition of several loads components, computed from either numerical, empirical or analytical models. A test case and its results are presented to show the reliability and efficiency of the developed numerical tool, by comparing response amplitude operators of a reference hull form with experimental and numerical data. Finally, the paper outlines two 6DOF dynamic simulations of an offshore trimaran. The first case shows a simple bearaway maneuver and compares two sail tuning strategies, while the second one presents the yacht evolution in unsteady wind demonstrating how in varying conditions the boat may reach attitudes that widely differ from the steady ones.

Recent Advances in Experimental Downwind Sail Aerodynamics

Jean-Baptiste R. G. Souppez, Warsash School of Maritime Science and Engineering, Solent University, Southampton, UK.
Abel Arredondo-Galeana, School of Engineering, Institute for Energy Systems, University of Edinburgh, Edinburgh, UK.
Ignazio Maria Viola, School of Engineering, Institute for Energy Systems, University of Edinburgh, Edinburgh, UK.

Over the past two decades, the numerical and experimental progresses made in the field of downwind sail aerodynamics have contributed to a new understanding of their behavior and improved design. Contemporary advances include the numerical and experimental evidence of the leading-edge vortex, as well as greater correlation between model and full-scale testing. Nevertheless, much remains to be understood on the aerodynamics of downwind sails and inherent flow structures. A detailed review of the different flow features, including the effect of separation bubbles and leading-edge vortices will be tackled, to provide a comprehensive presentation of the aerodynamics of downwind sails. New experimental measurements of the flow field around a highly cambered thin circular arc geometry with a sharp leading edge will also be presented. These results allow for the first time to interpret some apparently inconsistent data from past experiments and simulations, and to provide guidance for future model testing and sail design.

An Energy Aware Autopilot for Sailboats

Mathilde Tréhin, Université de Bretagne Sud/Madintec, Lorient, France
Johann Laurent, Université de Bretagne Sud, Lab-STICC UMR6285, Lorient, France
Hugo Kerhascoët, Madintec, Lorient, France
Jean-Philippe Diguet, CNRS, Lab-STICC UMR6285, Lorient, France

In this paper, we propose a new control method for the next generation of autopilots. These new systems will need to manage more actuators to control the hydrofoils, which is going to significantly increase the energy requirements. So, this method is aware of the autopilot power consumption. It uses a model predictive controller to manage the actuators (position control - appendage angle control). This controller uses a dynamic model of the actuator, running in real time, to anticipate the future behavior of the system. Once the predictions are made, it determines the future control sequence to apply in order to follow the reference trajectory. To do so, it minimizes a cost function which includes the quadratic error according to the behavior prediction and the associated energy consumption. So, it takes into account two criterions: the precision/rapidity of the system and the energy. With the proposed control method, skippers can weight each criterion in order to focus on one or the other depending on their goals and the boat's energy balance. We apply this method to one of the autopilot's subsystems, namely the rudder control. The electric actuator intervening in this control loop and the load representing the force opposed to its motion are modelled to design the control law. The first results of that method are compared with a standard autopilot. We increase by 40% the precision level and we are able to reduce the consumption by at least 20%. This work provides the first necessary components of a future autopilot that will control the whole appendages to a three-dimensional piloting. Moreover, this type of management is a first step towards possible fossil fuel free sailboats.

Sailboat Routing with Multiple Objectives for Sailing Races

Goulven Guillou, Lab-STICC, University of Brest, France
Laurent Lemarchand, Lab-STICC, University of Brest, France
Jean-Philippe Babau, Lab-STICC, University of Brest, France

Sailboat routing consists in computing the best route for a sailboat taking into account the characteristics of the ship and environmental data such as weather forecast. In the context of sailing races, the best route computation is usually based on the isochrone algorithm, a sub-optimal solution to optimize the time to destination (TtD) criterion by computing a route as a sequence of waypoints. In this paper, we propose to compute a set of possible routes by considering two criteria: the time to destination and the stress. The time to destination is evaluated according to weather forecast and boat polar diagrams. The stress function is a combination of human and environmental factors. The set of possible routes are then obtained by using an iterative multiple objective optimization algorithm. Isochrone algorithm is used for initializing the set of routes. Then mutation operators are used to explore alternative solutions. Applied to realistic test cases, our search strategy allows to obtain routes with very different characteristics in terms of time to destination and stress values, asserted by experimented sailors. Concerning the main objective of minimizing time to destination, we are competitive with commercial software such as MaxSea or Adrena.

Scoring Methods for Handicap Racing.

Jim Teeters

Two boats sailing in close proximity may inevitably compare their boat speeds, perhaps even "race" each other. Over the last few centuries there have been numerous handicap systems designed to equalize the competitive chances of racing sailboats. Corollary to handicapping is another arcane art; that of scoring races. Scoring methods have both technical options, in part determined by handicap rules, as well as "human engineering" options in the sense that different solutions can work best for different constituencies,

be they race organizers or sailors. Options may include single vs. multiple ratings, time on distance vs. time on time, pre/during/post-race handicapping, attempts to predict the environmental conditions on the race course, constructed courses, pursuit vs. staggered vs. fleet start racing, performance line and performance curve scoring. The underlying assumptions and motivations for these choices are presented along with the consequences of adopting them. The expectations of the competitors, and indeed their ability to intuitively grasp the fundamentals of how elapsed times are transformed into race rankings, are discussed with a view towards finding solutions that achieve a successful balance of fairness, transparency and acceptance.

A Case Study on the Effect of Sweep and Variations in Free-Surface Cross Section Geometry on the Lift and Drag of Transom-Hung Sailboat Rudders
Paul H. Miller, United States Coast Guard Academy

Conventional transom-hung rudders are often used on small sailboats because of their simplicity compared to rudders mounted under the hull; however, they present substantial performance penalties, including (1) the rudder is more likely to ventilate by drawing air down from the free surface, (2) the effective aspect ratio, and therefore the lift-to-drag ratio, is not increased by the mirror-plane of the hull bottom and (3) there is additional spray and wavemaking resistance that arises as a result of the rudder passing through the free surface. This case study focuses on a means to mitigate the last of these penalties, the increased spray and wavemaking resistance. While many transom-hung rudders are essentially parallel, or tapered with the maximum chord at the top where it meets the tiller handle, the reader will recognize that having the largest cross section of rudder at the free surface will generate significant spray and wavemaking resistance, especially when the rudder is turned. This study investigated the use of minimizing the rudder chord length where it passes through the free surface, demonstrating the findings by full-scale towing tests of a series of rudders designed for a *Fireball*-class dinghy. Running the tests at full-scale, therefore matching Reynolds number and Froude number, eliminated questions on scaling. Experimentation on the effects of sweep angle, section shape and chord length at varying angles of attack and velocities showed a noticeable increase in lift-to-drag ratio of foils with reduced chord length at the free surface and by sweeping the rudder forward. To complete the case study, a velocity prediction program was used to estimate the change in speed around a notional race course.

Science of 470 Sailing Performance
Yutaka Masuyama, Kanazawa Institute of Technology, Kanazawa, Japan
Munehiko Ogihara, SANYODENKI AMERICA, Torrance CA, USA

In order to sail 470 faster, authors consider the sailing performance of 470 from the various measured data and the simulated results of VPP (Velocity Prediction Program). This is a summary of TOBE-470 presented by the author.

Sailing Catamarans: Design for Cruising
Albert Nazarov, Albatross Marine Design, Thailand

Statistics is provided for number of sailing catamarans and approaches to craft dimensioning are reviewed. Styling trends and typical catamaran arrangements are featured. Weight components are studied for number of catamarans of different size and level of comfort on board. Effect of catamaran architecture on performance is studied by combining VPP predictions with CFD modeling of various deck/cabin configurations. Summary of safety requirements specific to catamarans is given. Case studies are presented number of cruising catamarans designed by AMD; new perspective concept of 44' catamaran featured.

Hydrodynamics of Three Slender Models Resembling Polynesian Canoe Hulls

Richard G.J. Flay, University of Auckland, Auckland, New Zealand
Ignazio Maria Viola, Institute for Energy Systems, University of Edinburgh, United Kingdom
Geoffrey J. Irwin, University of Auckland, Auckland, New Zealand

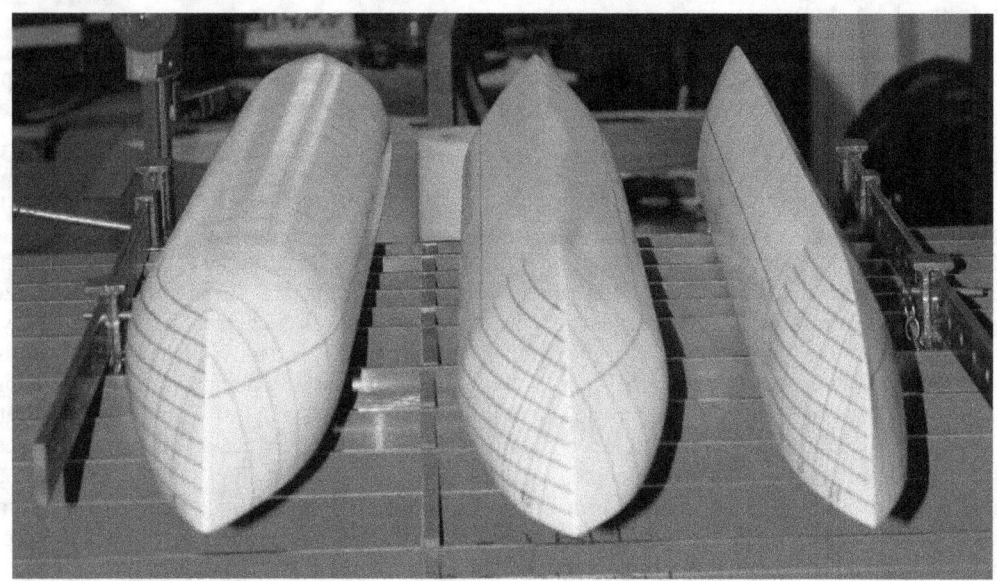

ABSTRACT

Towing tank tests were carried out on three slender models in the Towing Tank at Newcastle University at fixed sink and trim in order to obtain more information on the hydrodynamics of such shapes, and in particular, how the shapes could generate side force when operating at a leeway angle. The research was motivated by a study of ancient Polynesian multi-hull vessels which did not appear to have keels, and so the side-force had to be generated by the hulls. The authors speculate that the earliest vessels had rounded hulls (from trees) and were probably used mainly for sailing downwind. However, it appears that evolution has caused a change in shape from circular to Vee, presumably because such shapes are better able to generate side force to enable the vessels to also sail across the wind.

A CFD study with ANSYS-CFX using three different hulls was carried out as suggested by the first author and it showed that sharper Vee sections were better at generating side-force than a rounded hull. The purpose of the present tests was to investigate whether such behaviour could also be observed in physical testing. Three models were manufactured and were tested in the Towing Tank at Newcastle University in July and August 2013.

It was found that there was good agreement between the CFD and tank test results, and that indeed the hypothesis that narrower Vee-shaped hulls would generate more side-force when at leeway than a rounded hull was proved.

INTRODUCTION

This work was carried out while the first author was on Sabbatical Leave at Newcastle University during May to September 2013, and it was one of the Work Packages in the Sailing Fluids research project, funded by the Marie Curie International Research Staff Exchange Scheme (FP7 IRSES) and by the Royal Society of New Zealand.

Towing tank tests were carried out on three slender models in order to obtain more information on the hydrodynamics of such shapes, and in particular how the shapes could generate side force when operating at a leeway angle.

Slender hulls are of much interest for multi-hull vessels. There is increased interest in such vessels at present due to the use of wing-sail multi-hulls in the recent 2013 America's Cup races in San Francisco, although the AC yachts use foils to generate a lot of the side force, and only at low speeds are the hulls in the water. On the other hand, ancient Polynesian multi-hull vessels did not appear to have keels, and so the side-force had to be generated by the hulls. The authors speculate that the earliest vessels had rounded hulls (from trees) and were probably used mainly for sailing downwind. However, with time, it would have been realised that sail powered vessels could also manoeuvre across the wind, and then the importance of side force as well as drag would have begun to be appreciated. Modern Polynesian multi-hull vessels often have one or both hulls which are Vee-shaped in cross-section. Hence it appeared that evolution may have caused a change in shape from circular to Vee, presumably because such shapes are better able to generate side force.

A CFD study with ANSYS-CFX using three different hulls was carried out as suggested by the first author and it showed that sharper Vee sections were better at generating side-force than a rounded hull. The purpose of the present tests was to investigate whether such behaviour could also be observed in physical testing. Three models were manufactured and were tested in the Towing Tank at Newcastle University in July and August 2013. One set of tests were carried out with fixed sink and trim, and the August tests had free sink and trim, with a thrust moment applied which approximated the sail force located partially up the mast. This paper is restricted to discussing the fixed sink and trim tests only.

It was found that there was good agreement between the CFD and tank test results, and indeed that the hypothesis that narrower Vee shaped hulls would generate more side-force when at leeway than a rounded hull was proved.

This work was primarily motivated by two objectives: first, to obtain information on the hydrodynamics of slender hulls that could be used for helping to understand and predict the performance of Polynesian sailing vessels; second, to learn how to use a Towing Tank, as there is not one at the University of Auckland. Thus it was decided that the slender hull tank results would be analysed and written up in a report and paper for their own sake, to add to the literature on such shapes. Also the tests would be used to train University of Auckland academics, students and technicians in the techniques of tank testing. The tests were carried out by Richard Flay, Alexander Blakeley (UOA PhD student), Joshua Taylor (NU Summer student, and Nick Velychko (UOA Technician).

DESCRIPTION OF NEWCASTLE UNIVERSITY TOWING TANK

Since its construction in 1951 the Towing Tank at Newcastle, Figure 1, has been in almost continuous use. Since then, the tank has been regularly updated to keep abreast of modern trends. This includes the fitting of wave making and electronic recording equipment. Recent upgrades include the installation of a state-of-the-art motor control system to enhance very slow speed and high speed testing capabilities. A recent innovation is a modern telemetry system which will allow data to be sampled without any wired connections between the carriage and shore-based equipment. Further details are available in Newcastle Towing Tank, 2018 and Table 1.

Figure 1 Photograph of the Newcastle University Towing Tank

The Towing Tank is used mainly for calm water, wave resistance and sea-keeping experiments. Models are towed using a monorail carriage system that has a maximum speed of 3 m/s in its normal mode. The carriage can be remotely or manually controlled, while the 32-channel data retrieval system is on-line to a PC. The wave-makers can be used to generate regular waves of up to 0.12 m in height and wave periods in the range of 0.5 to 2 seconds. They are also capable of generating long crested random seas using a variety of wave spectra.

Table 1 Towing Tank Specifications

Specifications	
Tank length	37 m
Width	13.7 m
Water depth	1.25 m
Normal Carriage velocity	3 m/s
Wave capability	
Period range	0.5 - 2 Sec
Wave height	0.02 - 0.12 m (Period Dependent)

DESCRIPTION OF THE THREE MODELS

Model Manufacture

The models were based on those used for the CFD investigation by Boeck 2010 and 2012, using ANSYS-CFX, when he was studying in the Yacht Research Unit at the University of Auckland, and CAD images of the three models can be seen in Figure 2. It was decided to manufacture models for tank testing at the University of Auckland, and since the tank was small, and the models had to be shipped to the UK, a scale of 1:10 was used. The CFD investigation used hulls of length 12 m, as advised by Emeritus Professor Geoffrey Irwin, and so the models had an overall length of 1.2 m. Further details of the models are given in Table 2. They are rather small models, and it was realised once the tests were started that it would have been possible to test larger models at yaw, such as 50% longer, but they would have been more difficult to ship from NZ to the UK and back.

Table 2 Model-scale details of the three models used in the CFD and towing tank investigations, Boeck 2010. V1 – rounded hull, V2 – wide Vee, V3 – narrow Vee hull.

	V1	V2	V3
Draft amidships [m]	0.046	0.046	0.046
Waterline length [m]	1.16	1.16	1.16
Keel angle [deg]	180	100	70
Displacement [kg]	3.8	2.1	1.3
Wetted surface area [m^2]	0.163	0.129	0.111
Water plane area [m^2]	0.117	0.085	0.054
WL beam [m]	0.117	0.086	0.054
Prismatic coeff.	0.822	0.823	0.827
Block coeff.	0.492	0.341	0.263
Midship area coeff.	0.753	0.556	0.556
Water plane. Area coeff.	0.867	0.851	0.856
Projected side area [m^2]	0.0488	0.0488	0.0488

Igs files developed in Boeck 2010 were provided to the Faculty of Engineering Workshop, and these files were used to drive a numerically controlled milling machine to cut the shapes out of blue foam (Styrofoam). The foam models were then covered in a layer of fibre glass and painted, to strengthen them and to make them waterproof. A photograph of the models can be seen in Figure 3. The design waterline is shown in red and is the same for all three models. It was decided that the testing would be based on having constant draft, and hence the displacements of the models varied. The projected side area of all models is the same and is 4.88 m^2 in full-scale, or 0.0488 m^2 in model scale.

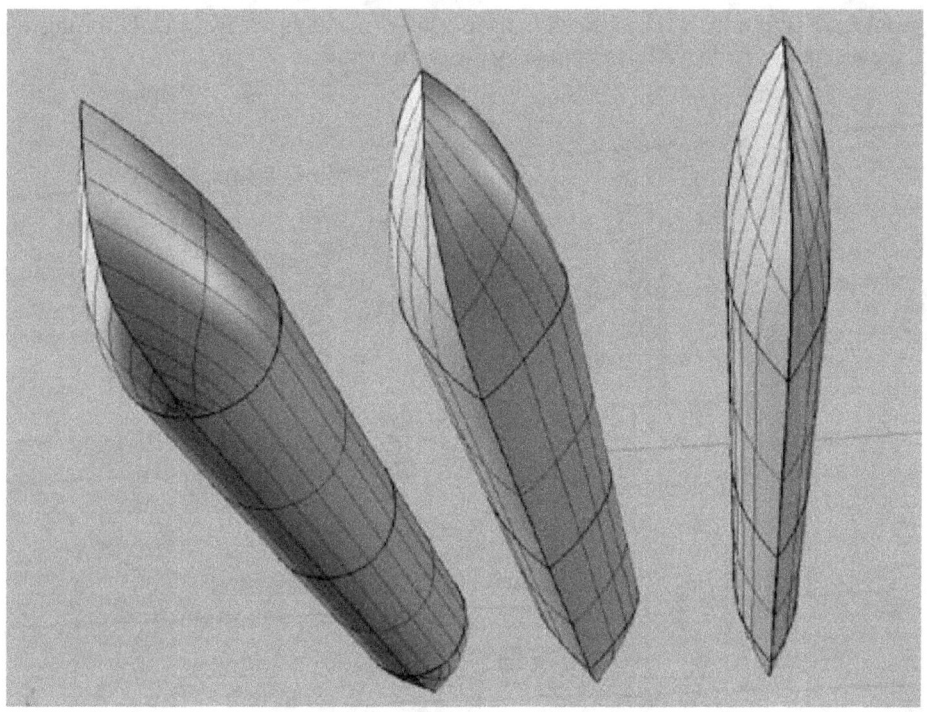

Figure 2 CAD image of the three models used in the CFD investigation from Boeck 2010. Left V1 – rounded hull, Centre V2 – wide Vee, Right V3 – narrow Vee hull

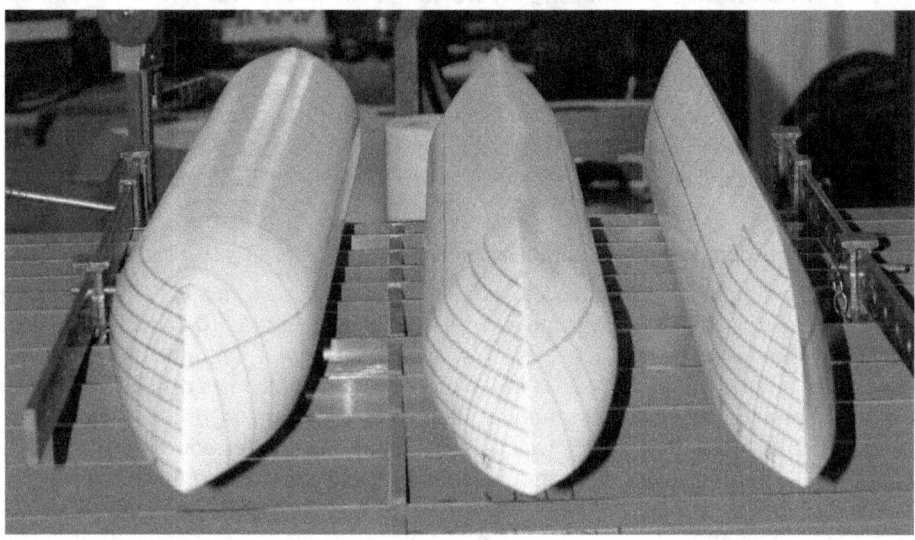

Figure 3 Photograph of the three models showing the underside and the painted grid pattern for assessing wave motion. The water line is shown in red. V1(left), V2(centre) and V3(right).

Model fixture to the Gifford dynamometer

A schematic diagram of the Gifford dynamometer is shown in Figure 4. The heave post is attached to the carriage. For fixed sink and trim tests it is clamped so that the model is immersed to the water line when at rest in still water. The pivot base is also secured with rectangular blocks so that it cannot rotate, and during setup the model attitude is adjusted to horizontal using shims. For free to sink and trim tests, the heave post is free to move vertically under the action of the force transmitted to it by the model and the pivot base is free to rotate. The towing plate is rigidly attached to the model, and its angular position is set by screws through holes at predetermined locations. The towing plates can be seen attached to the models in Figure 5. Holes were pre-drilled into the plates at 2° intervals, and enabled the larger two models to be tested at leeway angles from 0° up to +/- 24°. The narrower V3 model could be tested up to +/- 16°. It was expected that the narrower V3 model would generate more side force for a given leeway angle, which is why it was decided to have a lower maximum leeway angle for this model, although as can be seen in Figure 5, the maximum angle was also limited by the model width.

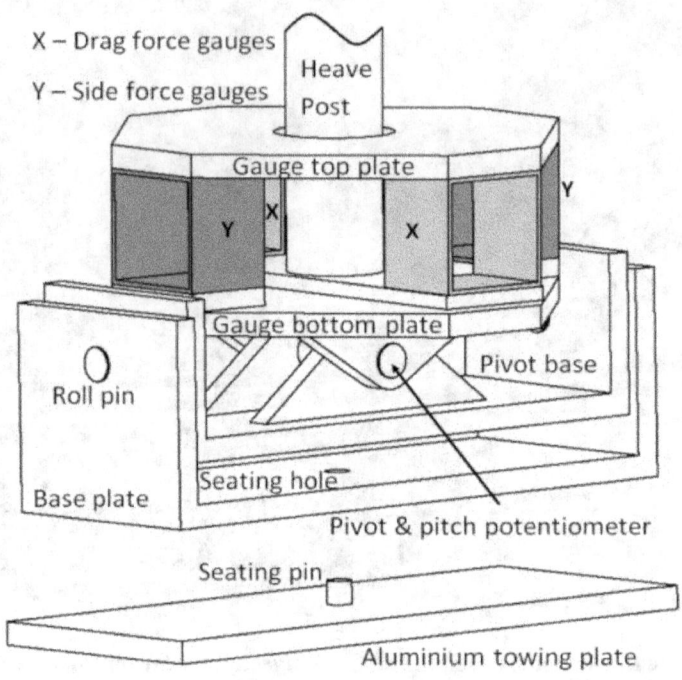

Figure 4 Schematic diagram showing the working principles of the Gifford dynamometer.

Trip wire placement

Trip wires are commonly used on tank test models to fix the location of transition, which would occur further aft at model scale compared to full scale, because of the lower Reynolds number (Re) in the former situation. If we assume that these vessels could travel at say 10 knots or 5 m/s, then based on their length of 12 m, the Reynolds number is: $Re_{fs} = 60\ million$, and the Froude number is: $Fr_{fs} = 0.461$.

If we assume that transition occurs at 0.5 million, then this is at $x_{tr} = 12 \times \frac{0.5}{60} = 0.1m$. As a fraction of the length, this is 0.83%. At 5 knots (half the speed) transition would occur at 0.2 m from the bow, which is 1.6% of the length. Thus in full-scale transition is expected to occur very near the bow. At model scale Re is much less. At 10 knots $Fr_{fs} = 0.461$, so for a model scale of $1/10^{th}$, the equivalent test speed is given by $V_m = 1.58\ m/s$, and thus $Re_m = 1.9$ million. Assuming as above that transition occurs at half a million means that it will occur at 26% of the model length, which is a distance of 0.32 m from the bow. Thus the nature of the flow would be very different, and would be likely to lead to errors in viscous drag.

The recommended ITTC 2002, Rev 01 approach was taken to determine the location of a trip wire to fix transition. ITTC recommends that a trip wire be located 5% of the length between perpendiculars (*Lpp*) aft of the forward perpendicular (*FP*), and that the wire diameter be between 0.5 and 1.0 mm. Thus for all three models, the wires were located a distance 58 mm aft of the *FP*, which itself is 20 mm aft of the foremost part of the model, so they were 78 mm aft of the foremost part of the model. Note that it is standard practice at Newcastle to use wire of diameter 1.2 mm, and this wire diameter was used on the three models. This slightly larger diameter than that recommended by ITTC could result in a slight increase in drag. An example of the trip wire on model V2 can be seen in Figure 6. The two other models had trip wires at the same location. Thus the trip wire was located a distance 6.5 % of the model length from the bow, i.e. aft of where one might expect transition to occur in full-scale, but well ahead of where it could be expected to occur naturally at model scale.

Definition of drag, lift, resistance, side-force coefficients

The main model dimensions are given in Table 2. Coefficients are made dimensionless using the product of area and dynamic pressure. The reference area, *A*, is normally the immersed projected side area, which is the same for all three models, and so changes in the coefficients relate directly to forces. For some of the analysis, especially for the drag, sometimes the reference area is the wetted surface area, with the model upright and immersed to its waterline. It is given the symbol *S*. Which reference area is used to form the coefficients is made clear in the relevant text. For example when applying the Prohaska-Hughes method to relate model to full-scale drag coefficients, it is convenient, and conventional to use the wetted surface area. ρ is the density, and V is the velocity of the model through the still tank water. The International Towing Tank Conference (ITTC) Recommended Procedures (2002) defines density for g = 9.81 m/s² as in Eq. (1).

Figure 1 Photograph of models showing attachment of the towing plate with holes pre-drilled at 2° intervals.

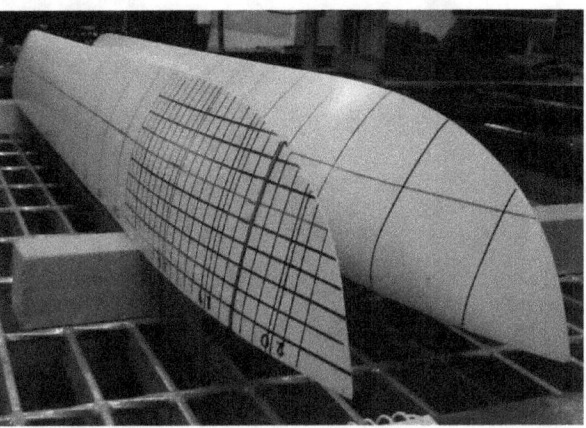

Figure 6 Photograph of model V2 showing the trip wire to initiate transition.

$$\rho = 1000.1 + 0.0552t + 0.0077t^2 + 0.00004t^4 \tag{1}$$

where t is temperature of water in °C.

The coefficients defined herewith are all located in the horizontal plane, normal to gravity.

Drag

Drag is the force in the direction parallel with the onset flow, so in the tank it is the force along the tank axis, and the drag force coefficient is defined in Eq. (2).

$$C_D = \frac{F_D}{\frac{1}{2}\rho A V^2}, C_D = \frac{F_D}{\frac{1}{2}\rho S V^2} \tag{2}$$

Lift

Lift is the force in a direction normal to the onset flow, so in the tank this is the force perpendicular to the tank axis. The lift coefficient is defined in Eq. (3).

$$C_L = \frac{F_L}{\frac{1}{2}\rho A V^2} \tag{3}$$

Resistance

The resistance coefficient is defined here as the force along the longitudinal axis of the model. When there is no leeway it is the same as the drag coefficient. When the model is at a leeway angle λ, the resistance coefficient is related to the drag and lift coefficients as in Eq. (4).

$$C_R = C_D cos\lambda - C_L sin\lambda \tag{4}$$

Side force

The side force is normal to the resistance, and the side force coefficient is related to the drag and lift coefficients as in Eq. (5).

$$C_S = C_D sin\lambda + C_L cos\lambda \tag{5}$$

PREDICTION OF DRAG AND SIDE FORCE FROM THE CFD INVESTIGATION

A CFD investigation using ANSYS-CFX was carried out by Boeck 2010 and 2012, for his master's thesis while studying in the Yacht Research Unit at the University of Auckland. The aim of the study was to investigate the generation of side force for hull shapes ranging from circular to a narrow Vee. Three model shapes were investigated, as depicted in Figure 2Figure .

Boeck initially modelled the so-called Wigley hull which has been well studied, and for which there are numerous experimental data available. He compared his predictions with the published Wigley results and verified that his CFD modelling appeared to be working well both in terms of the wave height along the hull, and the drag. For example, comparisons in Boeck, 2010 with experimental data from Wong 1994 show that resistance coefficient differences for Froude numbers of 0.225 and 0.267 are less than 0.005, and for a large Froude number of 0.316 the CFD under-predicts the experimental value by 0.03. He then modelled the three hull shapes which were the main subject of the investigation. Predictions of lift and drag were made at leeway angles of 0, 5, 10 and 15°, at a speed of 1.543 m/s (3 knots) which corresponds to a Froude number of 0.12. He also tried to obtain predictions at the higher speed of 5 m/s, but these runs did not converge to reliable solutions for reasons that are not entirely clear.

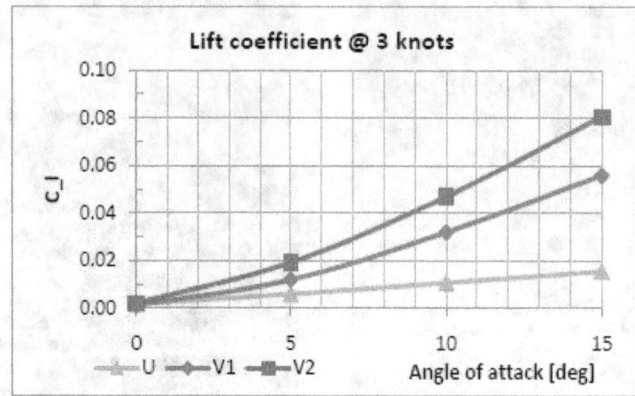

Hydrodynamic lift coefficient

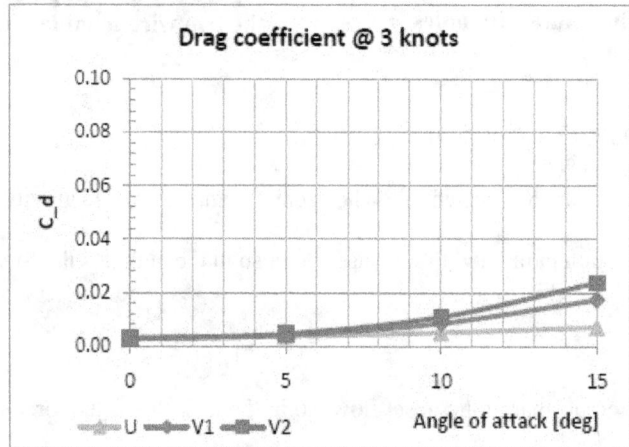

Hydrodynamic drag coefficient

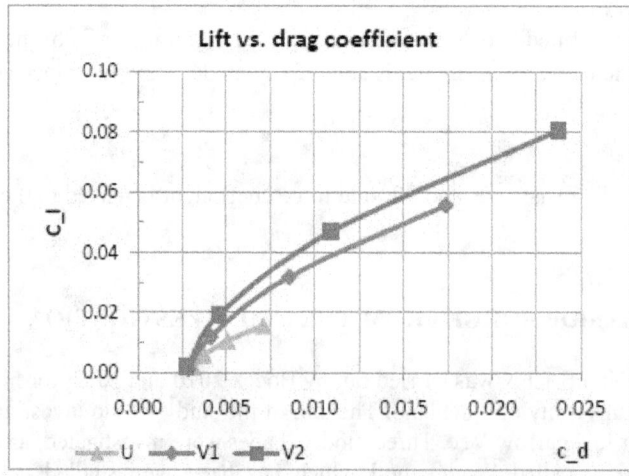

Hydrodynamic lift versus drag coefficient

Figure 7 Drag and lift coefficient predictions by Boeck 2010 using ANSYS-CFX. Note that in this figure U corresponds to the rounded hull, V1 to the wide Vee, and V2 to the narrow Vee. Wetted surface area was used as the reference areas

for these hulls. Top: lift coefficient versus angle of attack, centre: drag coefficient versus angle of attack, bottom: lift coefficient versus drag coefficient.

The results available in Boeck 2010 use the wetted surface area of the hull as the reference area, which is different for all the three models. In order to eliminate this variable, it was decided that for the present work the reference area would generally be the immersed projected side area, which is significantly smaller than the wetted area, and would therefore lead to higher force coefficients. The values of the various areas for the models are given in Table 2. The CFD predictions from Boeck 2010 are given in Figure 6 and Table 3.

Table 3 Force and moment predictions using CFD for the narrow Vee hull, from Boeck 2010.

speed [m/s]	AOA [°]	Fx [N]	Fy [N]	Fz [N]	Mx [Nm]	My [Nm]	Mz [Nm]
1.543	0	35.85	26.42	13621.61	18.50	-86.82	-94.27
1.543	5	35.00	250.00	13500.00	200.00	-175.00	-600.00
1.543	10	34.38	629.22	13384.93	451.94	-345.16	-1149.93
1.543	15	27.13	1099.48	13232.65	783.48	-660.20	-1972.66

FIXED SINK AND TRIM TESTS (JULY 2013)

A week of testing time in the tank was available during the period 1-5 July 2013. In order to make the best use of a relatively limited time, it was decided that these tests would be conducted with "fixed sink and trim". This means that the heave post is locked so that the model cannot move vertically, nor pitch. Such tests are simpler to conduct that "free to sink and trim" tests, as the former do not require the pitching moment to be calculated for a given towing force. In addition, these tests were used for the initial training of The University of Auckland staff and student in tank test techniques. The paper does not discuss the free to sink and trim tests.

Test schedule

Since the primarily objective of the test programme was to understand how the hull shape affected the lift or side force when the model had various amounts of leeway, a test schedule was prepared which was a matrix of with speeds ranging from 0.4 to 1.6 m/s, and leeway angles from 0 to 24°. As the test progressed, there were changes based on observations of tests, such as the speed and leeway which caused water to go spill onto deck. The actual tests that were undertaken are shown in Table 4. Note in Table 4 that V1 corresponds to the rounded hull, V2 to the wide Vee hull, and V3 to the narrow Vee hull.

Table 1 Tests done from 2 July – 5 July in the Newcastle Towing tank on slender hulls

Yaw/Speed	0.4 m/s			0.8 m/s			1.0 m/s	1.2 m/s			1.6 m/s	
0 degrees	V1	V2	V3	V1	V2	V3		V1	V2	V3	V1	V2
4 degrees	V1		V3	V1		V3		V1		V3	V1	
8 degrees	V1	V2	V3	V1	V2	V3		V1	V2	V3		V2
12 degrees			V3			V3				V3		
16 degrees	V1	V2	V3	V1	V2	V3		V1	V2	V3		V2
24 degrees	V1	V2		V1	V2		V2	V1	V2			

Analysis of fixed sink and trim results
Description of testing procedures

Normal tank testing procedures were used to carried out the tests, such as waiting for sufficient time for the water to settle between runs, checking the waterline height each day after skimming, zeroing the load cells, repeating the same test from the last run of the previous day to the first run of the following day, and repeating runs were the results looked questionable.

Drag, lift, resistance and side force as a function of speed

It is common to plot tank test results as a function of model speed, or Froude number. Here the force results are plotted against model speed. The test speeds were nominally 0.4, 0.8, 1.2 and 1.6 m/s. These correspond to Froude numbers of 0.12, 0.23, 0.35 and 0.47. Drag force is plotted against speed in Figure 8 for all three models. Several features are apparent in the figure. The narrowest model, V3 has the lowest drag for a given speed, followed by the wider Vee model (V2) and the rounded model (V1). The drag increases as speed increases, as expected, and the drag increases as the leeway angle is increased, although the V2 and V3 models have very similar drag when yawed at 16°. Drag coefficient is plotted against speed in Figure 9. Here it is evident that the coefficients are essentially constant over the speed range of the tests. It is more evident that yawing the model to produce leeway gives a significant increase in the drag coefficient as expected. The V3 model was not tested at a leeway angle of 24°. Note that the reference area here is the projected immersed side area, which is the same for the three models, and is 0.0488 m^2.

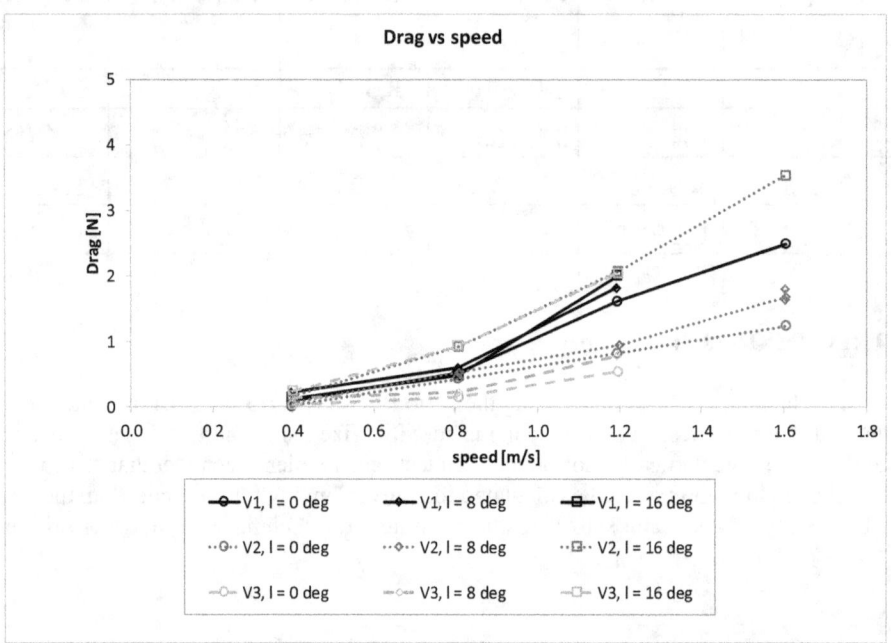

Figure 8 Fixed sink & trim, drag versus speed for all three models

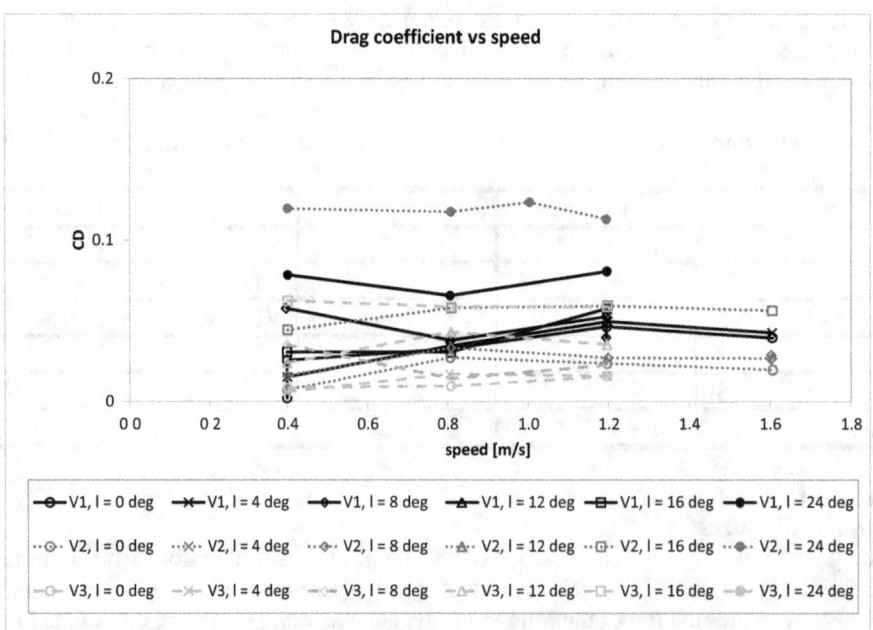

Figure 9 Fixed sink & trim, drag coefficient versus speed for all three models

Resistance coefficient is plotted against speed in Figure 10. One might expect the resistance and drag coefficient plots to look similar, and indeed this is the case for the low leeway angles, but for larger angles the significant amount of lift that is generated (of order twice the drag force) has a component that subtracts from the resistance, which lowers it significantly. This also has the effect of reordering the results. The rounded V1 model (black solid lines), which is not so good at generating lift, has the highest resistance, and conversely the narrow V3 model (yellow dashed lines) has the lowest resistance coefficient.

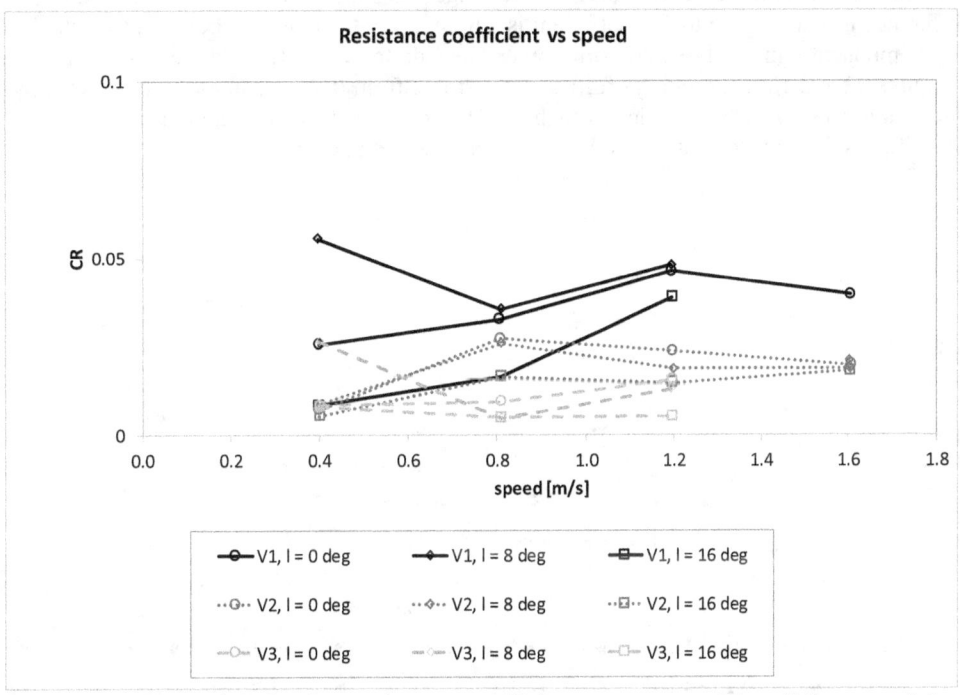

Figure 10 Fixed sink & trim, resistance coefficient versus speed for all three models

The lift and side-force coefficients are the primary foci of this work, and lift force is plotted against speed in Figure 11. It can be seen that the lift force is approximately twice the drag force, and that as the models become narrower, they become more efficient at generating lift. The V1 model at 0° shows a small negative lift force at a speed of 1.6 m/s. This is due to experimental error, possibly a very small angle misalignment of the model.

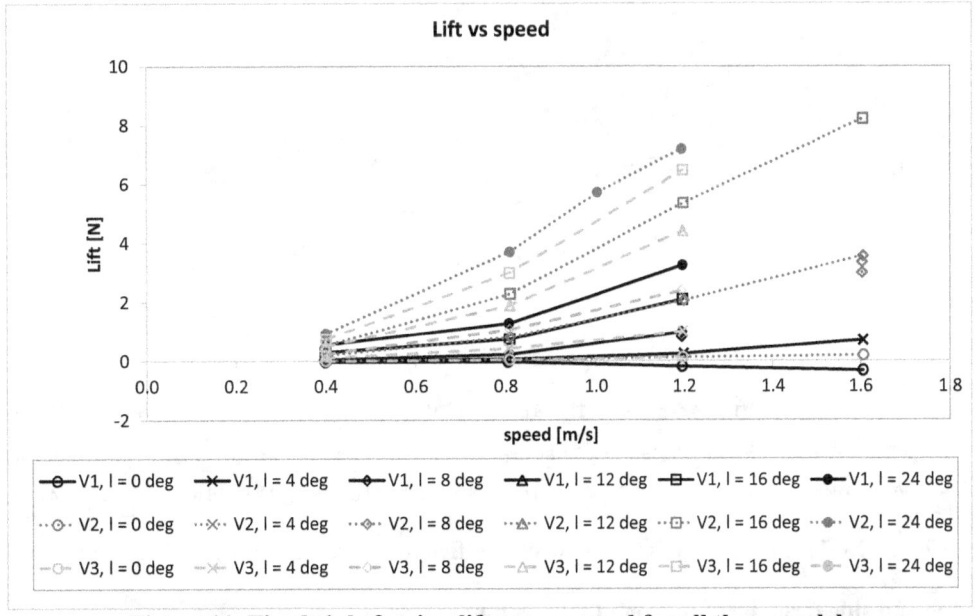

Figure 11 Fixed sink & trim, lift versus speed for all three models

The lift coefficients are shown in Figure 12, and as observed for the drag coefficients, the lift coefficients are also relatively unaffected by the speed over the range tested, although there is perhaps a slight trend to smaller coefficients as speed is increased. It is also evident that there is a much larger increase in lift coefficient between the rounded model V1 and the wide Vee model V2, with a smaller difference to the V3 model.

The side force coefficient is plotted against speed in Figure 13, where the reference area is A, the immersed projected side area which is the same for all models. It can be seen that this figure is similar to Figure 12. The results are ordered the same, and the main difference is that the side force coefficient is a little larger than the lift coefficient for the larger leeway angles, as the drag has a component which adds to the side force. The side force coefficients are also relatively unaffected by test speed. In both Figures 12 and 13, the ranking of lift and side-force from the three lines can been seen by studying the lines with the open square symbols (16° leeway). Similar to the CFD results, these experimental results show that having a narrower hull enables more lift of side-force to be generated for a given leeway angle.

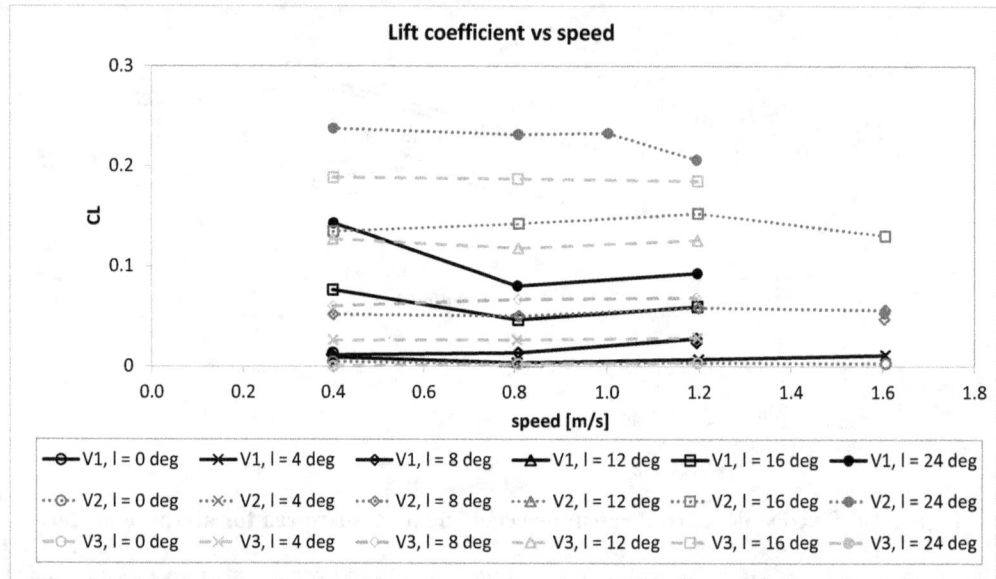

Figure 12 Fixed sink & trim, lift coefficient versus speed for all three models

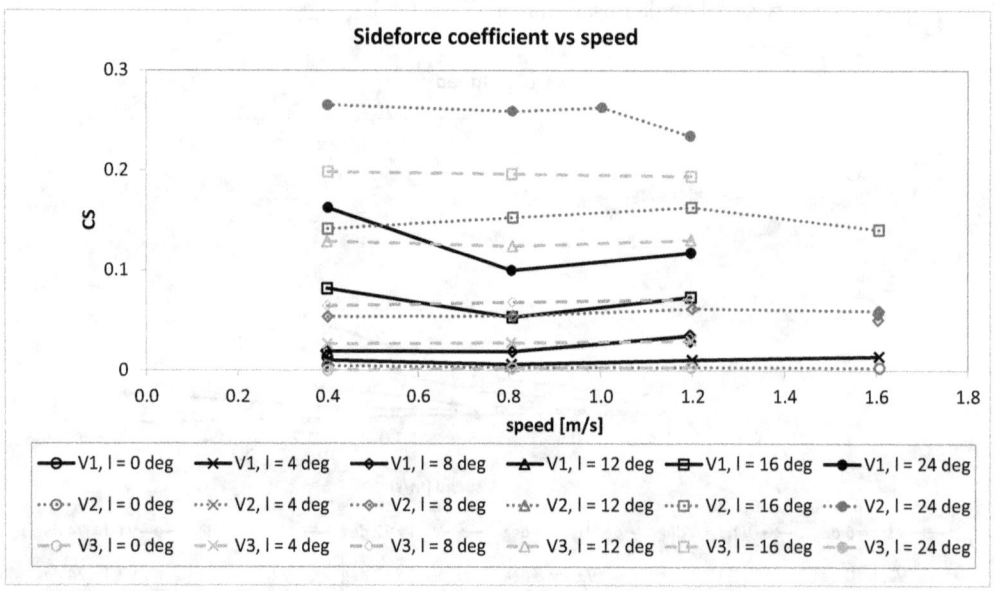

Figure 13 Fixed sink & trim, side force coefficient versus speed for all three models

Drag, lift, resistance and side force as a function of leeway angle

Since one of the objectives of carrying out this work was to obtain better estimates of side force for Polynesian sailing vessels as a function of leeway angle, the results are also shown in that manner. This presentation also enables them to be compared with the CFD predictions available in Boeck 2010, which were carried out at a speed of 3 knots (1.5433 m/s) for several leeway angles.

Drag force is plotted against yaw in Figure 14 for all three models. Here it can be seen that drag increases with increasing yaw angle for all three models, and that the rounded V1 model has the highest drag for a given yaw, followed by the moderate V2 model and by the narrowest V3 model. The drag coefficient is plotted against leeway angle in Figure 15. Although the variation in the results indicates some effect of experimental error, it is clearly evident that the three models show different trends. The drag coefficient for the rounded V1 model does not increase as quickly with yaw angle as does the drag coefficient for the V2 model, and the narrowest V3 model shows the most rapid increase in drag coefficient with yaw angle. Hence at a yaw angle of 0°, the drag coefficient is highest for the V1 model and lowest for the V3 model, but at a yaw angle of 16° the order is reversed. The V3 model was not tested at a yaw angle of 24°, which perhaps is unfortunate, as it might have shown up this effect even more clearly. All models had the same drag coefficient at an angle of about 7°.

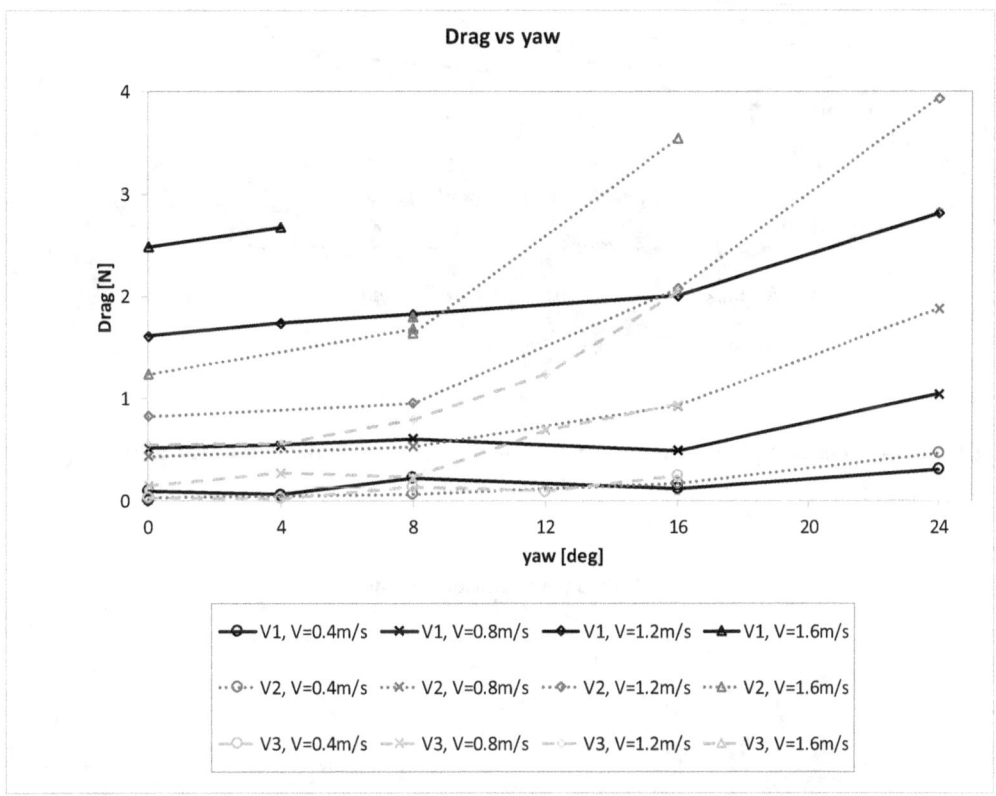

Figure 14 Fixed sink & trim, drag versus leeway angle for all three models

Figure 15 also shows the CFD predictions for drag coefficient as lines without symbols. Boeck 2010 and 2012 used CFD to predict the drag of a full-scale vessel at a speed of 1.543 m/s, and for coarse increments of leeway angles of 0, 5, 10 and 15°, for all the three hull shapes. Boeck's data were fitted by second order polynomial functions and used to plot the lines shown. Note that there is a disparity in the CFD and tank Reynolds numbers, and this point is returned to later when the tank coefficients are scaled to full-scale Re using the Prohaska-Hughes method. It can be seen that the CFD predictions are low for the rounded V1 model, but the trend of increase with yaw angle is represented in a similar fashion both in the CFD and model tests. There is reasonable agreement between the CFD predictions and the model drag coefficient values for the tests at a speed of 0.4 m/s for the V2 model. Interestingly, the Froude number corresponding to a model test speed of 0.4 m/s is 0.12, which converts to a speed of 1.26 m/s at full-scale. This model result is at the closest Froude number to the CFD predictions done at a speed of 1.543 m/s. There is fairly good agreement between the model and CFD results for the narrow V3 model over the range of leeway angles that were tested up to 16°.

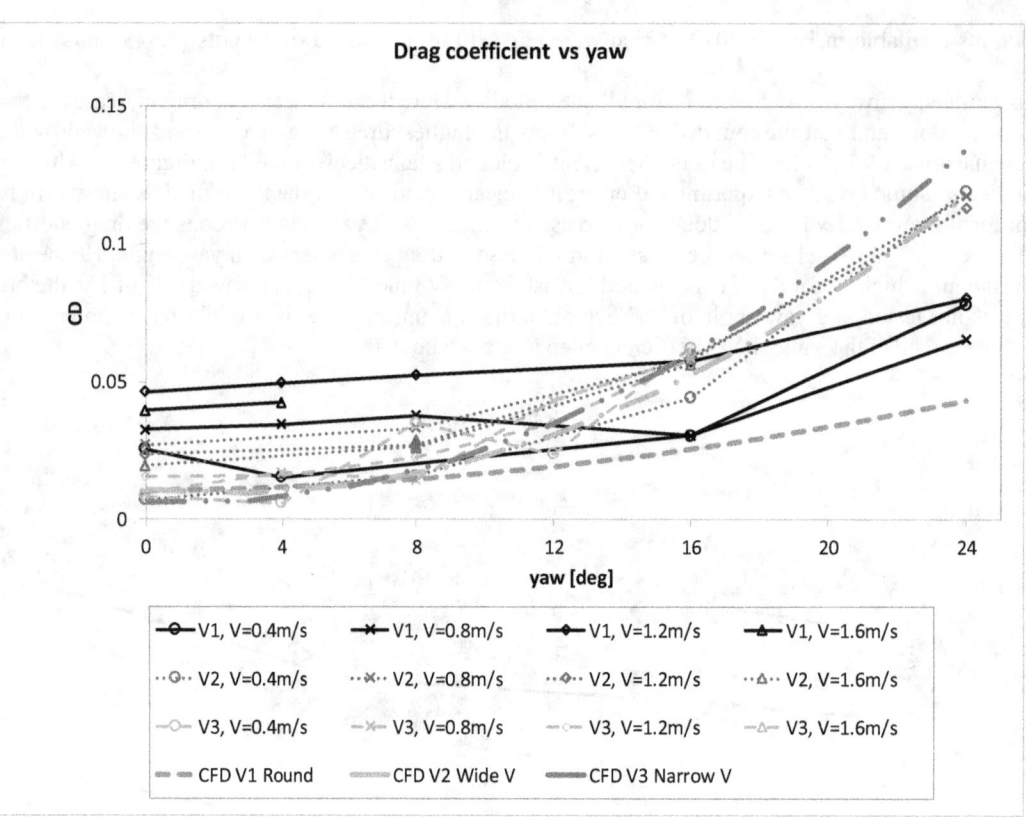

Figure 15 Fixed sink & trim, drag coefficient versus leeway angle for all three models, and comparison with CFD predictions.

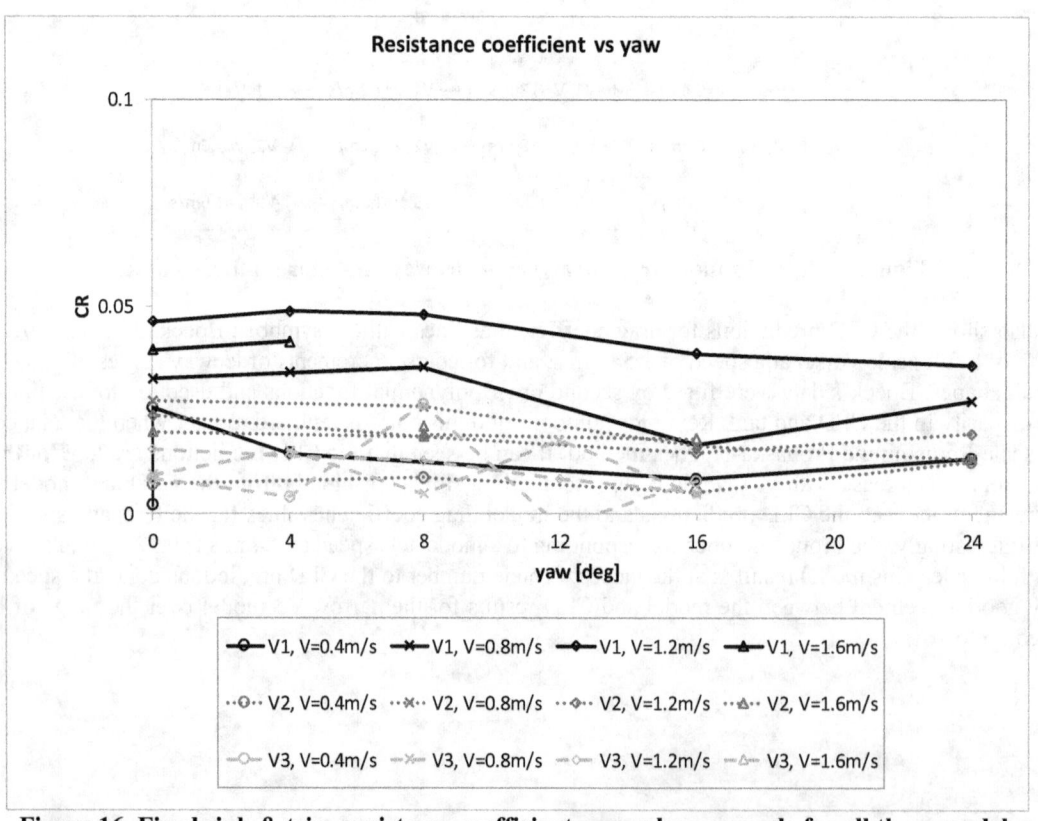

Figure 16 Fixed sink & trim, resistance coefficient versus leeway angle for all three models

The resistance coefficient is plotted against leeway angle in Figure 16. There is a rather striking difference between the drag and resistance coefficient plots, which may seem rather surprising. The resistance coefficients are generally lower than the drag coefficients, for the same reason as discussed in the previous section, that the lift force subtracts from the resistance, particularly at the higher leeway angles, where it can be seen in Figure 16 that the coefficients actually decrease with increasing yaw angle. The second major difference with the drag coefficient plots is that the resistance coefficient is always highest for the rounded V1 hull, and lowest for the narrow V3 hull, across the entire yaw angle range.

The lift force is plotted against leeway angle in Figure 17 for all three models. It is evident that lift force increases with increasing yaw angle, and that the narrowest V3 model is the best at producing lift, and the rounded V1 model is the worst at producing lift. Turning these lift forces into coefficients, as shown in Figure 18, puts the lines into clear groups, with it even more evident how a rounded hull is much worse at producing lift compared to having a narrow Vee. The effect of model speed can be seen to have a relatively minor effect on the coefficients for the V2 and narrow V3 model, but a slightly larger effect on the rounded V1 model.

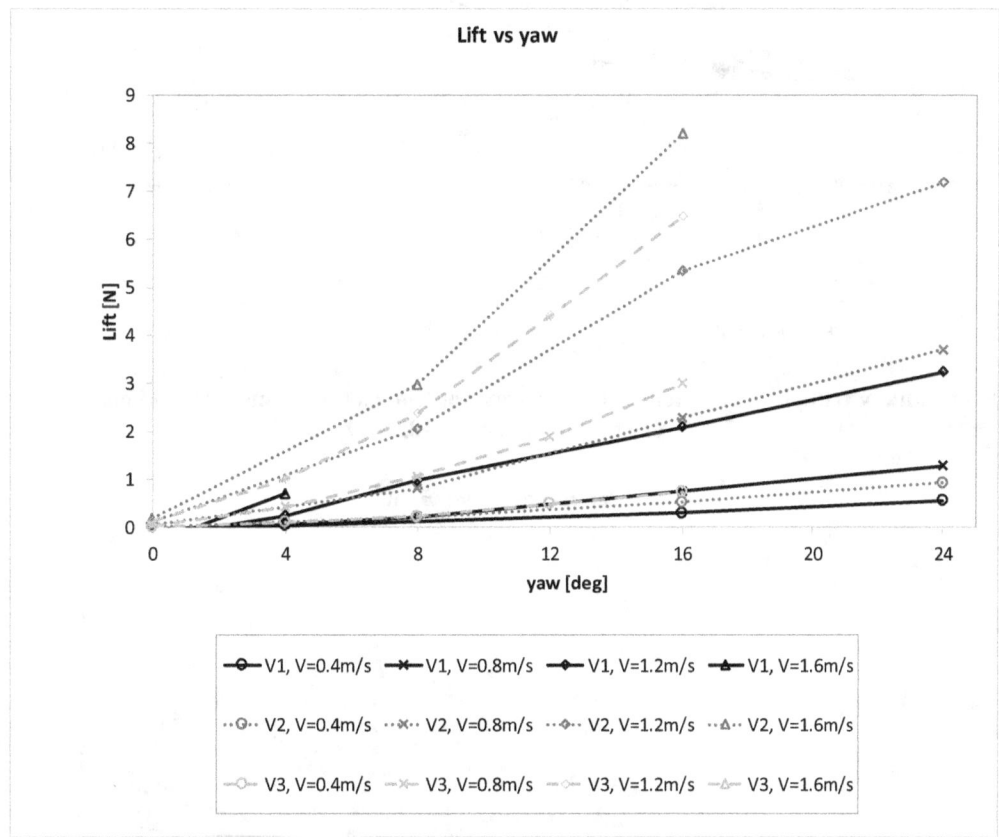

Figure 17 Fixed sink & trim, lift versus leeway angle for all three models

The CFD predictions of the lift coefficient from the work of Boeck 2010, 2012 are also plotted in Figure 18. Boeck carried out simulations on full-scale vessels of identical shape to the present tank test models, at a speed of 3 knots (1.5433 m/s), for a range of yaw angles. His data have been fitted to second order polynomials, and plotted as shown. It is clearly evident that the agreement is extremely good, and that the CFD predictions lie within the band of experimental results for each of the three models, up to a leeway angle of 16°, and beyond that angle, the predictions are still rather good.

The side force coefficients are plotted in Figure 19. These can be seen to look almost identical to the lift coefficients in Figure 18. There are groups of lines bunched together for each of the test models showing that the coefficients are only weakly dependent on speed.

14

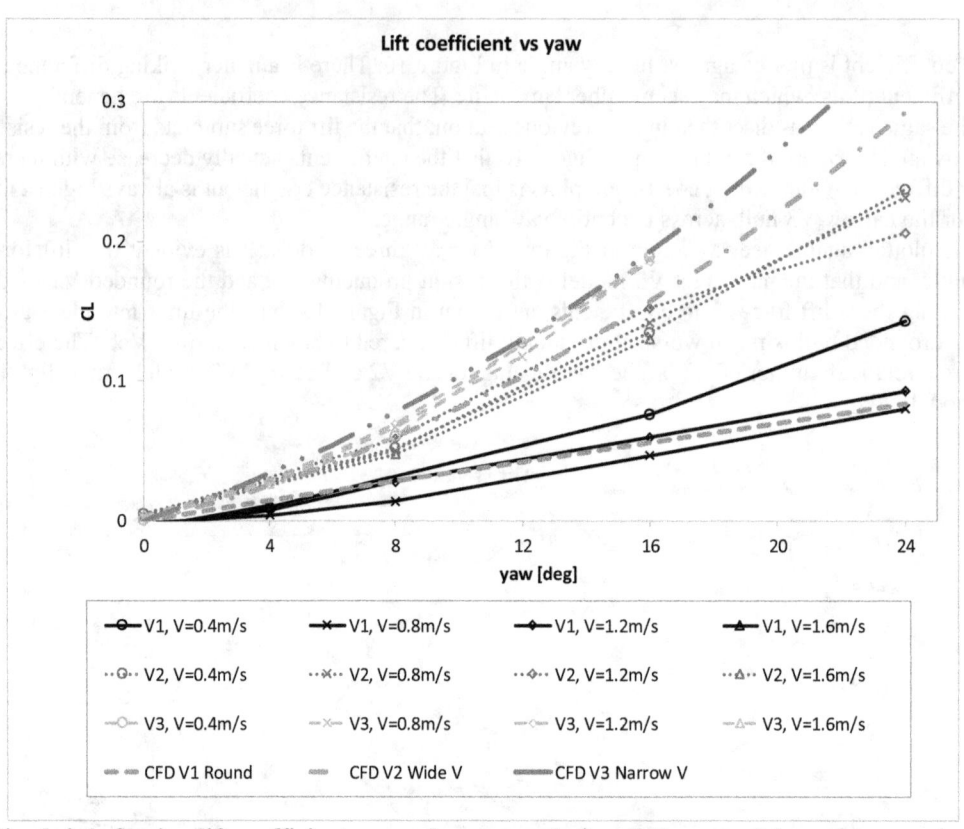

Figure 18 Fixed sink & trim, lift coefficient versus leeway angle for all three models, and comparison with CFD predictions Boeck 2010, 2012

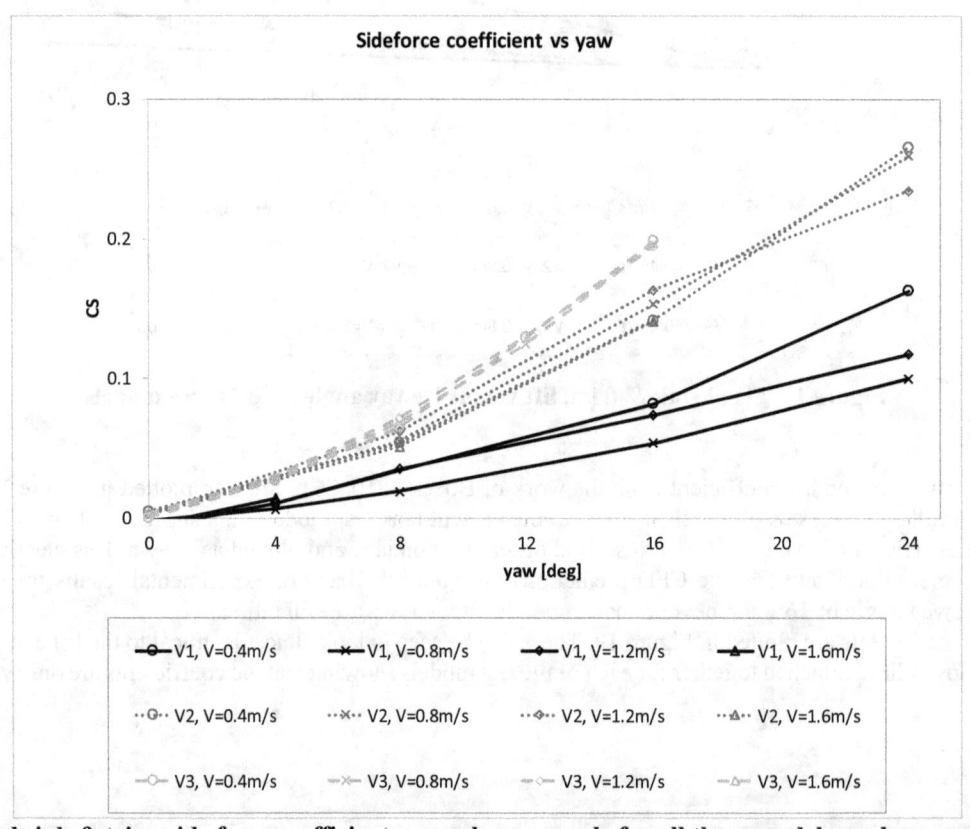

Figure 19 Fixed sink & trim, side force coefficient versus leeway angle for all three models, and comparison with CFD predictions Boeck 2010, 2012

It should be noted at this point that the CFD simulations were carried out at full-scale, i.e. at a Reynolds number of about 20 million, whereas the present tank results were carried out at a Reynolds numbers in the range 0.5 to 2 million, or more than 1 order of magnitude less. Whereas there are well developed procedures for scaling tank resistance results to full-scale Reynolds numbers, it is not clear that there are similar procedures for scaling the lift or side force results. Hence the comparisons shown in Figures 18 and 19 are for different CFD and tank model Reynolds numbers. No Reynolds number scaling corrections have been applied to the lift and side force results presented in the paper. The effect of Reynolds number on the drag is discussed in the next section.

Scaling of drag coefficients from model to full-scale

When predicting full-scale drag using model tests, the test results have to be scaled, or extrapolated, from model to full-scale. The method used in the present tests is due to Hughes, and was introduced in the 1950s, and later adopted by the International Towing Tank Conference (ITTC). Details on the method can be found in ITTC 7.5-02-01-03, 2002 and ITTC 7.5-02-02-01, 2002, Rev. 01. A summary description is provided herewith.

The basis of the approach is that the resistance is the sum of frictional resistance which scales with Re, and residuary resistance which scales with Fr, since it mainly consists of wave-making resistance, as proposed by Froude, and given in Eq. (6).

$$C_T = C_F(Re) + C_R(Fr).$$
(6)

Hughes proposed the form factor approach given by Eqs. (7) and (8),

$$C_T = C_V + C_W, \text{and}$$
(7)
$$C_V = (1 + k)C_{F0},$$
(8)

where $(1+k)$ is the form factor, which depends on hull form (i.e. shape), C_{F0} is the skin friction coefficient based on flat plate results, C_V is a viscous coefficient taking account of both skin friction and viscous pressure resistance, and C_W is the wave resistance coefficient.

k is generally small, say 0 to 0.2. C_{F0} is the "correlation line" adopted by the ITTC in 1957. It incorporates some three-dimensional effects and is defined in Eq. (9).

$$C_{F0} = \frac{0.075}{(log_{10}Re-2)^2}.$$
(9)

The full-scale drag coefficient is found by assuming that the wave drag coefficients are the same at model and full-scale at the same Reynold number, and that the form factor is the same in both model and full-scale. The skin friction coefficient terms is evaluated using the ITTC correlation line formula using an appropriate Reynolds number. In equation form, this can be written as in Eqs. (10) and (11).

$$C_{Tms} = C_{Vms} + C_{Wms} = (1 + k_{ms})C_{F0ms} + C_{Wms}$$
(10)
$$C_{Tfs} = C_{Vfs} + C_{Wfs} = (1 + k_{fs})C_{F0fs} + C_{Wfs}$$
(11)

The full-scale wave drag can be found from the model results using Eq. (12).

$$C_{Wms} = C_{Wfs} = C_{Tms} - (1 + k_{ms})C_{F0ms}$$
(12)

Therefore the total full-scale drag can be written as in Eq. (13).

$$C_{Tfs} = (1 + k_{fs})C_{F0fs} + C_{Tms} - (1 + k_{ms})C_{F0ms}$$
(13)

Since the model and full-scale form factor are the same, then we may simplify as given in Eq. (14).

$$C_{Tfs} = C_{Tms} - (1 + k_{ms})(C_{F0ms} - C_{F0fs}),$$
(14)

finally giving Eq. (15).

$$C_{Tfs} = C_{Tms} - (1 + k_{ms})\left(\frac{0.075}{(log_{10}Re_{ms}-2)^2} - \frac{0.075}{(log_{10}Re_{fs}-2)^2}\right)$$
(15)

The form factor is determined from the model results by postulating that the wave drag coefficient is proportional to Froude number to the power 4, and for conducting tests at low Fr say 0.1 to 0.2, then the straight line plot of C_{Tms}/C_{F0ms} versus Fr^4/C_{F0ms} will intersect the ordinate (Fr = 0) at $(1+k_{ms})$, and the slope, say A, can also be determined. This is called a Prohaska plot. It is described in Eq. (16).

$$C_{Tms} = (1 + k_{ms})C_{F0ms} + C_{Wms} = (1 + k_{ms})C_{F0ms} + A \times Fr^4, \qquad (16)$$

finally resulting in Eq. (17).

$$\frac{C_{Tms}}{C_{F0ms}} = (1 + k_{ms}) + A \times \frac{Fr^4}{C_{F0ms}} \qquad (17)$$

The measured tank drag data for all three models and for all leeway angles have been analysed using the method explained above. The Prohaska plots are shown in Figures 20, 21 and 22. It is evident that there is quite a lot of noise in the data. As mentioned previously, the drag values for these small models were quite low (for example at a towing speed of 0.4 m/s the drag was of order 0.1 N) so the accuracy was not as good as it would have been with larger models. It is also evident that the form factor appears to increase as the leeway angle increases. This is probably an artefact of the method, which assumes that there is no flow separation. Indeed, there is likely to be flow separation when the model is yawed, particularly for the V2 and V3 models.

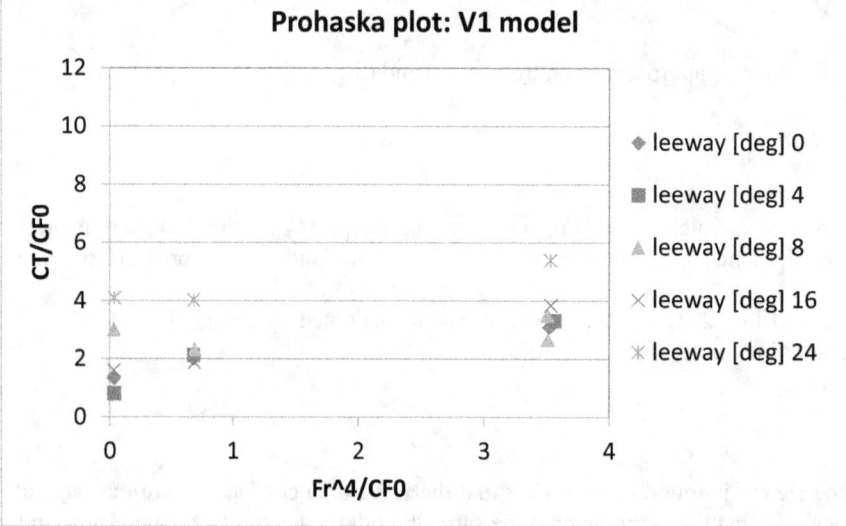

Figure 20 Prohaska plot to determine the form factor for the V1 model

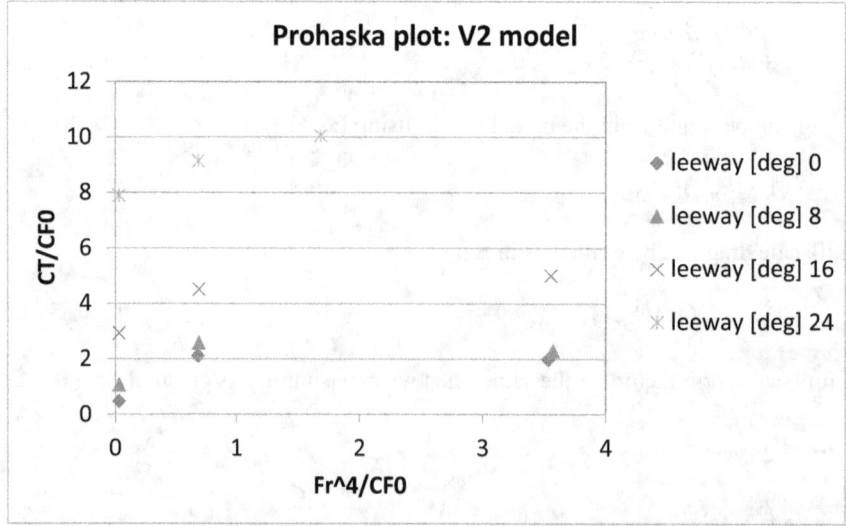

Figure 21 Prohaska plot to determine the form factor for the V2 model

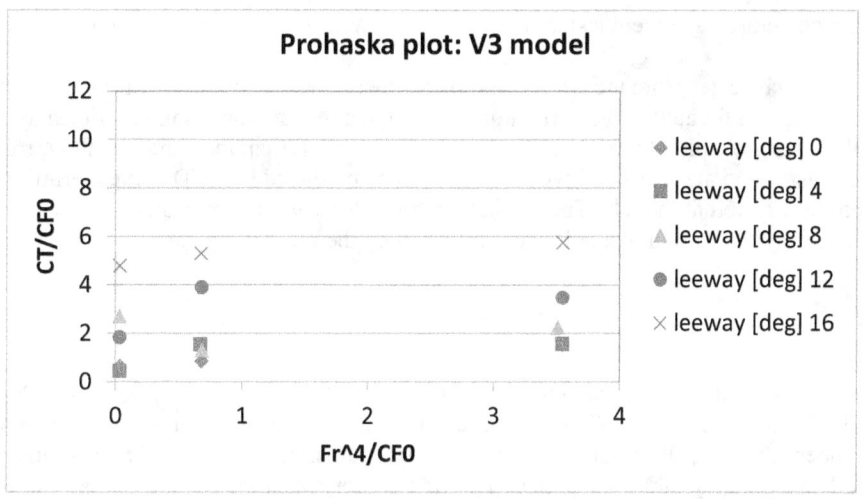

Figure 22 Prohaska plot to determine the form factor for the V3 model

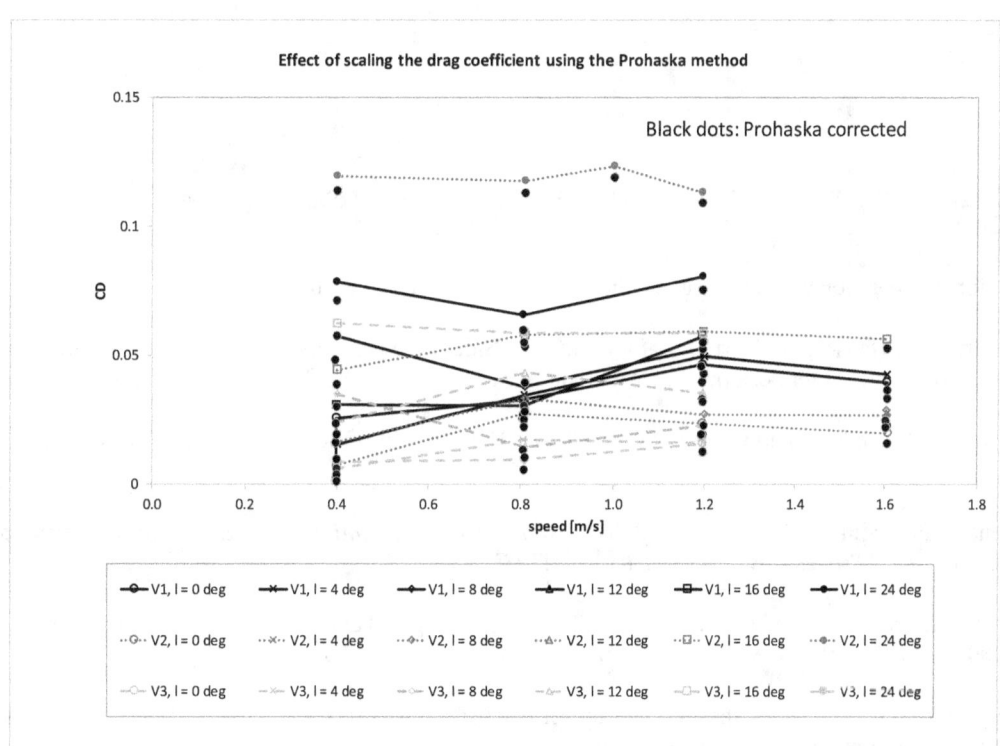

Figure 23 Plot of drag coefficients for all models showing the slight reduction in values (shown by the black dots) when the data are corrected using the Prohaska-Hughes method.

The form factor appears to decrease as the models become narrower. Estimates of the form factor for the V1, V2 and V3 models are 1.3, 1.1 and 1.0 respectively based on fitting straight lines to the data, although they cannot be determined with much accuracy. Nonetheless, these values have been used to recalculate the drag coefficients for full scale. These full-scale coefficients have been plotted on Figure as black dots. They all lie slightly below the model scale values. Although the value of the correction varied somewhat, it was found to result in a reduction of 0.005 on average, which although not large, is significant.

CONCLUSIONS

A CFD study with ANSYS-CFX using three different hulls was carried out as suggested by the first author and it showed that sharper Vee sections were better at generating lift and side-force than a rounded hull. The purpose of the present tests was to investigate whether such behaviour could also be observed in physical testing.

Three models were manufactured and were tested in the Towing Tank at Newcastle University in July 2013 with fixed sink and trim.

The drag coefficients were corrected from the tank Reynolds number of about 1 million to full-scale which is about 20 million using the Prohaska method. It was found that the correction resulted in a reduction of drag coefficient of the order of 0.005. this value is relatively small, and is probably of the same order, or in fact smaller, than the experimental error.

It was found that there was good agreement between the side-force predicted by CFD and experimentally determined values from tank test results on three different models. The predictions from the CFD that narrower Vee shaped hulls would generate more side-force when at leeway than a rounded hull was confirmed by the present tank test results.

ACKNOWLEDGEMENTS

This research was carried out as part of the Sailing Fluids project funded jointly by the EU and by NZ. The grant details are as follows. EU - Call: FP7-PEOPLE-2012-IRSES, Funding scheme: MC-IRSES (International Research Staff Exchange Scheme (IRSES)), Proposal number : 318924, Proposal acronym : SailingFluids. NZ - IRSES Application 2012-318920-SAILING FLUIDS funded by the Royal Society of New Zealand. The first author was assisted in carrying out the towing tank tests by Alexander Blakeley, Joshua Taylor, and Nick Velychko whose contributions are gratefully acknowledged. The technical support provided by the Newcastle University technical staff in the laboratory is also gratefully acknowledged.

REFERENCES

Newcastle towing tank, https://www.ncl.ac.uk/engineering/about/facilities/marineoffshoresubseatechnology/hydrodynamics/#towingtank accessed 20/12/2018.

Boeck, F., "Side force generation of slender hulls", *Study Thesis*, TU Berlin, Germany, 2010.

Boeck, F., Hochkirch, K., Hansen, H., Norris, S., Flay, R.G.J., "Side force generation of slender hulls – influencing Polynesian canoe performance", Proc. 4th High Performance Yacht Design Conference, Auckland, New Zealand 12-13 March 2012.

ITTC – Recommended Procedures and Guidelines 7.5-01 -01-01 Page 5 of 6 Model Manufacture Ship Models Effective Date 2002 Revision 01

ITTC Recommended Procedures 7.5-02-01-03, 2002. '*Testing and Extrapolation Methods, General Density and Viscosity of Water' ITTC Recommended Procedures 7.5-02-01-03*. s.l., ITTC.

ITTC – Recommended Procedures and Guidelines 7.5-02 -02-01 Testing and extrapolation methods resistance. Resistance Test. Effective Date 2002, Revision 01.

Wong, H.L., "Slender ship procedures that include the effects of yaw, vortex shedding and density stratification", *Thesis*, University of British Columbia, Vancouver, Canada, 1994.

THE 23RD CHESAPEAKE SAILING YACHT SYMPOSIUM
ANNAPOLIS, MARYLAND, MARCH 2019

A Comparative Study of Program FloSim Results against SYRF Wide Light Project Data

Brian Maskew, Computational Flow Simulations LLC, Winthrop, WA
Frank DeBord, Chesapeake Marine Technology LLC, Easton, MD

ABSTRACT

In November, 2015, the Sailing Yacht Research Foundation (SYRF) published the tank test data from their "Wide Light Yacht Project" for the hydrodynamics of a modern, high performance, semi-planing yacht. This comprehensive data set, comprising canoe body with and without appendages in upright, heeled and yawed conditions, provides an important validation base for CFD codes; previously, such data were not readily accessible mainly due to proprietary issues. The SYRF report includes a number of comparative results from commercial CFD codes, the RANS program Star-CCM+ results in general showing the best correlations with the measured data, albeit with some significant departures.

In this paper, we present results computed using an advanced Boundary Element Code, FloSim, these calculations being compared against the test matrix of "Wide Light" measured data and also Star-CCM+ calculated results. Times for computer model preparation and case execution are discussed together with computer requirements.

An outline of the FloSim method is presented; it is an unsteady, time-stepping code with a coupled integral boundary layer analysis for viscous effects such as skin friction resistance and boundary layer displacement effect. The free surface wave development uses a non-linear mixed Eulerian-Lagrangian treatment at each time step. FloSim has free convection and rollup of vortex wake elements providing non-linear lift characteristics and includes a number of modeling techniques for treating "real world" effects, such as flow separation and wave breaking. For the higher speed cases in the SYRF data set that have a breaking bow wave crest, FloSim's Wave-Breaker treatment is applied to convert excessive energy in the bow wave to a "dead-weight" pressure applied on the free surface; this effectively attenuates downstream wave amplitudes consistent with the loss of energy at the breaking crest. The empirical Foam Density Factor in the model has been established here as a simple function of Froude number based on the measured results at the Fn 0.8 upright case.

FloSim, already used in America's Cup and Volvo racing yacht analyses, was developed specifically to bridge the gap between basic potential flow panel methods and RANS codes, with the objective of providing accurate, practical solutions on a laptop computer within a reasonable turnaround time and cost. In essence, the results presented in this paper demonstrate these objectives have been achieved. The comparisons of FloSim's essentially "low-order" results against test data and Star-CCM+ RANS calculations, provide a measure of tradeoff between calculation accuracy versus cost and turnaround time for a case. Throughout the discussions presented below, the reader should keep in mind that this was not a "blind" comparison as it was for the original Wide Light Project participants who published their results *prior* to the tests.

NOTATION

CSYS   Chesapeake Sailing Yacht Symposium
SYRF   Sailing Yacht Research Foundation
BEM   Boundary Element Method
GFF   Ground-Fixed Frame of Reference
SFF   Ship-Fixed Frame of Reference
ρ Density of water (kg/m^3)
Fn   Froude Number
Rn   Reynolds Number based on reference length (hull length)
LCG   x-wise location of CG measured in the SFF (m)
Zfs   Free surface elevation (m)
θcrit   Limiting wave slope angle for FloSim's Wave-Breaker treatment (deg)
D_f Empirical *Foam Density Factor* for FloSim's Wave-Breaker model

INTRODUCTION

The Wide Light Project, conceived and sponsored by the Sailing Yacht Research Foundation (SYRF), was initiated in 2013 with the objective of extending the existing data base of yacht measured hydrodynamic characteristics to cover modern yacht design practices, the earlier *accessible* data bases, such as the Delft Systematic Yacht Hull Series (Keuning, 1998) and the nine-model series performed at the Canadian National Research Council (Teeters, 2003), etc., being no longer representative of today's high performance sailboats. Moreover, the data base from the Wide Light Project was to be made publicly available; although much tank testing of modern designs has been carried out, this has been mainly of a proprietary nature for specific projects and therefore not widely available to the yacht research community.

A broad data base of measured hydrodynamic characteristics is essential for CFD code validation and for handicappers trying to make equitable handicap rules for racing boats, and is useful for designers making initial decisions for new boat designs, since tank tests and even CFD analyses may be too time consuming for initial steps in the design process.

The measured data from the Wide Light Project were first presented at the 22nd CSYS in March 2016 (Prince and Claughton, 2016) and showed a comprehensive hydrodynamic data set from tank tests of a modern, semi-planing hull model over a range of speeds for canoe body alone and for an appended configuration in upright and in yaw and heel conditions. The paper also showed a number of comparative predictions by several CFD practitioners; in broad terms, the STAR-CCM+ RANS results by Cape Horn Engineering are perhaps closest to the measured data both in absolute terms and in trends, whilst the computationally less intensive panel code, FloLogic, by Ennova Technologies, though not as close in absolute terms, does capture the trends of the measured data.

The test matrix adopted for the Wide Light Project was extensive and necessitated significant computational resources; for example, to complete an individual run, RANS took anywhere from 3 hours (for upright canoe body cases) to 30 hours (for appended configurations in heel and yaw) on a multiple processor computer.

In this paper we present results calculated by the advanced Boundary Element Code FloSim, these results being compared against the test matrix of the Wide Light Project measured data and also the calculated results from the STAR-CCM+ RANS code. Comparisons of FloSim and STAR-CCM+ times needed to form a computer model, to prepare input files and to execute runs, are included, and computer requirements for the codes are discussed.

Development of the FloSim code began in 2003 but the method stems from a hierarchy of panel codes starting with a doublet panel code, QuadVort, (Maskew, 1970), developed in the 1960s, the steady code VSAERO (Maskew, 1987) in the late 1970s and the unsteady USAERO code (Maskew, 1991) developed in the late 1980s for submarine and surface piercing vessels in waves. A private (non-funded) extension of the USAERO program demonstrated the feasibility of a simultaneous hydro/aero/structural/analysis of a complete sailing yacht configuration in 6 DOF motion in waves (Maskew,1993).

FloSim was developed specifically to bridge the gap between RANS and Panel code approaches with the objective of leaning towards the "real world" accuracy of RANS, but with the speed advantage of the potential flow approach. The aim was to maintain the ease of use and low computer resource requirement of the panel code whilst avoiding the RANS method requirement for high level of user expertise and costly computational resources on multiple processor computers.

FloSim has already been used in America's Cup and Volvo racing yacht analyses. An outline of the method is given below; this includes discussions of some of the modeling tools installed in the code to achieve the FloSim development objectives such as treatment of "real-world" effects. A more detailed description of the basic method and formulation is given in Maskew and DeBord, 2009.

FLOSIM METHOD OUTLINE

Basic Procedure

FloSim is based on a non-linear Boundary Element Method (BEM) that treats the free surface deformations and finite amplitude vessel motions in a time-stepping procedure. The method uses uniform perturbation doublet and source singularity panels arranged on surface patches representing the free surface and all the wetted surfaces of the vessel and, if present, ocean floor, tank walls, etc. The patch panel grids are reformed to the changing waterline at each time step to maintain a "clean" panel relationship between intersecting patches as the waves develop and the vessel trims and heaves.

Similar singularity panels represent the transient mean surfaces of vortex wakes shed by lifting parts of the configuration. A new wake panel set is formed and shed from each trailing edge of the configuration at each time step, whilst wake panels shed earlier are convected at the locally computed flow velocity. The wake surfaces roll up in strong vortical regions, thus providing a non-linear lift capability

For viscous effects, a coupled integral method is used to calculate surface boundary layer characteristics along computed streamlines at each time step; the boundary layer displacement effect is represented by a source *transpiration* term in the boundary condition applied at each panel center, thereby affecting the current doublet solution and also the local external flow streamline paths and free surface geometry. The tangential force produced by the calculated boundary layer skin friction coefficient distribution is included in the force and moment integration over the vessel's surface.

The boundary layer calculation on each streamline starts at an upstream stagnation point with laminar flow conditions and proceeds either to laminar separation or to natural transition to turbulent flow, an empirical stability criterion being used in the transition step. Alternatively, transition can be *forced* either by specifying "transition strip" locations on the configuration surfaces, or by setting a transition flag "*as soon as possible*" when dealing with onset flows having a high turbulence level. When transition is established, the turbulent calculation proceeds until either a trailing edge is reached or conditions for turbulent separation are reached. If laminar separation occurs *before* transition the program looks for transition to turbulence *in* the separation bubble with possible turbulent reattachment, otherwise there is total separation.

The boundary layer calculation on a streamline stops when certain separation criteria are encountered. If the separated region is extensive then a separation model can be applied using either a source outflow model or a vortex sheet model (Maskew and Dvorak, 1978). Either approach provides a good prediction of separated base pressure.

The later cases in the Wide Light test matrix (HKR-091 to 110) involve rudder deflections. FloSim has a number of options for deflecting a control surface; the first is to define the control surface in its own reference frame and then specify the rotation of this relative to the Ship-Fixed Frame (SFF). There is a drawback with this approach if the control surface root intersects the hull surface, as it does here, since the local panel grid would need to be redone with each deflection angle to maintain panel matching. This is feasible but tedious to set up, especially when the deflection needs to be changed *during* a case run (as in a maneuver analysis). However, two alternative, and more convenient treatments are available that *simulate* the effect of control rotation; one uses a "*source transpiration*" distribution, the other uses a "*rotated normals*" option, whereby the vector *normal* to the surface at each panel center is rotated about an axis parallel to the control stock. Both options provide a *pseudo* rotation of the control, the control surface staying in its basic position with the initially established panel grids, however, the rotated normal option is currently preferred and was used for this project; it provides a reasonably accurate representation of control deflection effects well beyond the 5 degrees needed for this project.

The free surface wave development is treated using a mixed Eulerian/Lagrangian approach based on the 2D method by Longuet-Higgens and Cokelet, 1976. In this method, the free surface panel grid points are convected at the locally computed velocity at each time step. At speeds below about Fn=0.3 there is a problem of transient waves resulting from the impulsive start. This is dealt with using a "sieve" treatment whereby the vessel's disturbance to the flow is applied in a very gradual manner by varying the hull surface *porosity* from 1.0 to 0.0 over the first 300 steps, or so; initially the flow passes straight through the hull. Wave absorber beaches placed along the free surface patch edges dampen any residual transients.

At high speeds, there is an issue over wave energy in conditions where wave breaking occurs; whereas with the basic 2D method high panel densities can capture the development of breaking crests and the subsequent trajectory of the jet of broken-off material, in the 3D case such panel densities would be impractical as they would demand very high computer resources and run times. Clearly, this would be incompatible with the stated basic objectives for FloSim's development, namely, rapid turn-around of cases on a laptop computer. With *practical* panel densities the basic free surface procedure cannot resolve the wave crests, and amplitudes become excessive when wave breaking should have occurred. When wave breaking conditions do exist, ways of limiting the wave energy, and hence amplitude, have to be applied very carefully otherwise resulting downstream wave amplitudes and phasing may be incorrect; although practical limits may be set for the perturbation velocity magnitude to guard against "*excessive* energy" in generated waves, this does not fully represent the effects of energy loss when breaking-wave conditions occur. In particular, when the generated waves *are* breaking, the downstream attenuation of wave amplitude in a vessel's wave train is not adequately represented in the analysis with just a simple energy limit applied. Predictions of resistance, heave and trim values, for vessels at speeds above the onset of wave-breaking suffer as a consequence.

From the tank test photos in the Wide Light Project, wave breaking is apparent in the bow wave for speeds as low as 2.4 m/sec or Fn 0.35, hence, for the present correlation study, wave breaking treatment is clearly needed over a significant portion of the speed range that extends to Fn 0.8. A breaking-wave model that is more representative of the physics of the problem was developed earlier within the free surface treatment in FloSim; since this treatment has not yet been published, and since it is clearly needed for this project, a broader discussion of the model is included here.

Wave-Breaker Model

FloSim's Wave-Breaker model rests heavily on Duncan's observations from his hydrofoil-in-tank experiments, (Duncan, 1981 and Duncan, 1983) and also the theoretical treatment by Cointe and Tulin, 1994. In addition, the breaking-wave model presented by Muscari and Di Mascio, 2003, for their RANSE method was helpful though the implementation of the model in the context of the BEM is a little different.

To establish a breaker model in the context of the FloSim BEM approach, consider the "simple" situation of a steady spilling breaking wave as produced in the essentially two-dimensional situation of a towed hydrofoil moving steadily below the free surface in a tank, the hydrofoil set at, say, 5° and fully spanning the tank width. This is basically the setup in Duncan's 1981 experiment. A trough develops in the free surface above the hydrofoil and is followed downstream by a uniform wave-train, (see Figure 2 below).

For submergence depths greater than the foil chord, the wave amplitude is below that for breaking conditions and the wave energy resulting from the foil motion is fully represented by the kinetic and potential energies derived from the wave profile data. As the hydrofoil is moved closer to the free surface, the waves become steeper and eventually the first crest breaks. The downstream wave train then has reduced amplitude and the kinetic and potential energies derived from the residual wave profile no longer represent the *total* energy being put into the system by the towed hydrofoil; the difference is accounted for by the energy "lost" in the breaking action.

At the first wave, the broken-crest fluid falls forward onto the upstream face of the wave where it is first deflected before settling back on the free surface. During the deflection it entrains air and so a reduced-density "foamy" layer representing the broken crest material is carried back up to the crest on the free surface. The impact of the crest material and its "dead weight" following impact, essentially apply a pressure load on the forward face of the wave and this reduces the total head in the surface streamlines; consequently a "momentum-deficit layer" convects downstream just below the free surface.

One possible approach to modeling the effect of the breaking crest is to regard the broken-off material as a dead-weight layer of reduced density applying a hydrostatic pressure, Δp, on the free surface upstream of the crest. This pressure term can be included in Bernoulli's equation applied in the free surface boundary conditions in the FloSim numerical model: the expression for the velocity potential time derivative, $\frac{\partial \Phi}{\partial t}$, at a point in the free surface then has an additional term, $\frac{\Delta p}{\rho}$:

$$\frac{\partial \Phi}{\partial t} = gz + \frac{1}{2}v^2 + \frac{\Delta p}{\rho}$$

In FloSim the velocity potential, Φ, on the free surface is obtained by integrating $\frac{\partial \Phi}{\partial t}$, over each time step, hence the influence of the Δp term appears in the velocity potential distribution, and the effect of this in reducing the perturbation velocity magnitudes follows automatically from the singularity solution.

The quantity $\frac{\Delta p}{\rho}$ is a function of longitudinal position, x, and is determined as follows:

$$\frac{\Delta p(x)}{\rho} = g * D_f * h(x)$$

Where D_f is the *foam density factor* for the "foamy layer" relative to the water density, and $h(x)$ is the height of the downstream crest above the local free surface elevation, see Figure 1:

$$h(x) = z_{crest} - z(x)$$

The pressure is applied over the forward-face region where $h(x)$ is less than H; H being the "foamy height", i.e., the vertical extent of the breaker region.

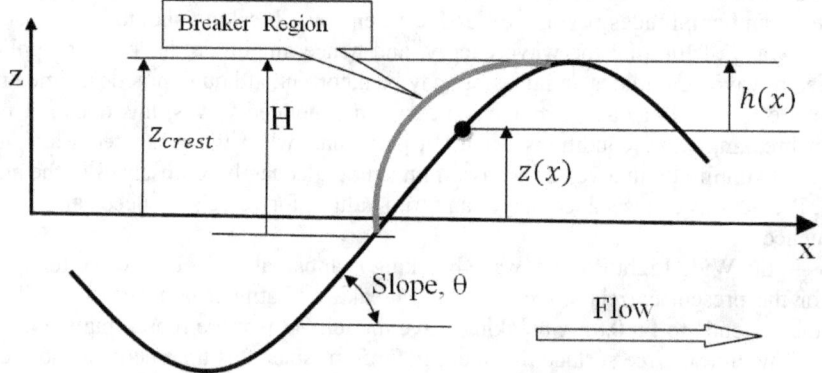

Figure 1 – Wave-Breaker Model

The wave-breaker model therefore has two parameters, D_f and H, that need "empirical help" for their evaluation: Cointe and Tulin, (1994), estimated the density factor, D_f to be in the range 0.2 to 0.6 *depending on the severity of the breaking*; for now we use 0.5 as a basic default value. This will be considered later when treating the SYRF data.
Muscari and Di Mascio, (2003), established an empirical relationship for H based on Duncan's experimental measurements:

$$H = H_f * z_{crest} \qquad \text{where } H_f = 1.28$$

In addition, we need a criterion to establish *when* the model should be applied. Based on Duncan's measurements it appears the model should be activated when the wave slope, θ, approaching a crest, exceeds a critical value of about $\theta_{crit} = 15^\circ$.

In testing the performance of the breaker model, results from FloSim calculations were compared with Duncan's hydrofoil-in-tank data for different submergence depths and angles of attack of the hydrofoil. This demonstrated the effectiveness of the breaker treatment in dissipating wave energy and in reducing the downstream wave amplitude.

To provide an example of the wave-breaker model here, a comparison is shown below for a particular condition in which the waves do not break *spontaneously*, but do so if a *temporary* disturbance is applied on the upstream free surface; the breaking then persists. For the FloSim test, the breaker model was activated after the wave calculation had become steady (at least for the first two crests; crests downstream of that takes longer to converge).

A view of the calculated wave train before and after the breaker treatment is shown in Figure 2; for clarity this just shows FloSim's "working section" with the upper part of the tank wall removed (for the FloSim calculation the tank was extended upstream and downstream of the "working section" to provide wave-absorbing buffer zones for transient waves resulting from an impulsive start.). Figure 2a shows the calculated wave contours just before activating the breaker model; at that point all the crests have upstream slopes greater than θ_{crit} and so when the breaker model is activated all the crests "break" simultaneously and the Δp term is applied at *all* the crests. As time progresses, the lower energy level from the first crest convects downstream thus reducing the amplitude of the following waves until finally all the crests have similar height at the lower level, as shown in Figure 2b. From that point on, the Δp term is applied only at the first crest, the approach slopes to all the *downstream* crests now being below θ_{crit}.

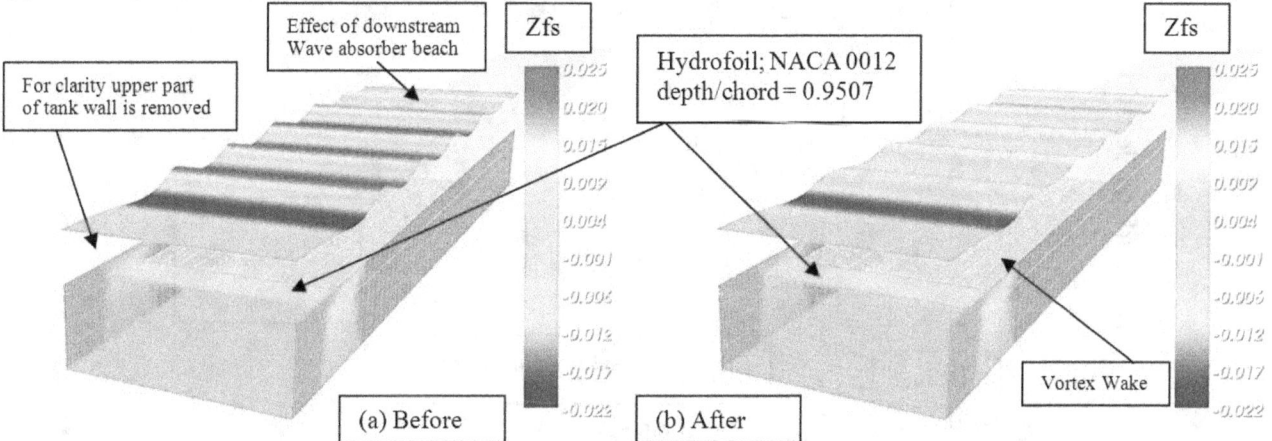

Figure 2 – FloSim Calculated Wave Contours for a Submerged Hydrofoil; Before and After Breaker Model Activation

The calculated first crest elevation history is given in Figure 3 and shows a dramatic drop in value immediately the Wave-Breaker model is activated. This is followed by a few oscillations as the Δp term adjusts, but the elevation eventually settles down to a value less than half the non-breaking level.

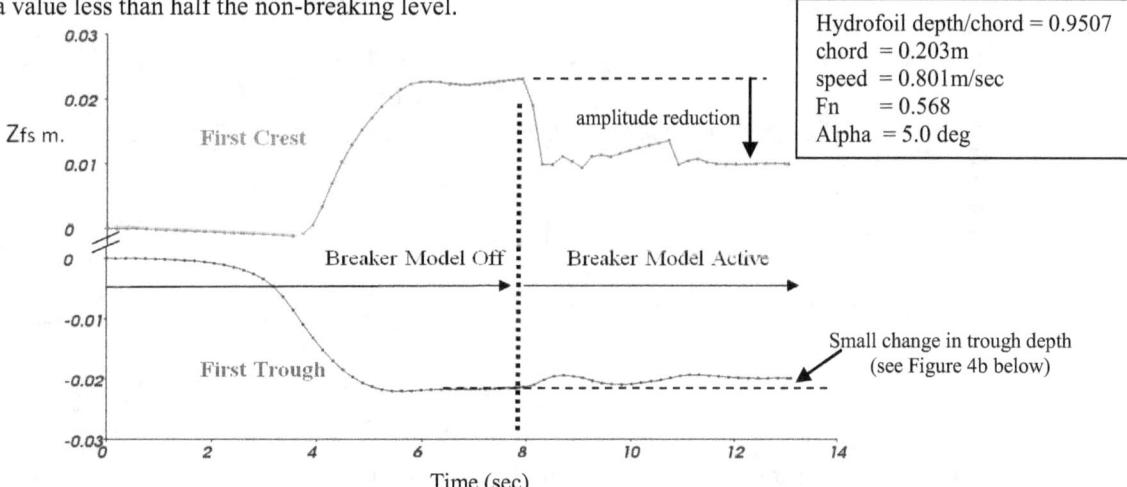

Figure 3 - History of Calculated Free Surface Elevation at first Trough and Crest

Comparisons between calculated and measured wave profiles for the depth/chord = 0.9507 submergence case are shown in Figure 4; Figure 4a shows the profiles before breaking and Figure 4b after breaking. The calculated before-breaking profiles are in excellent agreement with the Duncan's measurement, at least until the third crest where the calculation is somewhat high at this time step; (again, the downstream crests tend to take longer to settle down because of influence from the "beach" in the downstream buffer zone in the FloSim model). For comparison, the profiles from the RANSE calculation (Muscari and Di Mascio, 2003) are also shown; they reported some problem in reaching the experimental trough values, mentioning a similar trend seen by other researchers with Navier Stokes simulations.

The agreement in the profiles "with breaking" is good, but not as close as in the non-breaking case. In particular, the calculated elevation at the first trough increases by about 10% relative to the non-breaking value whereas the measurement shows no significant change; possibly, the H_f empirical factor in the breaker model needs to be adjusted to reduce the forward influence of the Δp term. Interestingly, the FloSim calculation *with breaking* follows Muscari's RANSE profile very closely up to the second crest; both calculations use $H_f = 1.28$ and both have the reduced first trough depth.

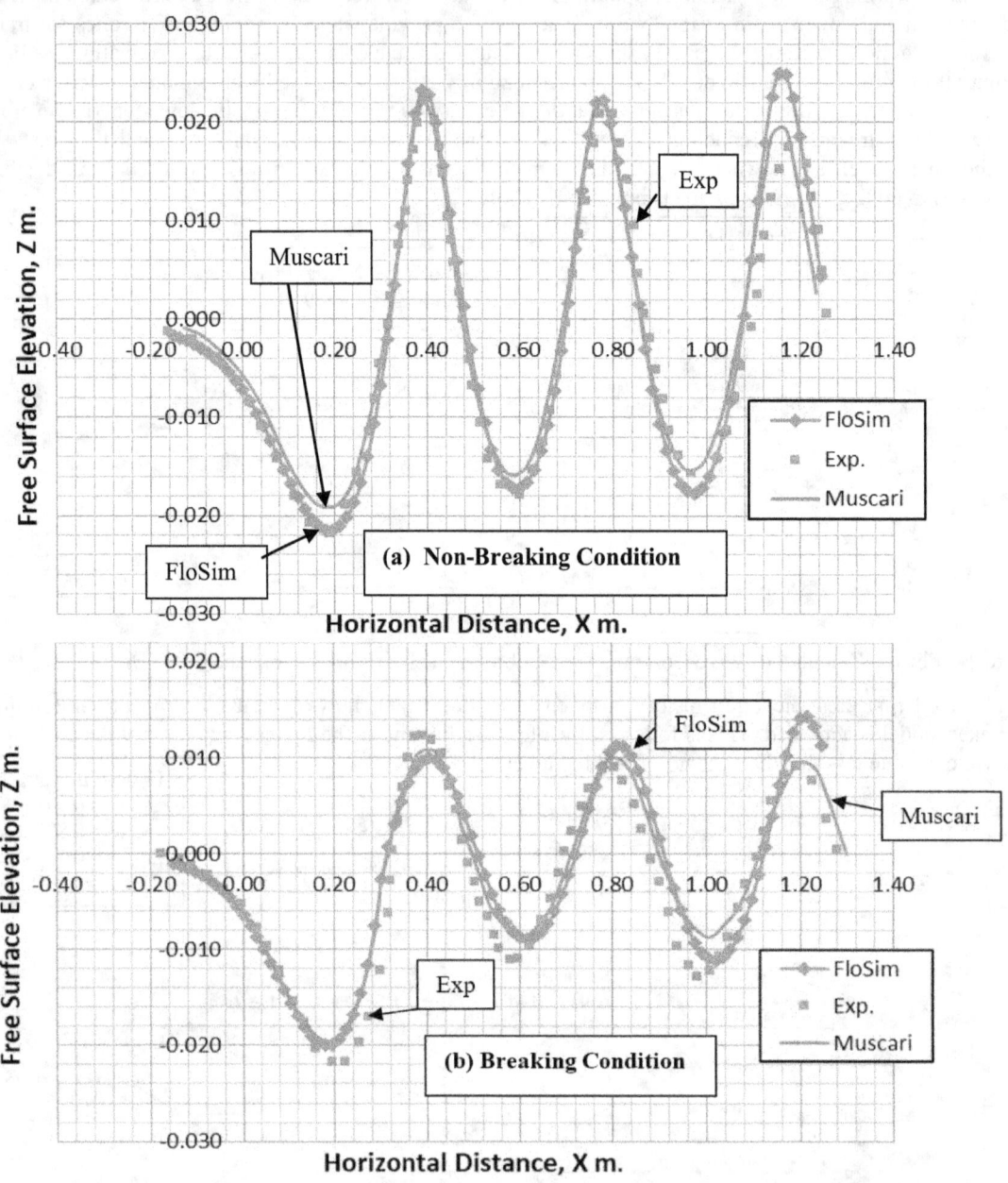

Figure 4 - Comparison of Calculated and Measured Wave Profiles for depth/chord = 0.9507

For the present application where we are treating a breaking *bow* wave the conditions are clearly more 3-dimensional, however, the basic principle in which we are following the path of a flow particle in the free surface is essentially the same, i.e., the particle experiences a disturbance that effects its energy level as it proceeds downstream, however, the breaker model parameters discussed earlier will probably need adjusting for the 3D situation; certainly, the H_f factor must be less than 1.0 since there is no upstream trough, and the critical wave slope may need adjusting due to the wave crest being swept relative to the flow direction. Initially, for the current 3D situation, we used $H_f = 0.8$, $D_f = 0.5$ and $\theta_{crit} = 12^{O}$, however, when the calculated results were compared with the SYRF data, high trim predictions at Fn 0.6 and above indicated that more energy needed to be extracted from the bow wave. Higher values for the D_f parameter were then explored using the upright canoe body test data at Fn 0.8, as described later in the DISCUSSION section.

WIDE LIGHT PROJECT BASICS

The model and tank testing procedure are given in Prince and Claughton, 2016; for completeness some details are repeated here.

Model

The model principal dimensions are shown in Table 1 and the hull lines and side view of the appendages are presented in Figure 5. Turbulence inducing studs were fitted on the hull at 0.3m from the bow and on the keel fin and rudder at 25% of the chord. On the bulb a carborundum grain strip was placed at 20% of the length.

The tow point location is at 2.484m aft of the hull bow and 0.172m above DWL. This is the reference point for heave measurements.

M1108		
Overall length	m	4.88
Design waterline length	m	4.60
Displacement (appended)	kg	215
Displacement (canoe body)	kg	197
Maximum beam	m	1.28
Draught to datum	m	1.15

Table 1 - Model Principal Dimensions

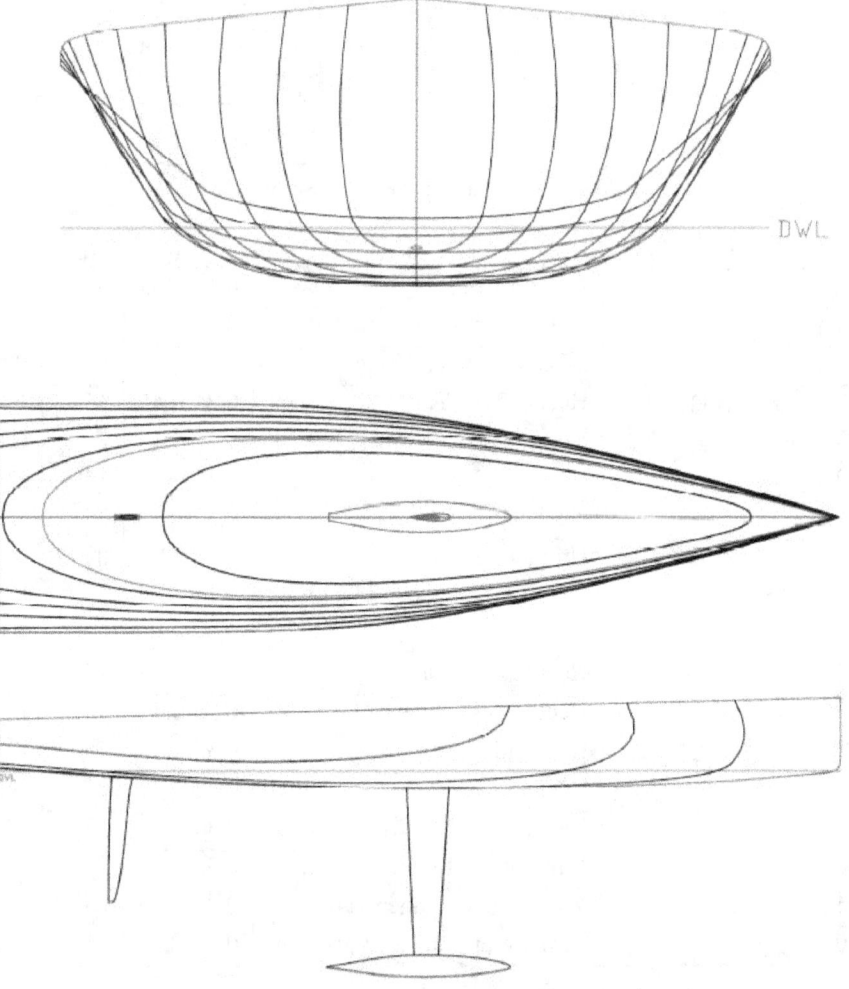

Figure 5 - Hull Lines

Test Matrix

The speed and LCG index settings are given in Table 2. Note that in the tank tests the x axis is positive forwards whereas in FloSim the x axis is positive aft, hence the LCG x values for FloSim are reversed in sign relative to those in Table 2.

The schedule of runs is given in Table 3. The FloSim runs follow the same numbering system but with the Froude number included, e.g., CB4-030-25, the last two digits indicate the Fn 0.25.

Speed Index	Fn	Vs m/s
V_001	0.1	0.681
V_002	0.15	1.022
V_003	0.2	1.363
V_004	0.25	1.703
V_005	0.3	2.044
V_006	0.35	2.385
V_007	0.4	2.725
V_008	0.45	3.066
V_009	0.5	3.407
V_010	0.55	3.747
V_011	0.6	4.088
V_012	0.65	4.429
V_013	0.7	4.769
V_014	0.75	5.110
V_015	0.8	5.451

LCG Index	LCG m
LCG_001	-2.488
LCG_002	-2.476
LCG_003	-2.469
LCG_004	-2.459
LCG_005	-2.447
LCG_006	-2.431
LCG_007	-2.403
LCG_008	-2.367
LCG_009	-2.330
LCG_010	-2.294
LCG_011	-2.262
LCG_012	-2.235
LCG_013	-2.206
LCG_014	-2.172
LCG_015	-2.127
LCG_016	-2.784
LCG_017	-2.634
LCG_018	-2.484
LCG_019	-2.334
LCG_020	-2.184

LCG Variation (LCG_016 to LCG_020)

Table 2 - Speed and LCG Index Settings

ID	Runs	Configuration	Test	Fn	Heel	Yaw	Rudder
CB-1	001-015	Canoe Body Only	Upright Resistance	0.1 to 0.8	0	0	0
CB-2	016-020		LCG Variation, 016 to 020	0.35	0	0	0
CB-3	021-025		LCG Variation, 016 to 020	0.5	0	0	0
CB-4	026-030		Heel at Zero Yaw	0.25-0.45	15	0	0
CB-5	031-035		Heel at Zero Yaw	0.25-0.45	25	0	0
CB-6	036-040		Heel with Yaw	0.35	15	-2 to 3	0
CB-7	041-045		Heel with Yaw	0.5	25	-2 to 3	0
HKR-1	046-060	Hull, Keel & Rudder	Upright Resistance	0.1 to 0.8	0	0	0
HKR-2	061-065		LCG Variation, 016 to 020	0.35	0	0	0
HKR-3	066-070		LCG Variation, 016 to 020	0.5	0	0	0
HKR-4	071-075		Heel at Zero Yaw	0.25-0.45	15	0	0
HKR-5	076-080		Heel at Zero Yaw	0.25-0.45	25	0	0
HKR-6	081-085		Heel with Yaw	0.35	15	-2 to 3	0
HKR-7	086-090		Heel with Yaw	0.5	25	-2 to 3	0
HKR-8	091-095		Yaw sweep, Rudder Variation	0.35	15	1, 3, 5	0, 2, 4
HKR-9	096-100		Yaw sweep, Rudder Variation	0.5	15	0, 2, 4	0, 2, 4
HKR-10	101-105		Yaw sweep, Rudder Variation	0.35	25	1, 3, 5	1, 3, 5
HKR-11	106-110		Yaw sweep, Rudder Variation	0.5	25	0, 2, 4	1, 3, 5

Table 3 - Run Schedule

FLOSIM COMPUTER MODEL PREPARATION

Panel Model

The surface geometry for the model hull, keelfin, bulb and rudder was provided in Rhino ~.3dm files from which *dense* sets of surface points were extracted in text files for direct input to FloSim. These points are independent of the panel grid; the user provides the desired panel *arrangement* and the program forms the panel grids automatically, given the longitudinal and lateral panel counts and *form* of distribution (i.e., equal, cosine or geometric progression, etc.). Similarly, the free surface patch panel grids are formed automatically to match the hull longitudinal grid. Trimming instructions between the hull and free surface patches are assembled within the program; as mentioned earlier, at each time step the free surface patch is automatically trimmed to the hull patches and all the affected patches are regridded to the changing waterline along the hull side.

Figure 6 shows a general view of the appended configuration with the basic paneling used for most of the runs here; the very low speed cases required higher longitudinal density along the hull in order to maintain a reasonable number of panel intervals (>20 recommended) per wavelength ($\lambda = 2\pi V/g$) at the lower Froude Numbers.

A general view of the hull in free surface is shown in Figure 7; again, for the very low speed cases (Fn < 0.25) a larger free surface patch was needed to deal with the transient waves after impulsive start; also, wave absorber "beaches" were needed along the free surface edges in these cases to dampen any transient waves.

The longitudinal grids on the free surface and hull are established based on the distance moved in one time step, DxMove. The majority of the runs here use a DxMove value of 0.1m per step, this maintains the number of panel-intervals per-wavelength greater than the recommended 20 down to a Fn of 0.3. We can still go down to Fn 0.25 with this grid but the number of intervals/wavelength is now about 15; this is below the recommended value for stable solutions, however, for the present configuration we found the cases converged smoothly and so the later cases at Fn 0.25 with appendages and with heel and yaw use the smaller free surface model rather than the larger one used for the upright runs at this speed. For lower speeds we use DxMove = 0.05 for Fn value of 0.2, and 0.03 for Fn values of 0.15 and 0.1.

The starboard half of the bulb surface is divided into an upper and a lower patch; the upper patch has 3 panels across and the lower one has 17; the distortion in the lateral grid due to the keelfin tip intersection is thereby confined to a smaller area, the lower patch panel grid is then undistorted by the intersection. The upper patch has a hole cut in it to accept the keelfin tip; actually, the hole is larger then the tip section so that a junction *filler* patch can be inserted to help blend the keelfin chordwise grid with the bulb longitudinal grid, see inset in Figure 6. A similar treatment is applied at the intersections of keelfin root and hull, and rudder root and hull.

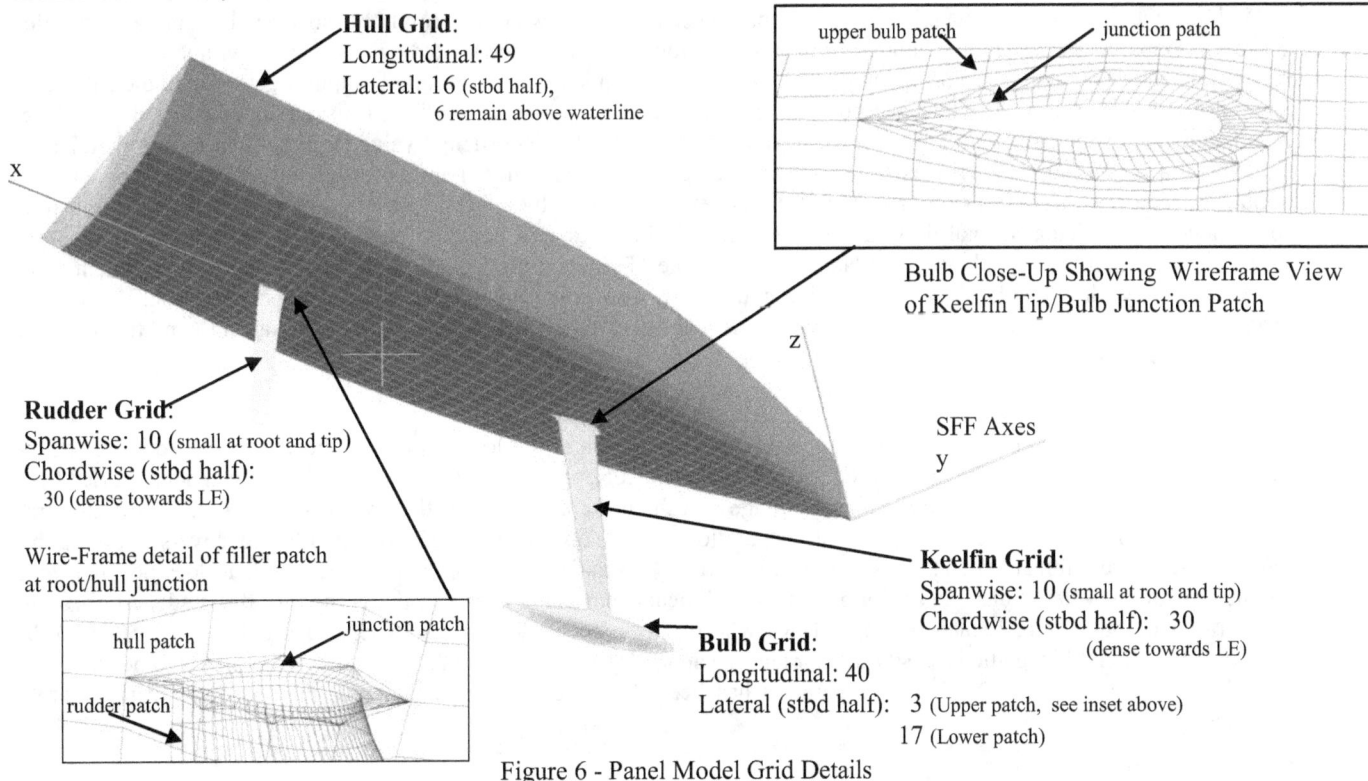

Hull Grid:
Longitudinal: 49
Lateral: 16 (stbd half),
   6 remain above waterline

upper bulb patch   junction patch

Bulb Close-Up Showing Wireframe View
of Keelfin Tip/Bulb Junction Patch

x

z

SFF Axes
y

Rudder Grid:
Spanwise: 10 (small at root and tip)
Chordwise (stbd half):
   30 (dense towards LE)

Wire-Frame detail of filler patch
at root/hull junction

hull patch   junction patch

rudder patch

Keelfin Grid:
Spanwise: 10 (small at root and tip)
Chordwise (stbd half): 30
   (dense towards LE)

Bulb Grid:
Longitudinal: 40
Lateral (stbd half): 3 (Upper patch, see inset above)
   17 (Lower patch)

Figure 6 - Panel Model Grid Details

Free surface has a port and a
starboard patch to allow a hole
to be cut for the hull intersection:

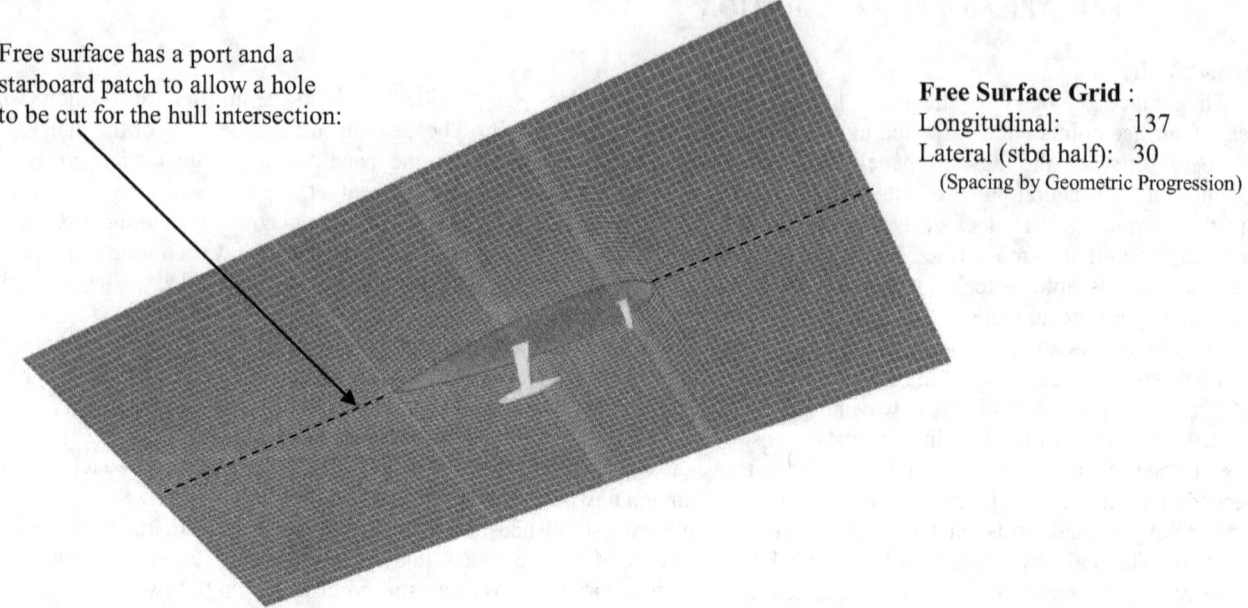

Free Surface Grid :
Longitudinal: 137
Lateral (stbd half): 30
(Spacing by Geometric Progression)

Figure 7 – Free Surface Panel Model

Special Treatments in Regard to the Measured Data
Laminar Flow Drag Correction

It is reported in Wolfson Unit MTIA, 2015, that the measured resistance data were corrected to allow for the resistance of the transition inducing studs *and for the region of laminar flow ahead of them*. So far we have not obtained the magnitudes of these corrections; certainly, initial FloSim calculations for the upright canoe body gave drag numbers *below* the test values and this was inconsistent with previous experience. The low prediction *is* consistent with the fact that FloSim's number has the laminar flow skin friction ahead of transition and should therefore be compared against the measured value with just the stud drags removed and not the value adjusted for completely turbulent flow.

With the actual adjustment magnitudes unknown, we needed to find what they might be, so special coding was installed in FloSim to extract the *calculated* laminar flow drag contribution from all the wetted panels flagged with laminar condition in the boundary layer analysis. This quantity, say, D_{LF}, is now displayed in the program output together with the calculated force and moment information for each component (hull, keelfin, rudder and bulb). From details of the boundary layer calculations along each streamline the change in calculated skin friction coefficient value from laminar to turbulent across the transition zone is automatically extracted and an average ratio of Turbulent/Laminar evaluated for each component as a turbulence factor, F_T, say. Typically, the skin friction coefficient jumps by a factor of between 3 and 6 across the transition zone. Hence, the calculated total drag number, that already includes D_{LF}, needs to have $(1-F_T)*D_{LF}$ for each component added to it to obtain the overall total calculated value with the "Turbulent Flow Correction" applied for fully turbulent flow.

Alternatively, the boundary layer calculation could have been forced to start as turbulent, but that would require an *artificial* shape factor, H, and momentum thickness Reynolds Number, R_θ, to be specified as starting conditions for the turbulent boundary layer analysis.

Laminar Flow Separation

Initial FloSim calculations on the yawed appended configuration at the lower speeds indicated extensive laminar flow separation regions on the keelfin and rudder. Actually, the *chord* Reynolds number, Rnc, for these appendages varies from about 33,000 to 263,000 over the current speed range and therefore enters the critical range where complex lift and drag characteristics are likely, due to long bubble separations, see Yongshong, 2006. Whilst FloSim covers short bubble separations, where laminar separation is followed quickly by transition and turbulent reattachment, it does not treat *long* bubble separations where transition and turbulent reattachment occur much later. For the purpose of this study, *catastrophic* laminar separations are restrained on appendages whose chord Reynolds number, Rnc, drops below, say, 500,000, by applying an artificial kinematic viscosity to maintain the appendage's Rnc at 500,000. This treatment needs to be reviewed; a preliminary treatment of long bubble separation has been developed in FloSim but this needs further work before it becomes operational.

RESULTS COMPARISONS

Upright Cases

The trim and heave curves comparing the tank test upright data and the FloSim and Star-CCM+ calculations are shown in Figures 8 and 9, respectively; in each figure the canoe body results are shown in part (a) and the appended results in part (b). In initial calculations, the FloSim trim values compared well with the measured data for low Fn but above about Fn= 0.55 the FloSim trim trend appeared to diverge upwards. This suggested that the trough following the bow wave was too deep and therefore still carrying too much energy; clearly, the Wave-Breaker model discussed earlier needed to extract more energy from the breaking bow wave than in the basic setup with $D_f=0.5$. A study was conducted based on the measured data at Fn 0.8 and from this a new model developed for the Wave-Breaker Foam Factor, D_f; this was used for the Fn cases 0.6 and above in the following Figures. Details of the study are included in the <u>DISCUSSION</u> section later.

In general, for both canoe body and appended cases, the calculated trim values fall below the measured results, FloSim (up to 0.3^O below) slightly more so than Star-CCM+ (at 0.2^O below), at least up to about Fn 0.55; above that speed FloSim values lie closer to the measured data than those from Star-CCM+.

For heave, Figure 9, the calculations follow the measurements closely for speeds up to about Fn = 0.5, FloSim being slightly closer than Star-CCM+. Above Fn=0.5 both calculations depart from the measured curve, Star-CCM+ above by about 0.3cm and FloSim below by about 0.2cm. At least for the canoe body, FloSim results follow the trend of the measurement very closely.

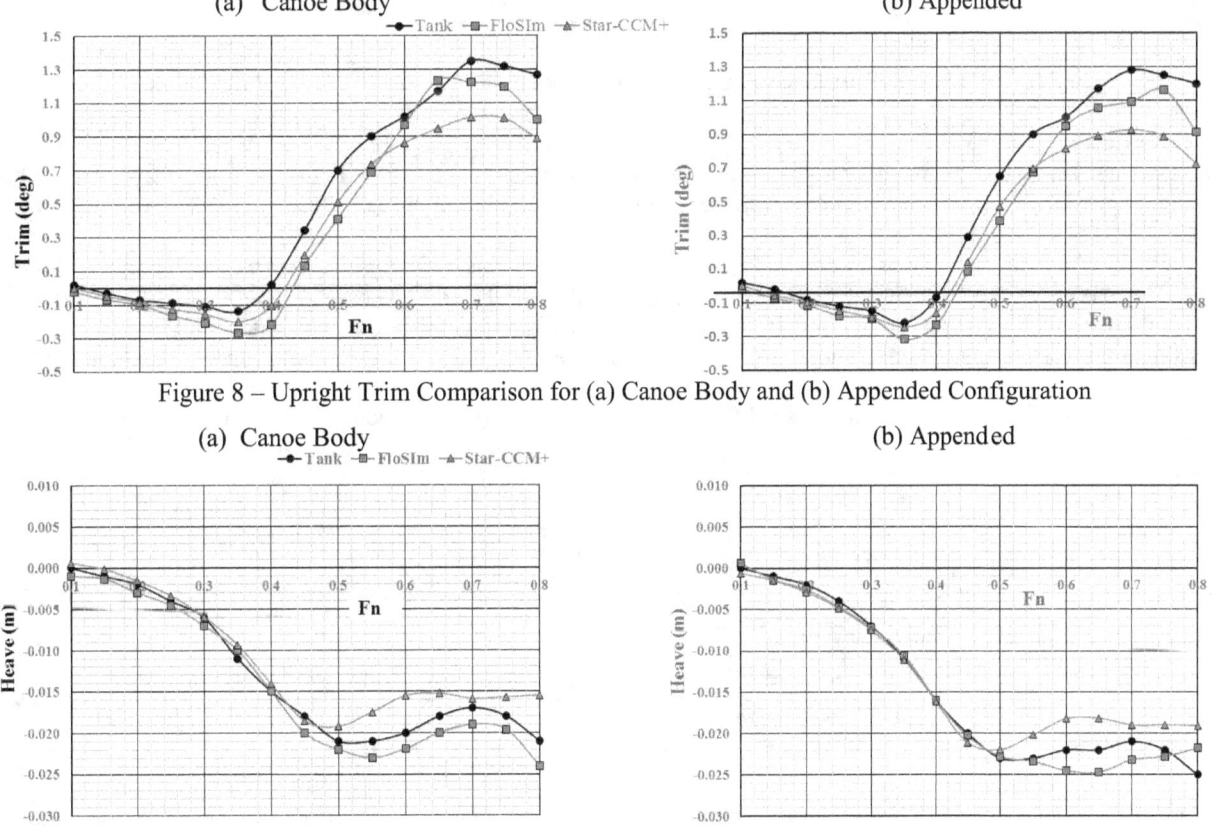

Figure 8 – Upright Trim Comparison for (a) Canoe Body and (b) Appended Configuration

Figure 9 – Upright Heave Comparison for (a) Canoe Body and (b) Appended Configuration

The upright resistance curves from the tank results and the computed results are compared in Figure 10 for both the canoe body and appended configurations. CB1 defines the canoe body results and HKR1 defines the results for the appended configuration comprising hull, keelfin, bulb and rudder. The calculated values for the canoe body are in close agreement with the measurements up to Fn 0.55 then they fall below, FloSim slightly more so than Star-CCM+; at Fn 0.8 the error is 10% for Star-CCM + and 12% for FloSim. FloSim's departure around Fn=0.6 is largely due to a 17% reduction in the computed wetted area, the transom edge being now slightly above water, hence the waterline has moved forward relative to the lower speed conditions where the transom was immersed. For the appended case both calculations follow the measured line very closely, FloSim being slightly closer than Star-CCM+; at Fn=0.8 the errors are 2% and 5% respectively. The appended case has a similar reduction in hull wetted area for speeds above Fn= 0.55, however, the turbulence correction term with appendages is now around 5 to 6% whereas for the hull alone the correction was typically around 2%.

Upright Resistance

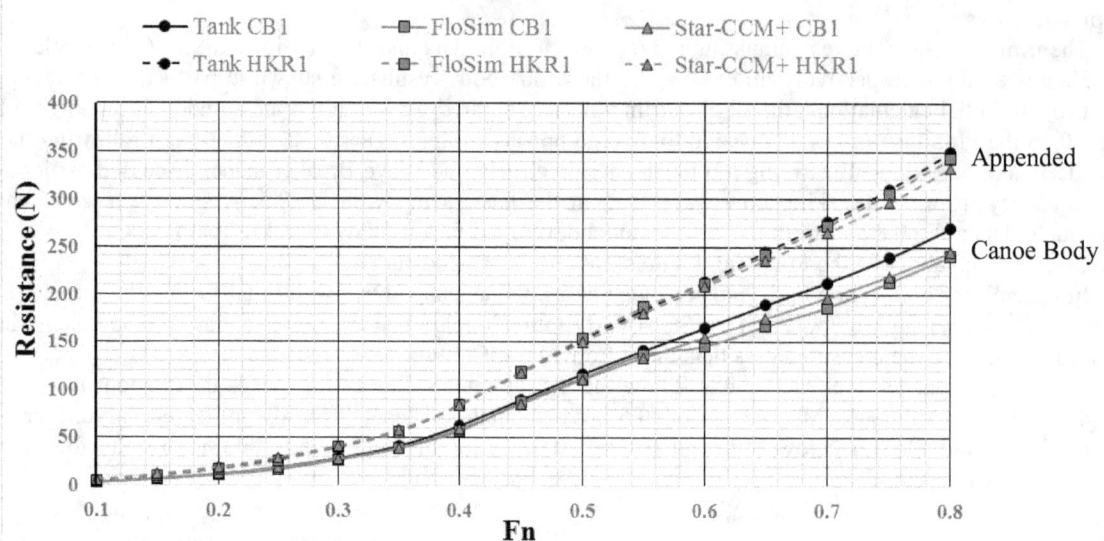

Figure 10 – Upright Resistance Comparison for Canoe Body and Appended Configurations

LCG Variation

The trim and heave curves for the upright cases with LCG variation are shown in Figures 11 and 12; the canoe body results are shown in part (a) of each figure and the appended configuration in part (b). Results are shown for two speeds, Fn = 0.35 and 0.5.

At Fn=0.35, the trim calculations in Figure 11 follow the measurements very closely, Star-CCM+ being the closest in general. The maximum departure for the FloSim curve is about 0.25^O for the aft LCG case.

At Fn 0.5 the calculated trim values slowly depart from the measurement as LCG moves aft, FloSim results slightly more so than those from Star-CCM+. The Star-CCM+ departure reaches about 0.25^O by the aft LCG position and FloSim about 0.6^O.

(a) Canoe Body Upright Trim Versus LCG   (b) Appended Upright Trim Versus LCG

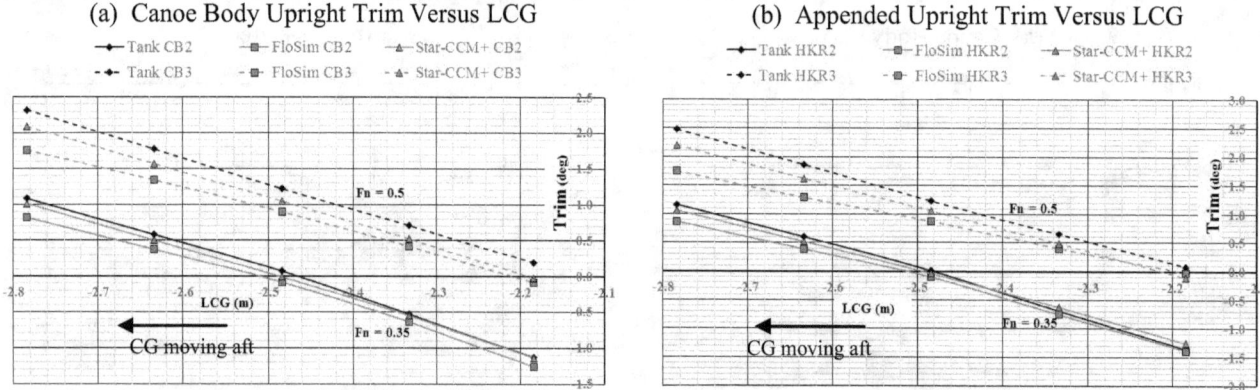

Figure 11 – Comparison of Trim Variation with LCG Position for (a) Canoe Body and (b) Appended Configuration

For the canoe body heave comparisons in Figure 12a, FloSim's value is within 0.2 cm (0.04% of hull length) above the measured value for the most forward LCG position, but crosses over and diverges gradually as LCG moves aft reaching about 0.5 cm below the measurement at the most aft LCG for the Fn= 0.35 case; Star-CCM+ is about 0.2cm above the measured curve throughout. For Fn=0.5, FloSim starts about 0.2cm below the measured value at the forward LCG position but gradually diverges to about 1cm below at the aft position. Star-CCM+ has a similar trend but with a greater departure of 1.4 cm at the end.

The FloSim heave values for the appended configuration in Figure 12b show a similar trend as before but with smaller departure, whereas Star-CCM+ remains close to the measured values throughout.

The comparisons of resistance variation with LCG position are given in Figure 13 for both the canoe body and appended configurations and for two speeds. At the lower speed the resistance curves are very close together but at the higher speed the curves tend to depart as the LCG moves aft, Star-CCM+ difference is about 10% and FloSim almost double that.

For the aft LCG case we have positive trim and the turbulence inducing studs located at 0.3m from the bow are completely above water and therefore ineffective. For FloSim's first run on this case we got a very low resistance value; since the "trip strip" was not encountered, the surface flow proceeded with laminar condition in a favorable pressure gradient ultimately going through natural transition at about 70% of the hull length. The case was rerun with the transition flag set to "as soon as possible" (this being generally used when the onset flow has a high level of turbulence); transition to turbulent flow conditions then started at about x= 0.7m; this is 0.3m aft of the current forward wetted point and well aft of the transition strip location.

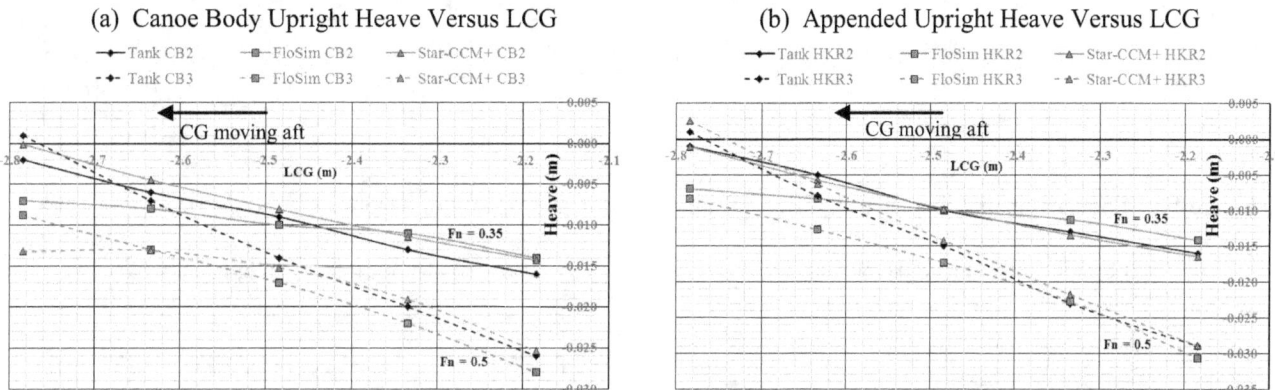

Figure 12 – Comparison of Heave Variation versus LCG Position for (a) Canoe Body and (b) Appended Configuration

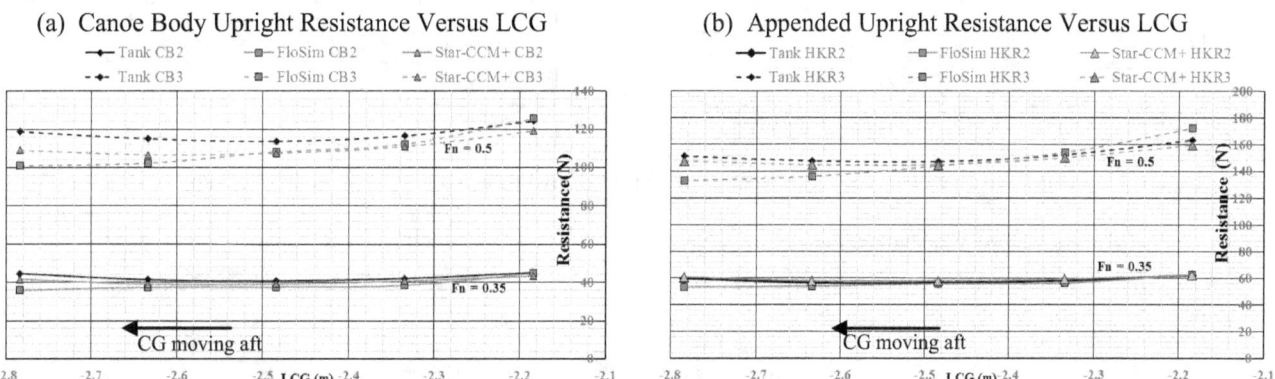

Figure 13 – Comparison of Upright Resistance versus LCG Position for Canoe Body and Appended Configuration

Cases with Heel at Zero Yaw

The trim versus speed curves for the canoe body heeled at 15° and 25° at zero yaw, are shown in Figures 14a and 14b, respectively. At heel= 15° the FloSim trim curve is about 0.1^O below the measurement at low speed but the discrepancy grows slowly to about 0.3^O by Fn 0.45, however, the calculation follows the same trend as the measurements.

At heel= 25° the FloSim trim curve essentially follows the experimental line but at about 0.2^O more bow down; it is similar to the Star-CCM+ line but this has less than 0.1^O discrepancy.

Figure 15 shows the heave versus speed results for the canoe body, again for heel values of 15° and 25° and at zero yaw. Here, both calculated lines run essentially parallel to the experimental line with the Star-CCM+ line above the experiment and FloSim below; the offsets being +0.5 cm for Star-CCM+ and -0.9 cm for FloSim at heel= 15° and 1 cm and -2.4 cm, respectively, at 25° heel.

The variation of resistance versus speed for the heeled canoe body at zero yaw condition is shown in Figure 16; part (a) is for heel= 15° and part (b) for heel= 25°. Here, for both heel values, the two calculations follow the trend of the experimental line very closely, however, in absolute terms there is a discrepancy in resistance value reaching about 8% for Star-CCM+ and about 10% for FloSim.

The corresponding plots for the appended configuration are given in Figures 17, 18 and 19. The trends and error values are similar to those for the canoe body except the absolute discrepancy in the resistance is perhaps smaller, about 7%. In Figure 17 at heel=15° around Fn 0.35, the FloSim trim result has 0.14^O error producing a slightly more pronounced bulge than the one on the measured curve; the error grows to 0.2^O by Fn 0.45 compared with the Star-CCM+ error of 0.1^O.

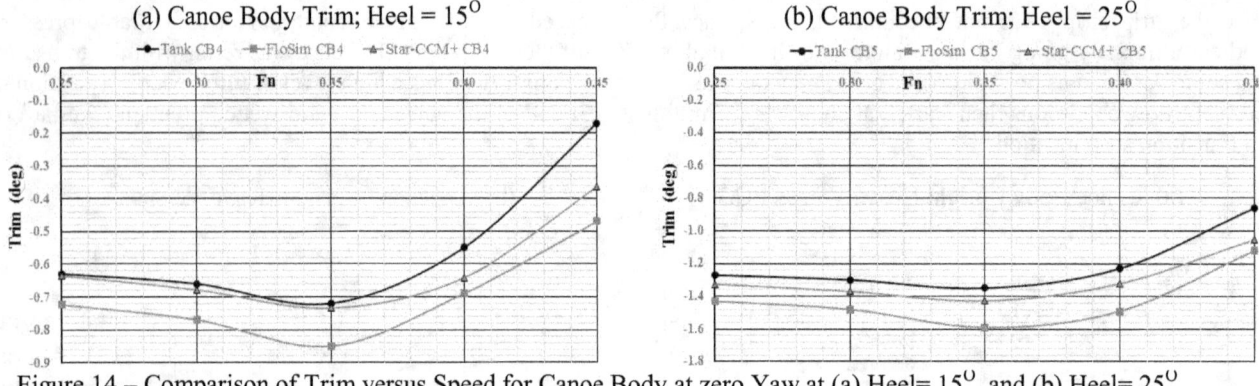

Figure 14 – Comparison of Trim versus Speed for Canoe Body at zero Yaw at (a) Heel= 15° and (b) Heel= 25°

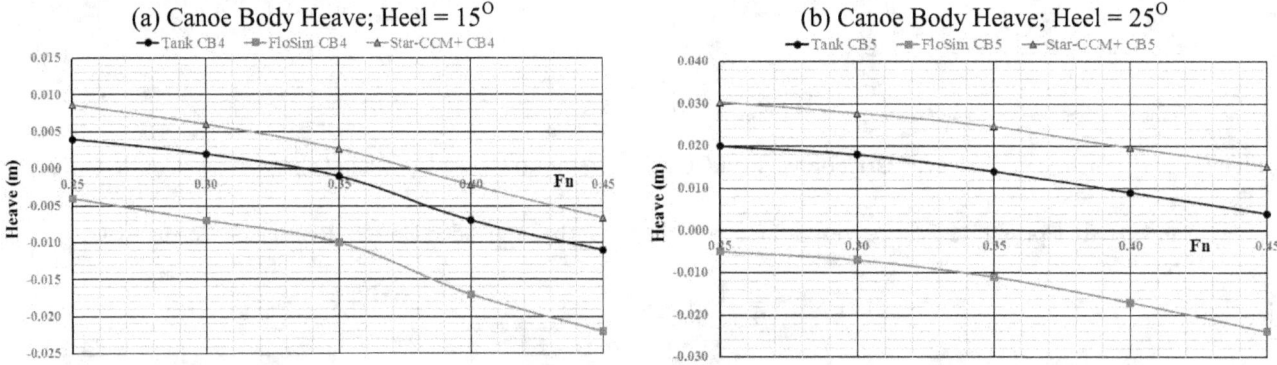

Figure 15 – Comparison of Heave versus Speed for Canoe Body at Zero Yaw at (a) Heel= 15° and (b) Heel= 25°

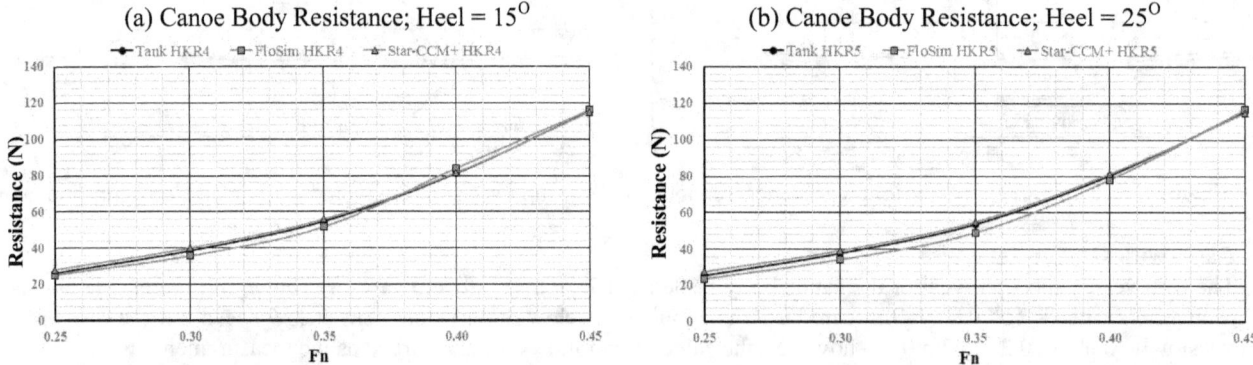

Figure 16 – Comparison of Resistance versus Speed for Canoe Body at Zero Yaw at (a) Heel= 15° and (b) Heel= 25°

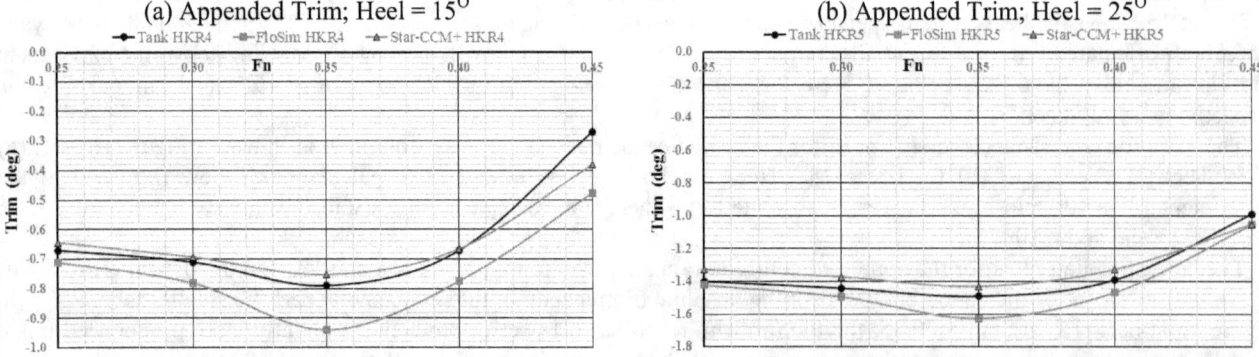

Figure 17 – Comparison of Appended Trim versus Speed at Zero Yaw at (a) Heel= 15° and (b) Heel= 25°

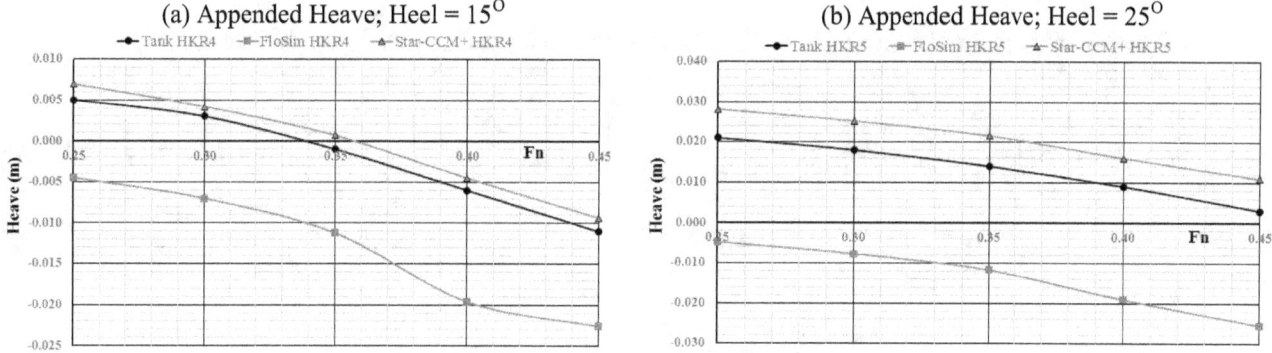

Figure 18 – Comparison of Appended Heave versus Speed at Zero Yaw at (a) Heel= 15° and (b) Heel= 25°

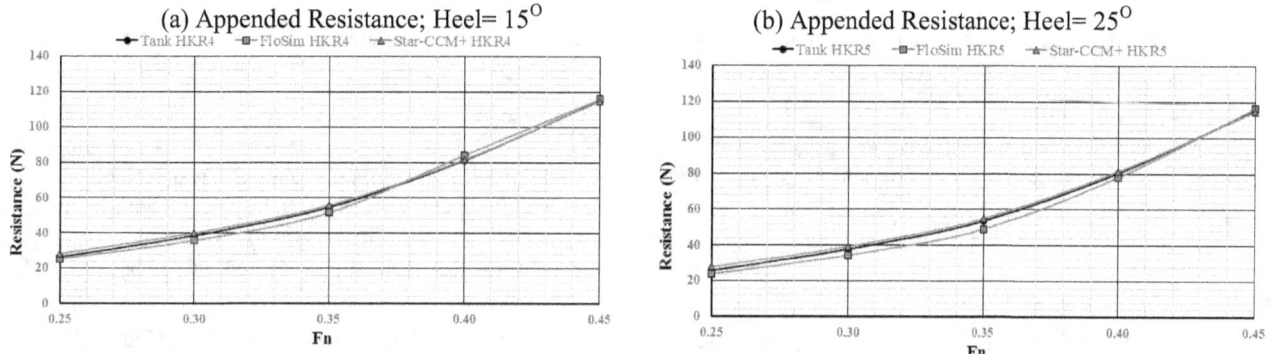

*Figure 19 – Com*parison of Appended Resistance versus Speed at Zero Yaw at (a) Heel= 15° and (b) Heel= 25°

Cases with Heel and Yaw

Figure 20 compares the resistance variation with yaw for (a) the canoe body and (b) the appended configuration. Results are shown for the configurations with heel= 15° at Fn 0.35 and for heel 25° at Fn 0.5. For the low speed case at heel= 15° the resistance varies very little for both the canoe body and the appended configuration. At the higher speed at heel= 25° the FloSim canoe body values have an error of about 12%; this is about double the Star-CCM+ error and moreover, the line has the opposite trend from that of the measurement; this is inconsistent with the heel 25° appended case result in Figure 20b, where, except for the value at -2.0°, the FloSim and Star-CCM+ lines follow the measured data closely with FloSim 3% above and Star-CCM+ 4% below the measured line.

The side force variation with yaw is shown in Figure 21; part (a) is for the canoe body and part (b) for the appended configuration. Again, results are shown with heel= 15° at Fn= 0.35 and for heel= 25° at Fn= 0.5. For the canoe body alone at the lower speed, the side force is quite low and is well predicted by Star-CCM+; the FloSim line is very close to the measurement at the start and end of the line but has an error of about 7N at yaw = 2°. For the canoe body at heel =25°, the FloSim result shows a wavering trend with a maximum error of about 15N. In absolute terms the wavering is small and is not apparent in the appended case in Figure 21b, where, at heel = 25°, the FloSim and Star-CCM+ lines run almost parallel but at a slightly different slope to the measured line; the FloSim line essentially coincides with the measured line around yaw = 3° and the Star-CCM+ line at yaw = -2°.

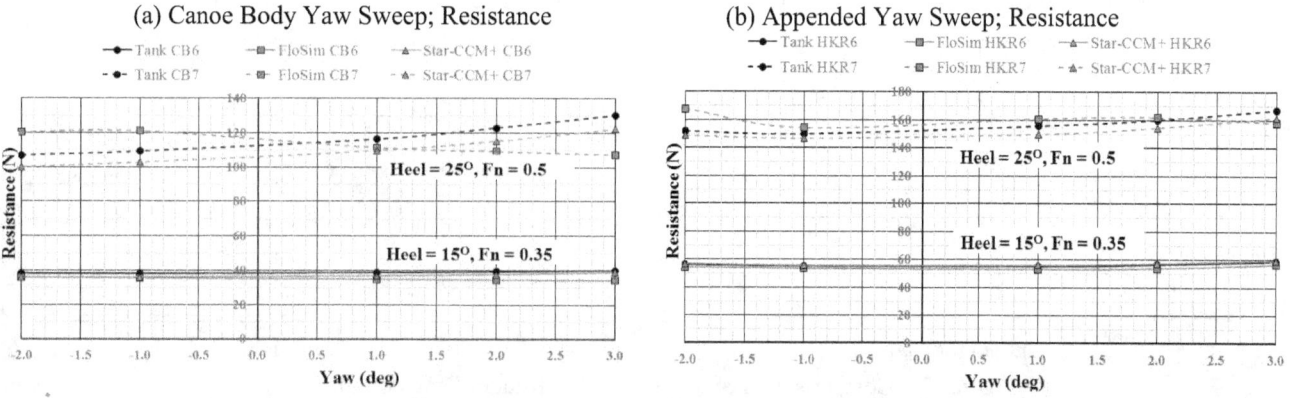

Figure 20 – Comparison of Resistance versus Yaw for (a) Canoe Body and (b) Appended Configuration in Heel

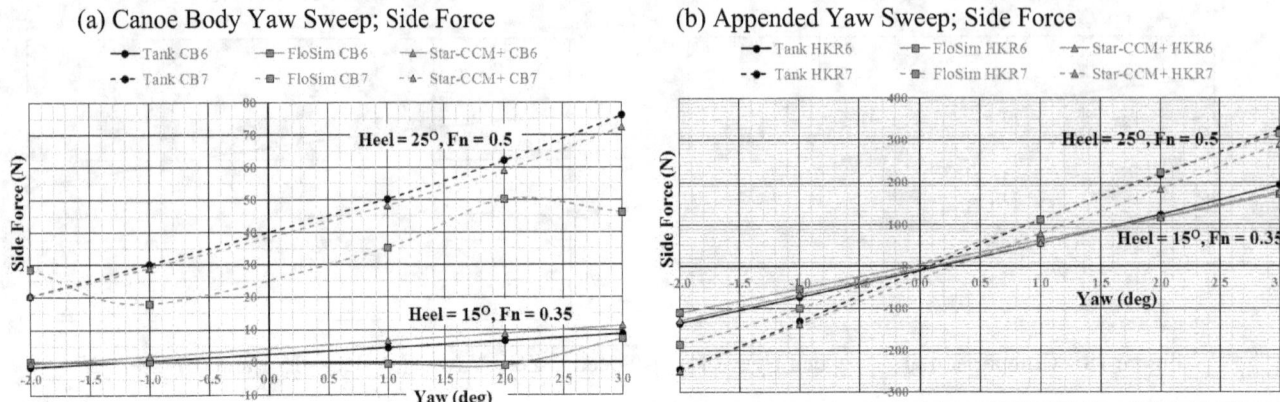

(a) Canoe Body Yaw Sweep; Side Force   (b) Appended Yaw Sweep; Side Force

Figure 21 – Comparison of Side Force versus Yaw for (a) Canoe Body and (b) Appended Configuration in Heel

Rudder Deflection with Yaw

The results with rudder deflection are presented in Figures 22 and 23. These are for the appended configuration at heel= 15° in part (a) and 25° in part (b), and for speeds of Fn = 0.35 and 0.5 with yaw angles 3° and 2^O, respectively.

Figure 22 shows the variation of resistance with rudder angle; for the low speed case at Fn 0.35 and for both heel= 15° and 25°, the values for tank, FloSim and Star-CCM+ are essentially on one line with little variation, the values varying at the most by about 2%. At the higher speed the calculated resistance values differ from the measured values by about 4%, the Star-CCM+ being below the measurement and FloSim above.

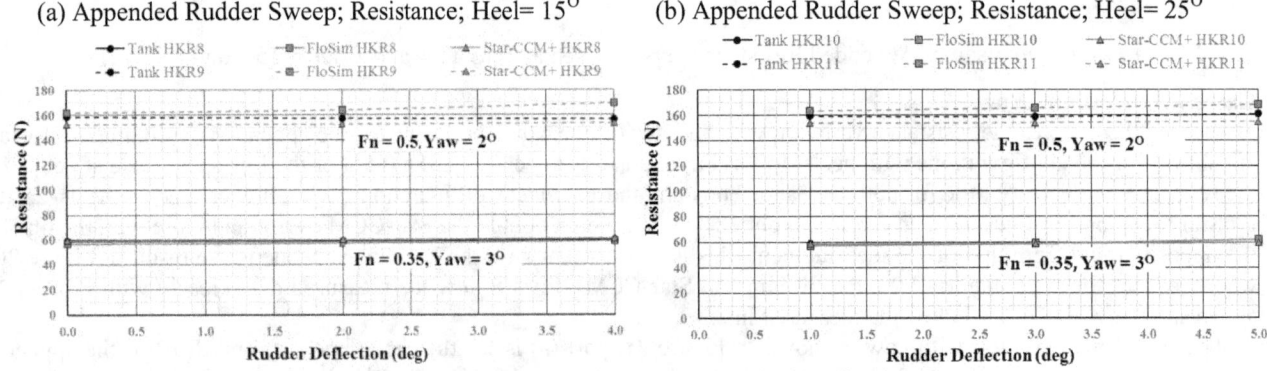

(a) Appended Rudder Sweep; Resistance; Heel= 15° (b) Appended Rudder Sweep; Resistance; Heel= 25°

Figure 22 – Comparison of Resistance versus Rudder Deflection for
Appended Configuration at (a) Heel = 15° and (b) Heel = 25°

The variation of side force with rudder angle is shown in Figure 23; part (a) shows the results for low speed conditions at Fn= 0.35 for heel= 15° and 25° and yaw 3°; both calculations show the same trend against rudder deflection; this matches the measured trend at low speed, Fn 0.35, albeit with a 17% error, Star-CCM+ being below the measurement and FloSim above for the 25° heel case and both calculations being below measurement at heel = 15°.

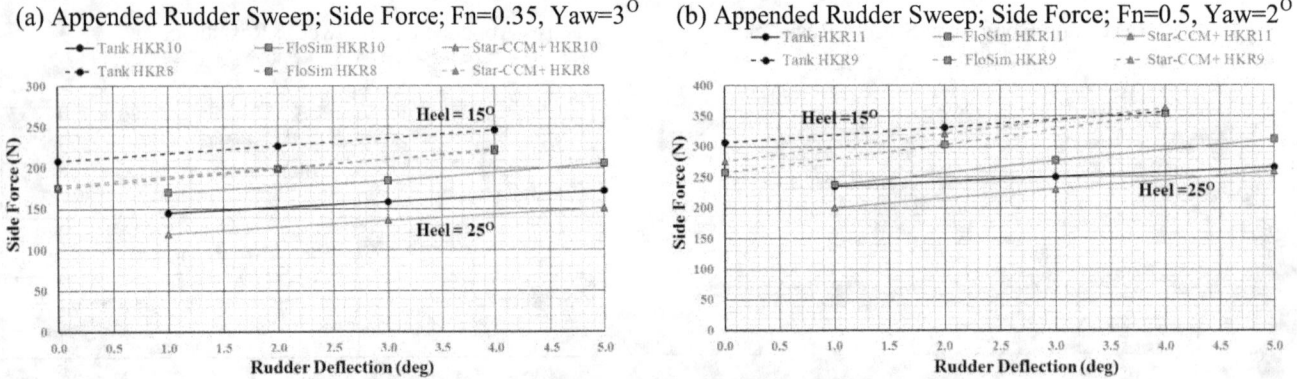

(a) Appended Rudder Sweep; Side Force; Fn=0.35, Yaw=3° (b) Appended Rudder Sweep; Side Force; Fn=0.5, Yaw=2^O

Figure 23 – Comparison of Side Force versus Rudder Deflection for
Appended Configuration in Heel at (a) Fn = 0.35 and (b) Fn = 0.5

At the higher speed condition in Figure 23b, the side force trend from the calculations is at a slightly higher slope than that of the measurements, FloSim's value matching the data at 1^O rudder deflection and Star-CCM+ at deflection of 5°. The maximum error is again about 17%. Potentially, the error is due to low Reynolds Number issues as noted earlier under MODEL PREPARATION; in the speeds covered in this rudder sweep, the rudder chord-Reynolds number, Rnc, is in the range 100,000 to 150,000 and so will be prone to laminar separations and possible long bubble events.

CASE COMPUTER REQUIREMENTS

Table 4 compares the FloSim and Star-CCM+ execution times for typical Wide Light cases.

Flosim execution times are on a HP Spectre-360 laptop computer with Intel dual core i7-7500 CPU at 2.7 GHz with 16 GB RAM, the latter being sufficient for cases up to at least 18,000 panels. With this RAM and dual core CPU, it is possible to run 2 cases *simultaneously* with no slow down.

Star-CCM+ times are on a multi processor computer; the single case values were obtained from the returned CFD Participant Questionnaire in the Wide Light final report documentation.

Configuration		FloSim	Star-CCM+	
	Condition	Time/case (hours)	Time/case (hours)	Number of Processors
Canoe Body	Symmetrical	0.2	4.67	8
	Asymmetrical	1.0	9.25	8
Appended	Symmetrical	0.5	5.75	16
	Asymmetrical	2.0	11.5	16

Table 4 – Comparison of Execution Times

DISCUSSION

The results presented here show the FloSim prediction of upright trim, heave and resistance compare well with measurement. FloSim's Wave-Breaker treatment described herein is clearly effective in attenuating downstream wave amplitude. As described earlier in the Method Outline, the wave-breaker treatment has three empirical parameters that were set based on measurements by Duncan, 1981. In particular, the *foam density* parameter, D_f, that was held constant initially, was later re-evaluated based on measured results at Fn = 0.8. An investigation was carried out, as described below, after initial results with the constant D_f value indicated that more energy needed to be extracted from the bow wave at the higher speeds; certainly, the investigation by Cointe and Tulin, 1994, suggested that D_f is dependent on the *severity* of the breaking crest and this will certainly increase with speed. In addition, for the present calculations we have the issue of 3D conditions with the sideways breaking bow wave and so modification of the wave-breaker parameters is to be expected.

Concentrating on the upright canoe body case at Fn 0.8, the Wave-Breaker Foam Factor, D_f, that was originally set at 0.5, was increased to apply more pressure on the approach to the crest. A D_f value was found that produced the measured trim value however, the corresponding heave value did not match the measurement. Another Df factor was then found that gave a match on heave but now trim was off. In all this, the resistance value did not vary significantly. Ultimately, a Df value of 0.9 was selected that gave a compromise trim and heave relative to the measured data. This factor is higher than the 0.6 upper limit suggested by Cointe and Tulin (1994), however, the situation here is far from the 2D case they considered; for one thing, with the swept bow wave the crest material falls more to the side than directly forwards.

The fact that the trim and heave values could not be matched with the same D_f setting implies the locations of the calculated trough and peak downstream from the bow wave crest are incorrect, i.e., a phasing problem. This is probably related to how the pressure is applied; currently we apply the pressure directly according to the distance h(x) of a point below the wave crest as shown in Fig. 1, and the total foam "height", H= H_f*Zcrest, is set with H_f = 0.8. These parameters would change the form of the pressure application but in the absence of measured wave profiles for comparison this would be an extensive trial and error exercise to reach a conclusion with just heave and trim data for comparison.

For the present results we have used the compromise 0.9 D_f value at Fn 0.8 and connected that with the earlier constant line at D_f= 0.5 up to Fn 0.5 using a parabolic curve to match slope and two values. For intermediate Fn values in the test matrix, the D_f value is obtained from this curve by interpolation on the basis of the case Fn value. Later, when further data cases can be found it is hoped the basis can be changed from Fn to that of *particle energy* value in the flow approaching a crest; physically, the incoming energy level must be a key factor in the *severity* of the breaking wave as it depends on the level of disturbance produced by the bow entry, i.e., this is a function of bow shape (bluffness) and state of trim, yaw and heave and also on wave encounters. Such test data must include measurements of wave profiles along the body side and along cuts away from the body, ideally, including velocity measurements.

For the present results, the FloSim predictions of trim, heave and resistance with LCG variation, are generally on par with the Star-CCM+ results but tend to diverge from the measured values as LCG moves aft. Certainly, for the most aft LCG case, the turbulence inducing studs located at 0.3m from the bow are completely above water and therefore ineffective. Since there is a long favorable pressure gradient in this bow-up condition, transition to turbulent flow had to be forced for the FloSim calculation.

It had been hoped that the actual magnitudes of the "Turbulence Correction" values that were applied to the measured results could be obtained, and then applied for the final manuscript, however, these were not obtained and so special coding was installed in FloSim to extract a theoretical value based on calculated boundary layer properties; details were described earlier in the Model Preparation section. The procedure evaluates a contribution for the hull and each appendage separately. Typically, the increments amount to about 2% of total drag for canoe body cases and up to 6% for appended configurations.

On the whole, FloSim's trim, and resistance predictions for the heeled cases compare well with measurement and with Star-CCM+, but the heave values are somewhat lower than the data implying the calculated trough is still too deep.

In general, FloSim's resistance and side force predictions for the heeled configurations in a yaw sweep compare favorably with the measured data and with Star-CCM+.

The calculated resistance predictions from FloSim and from Star-CCM+ versus rudder deflection on the heeled and yawed appended configuration follow the experimental lines closely; the maximum deviation being about 4% at the higher speed of Fn 0.5, Star-CCM+ being below experiment and FloSim above. For the side force predictions however, though both programs give essentially the same *trend* as in the experiment, the absolute side force error for both programs is about 17%, Star-CCM+ being below experiment and FloSim above. This side force discrepancy is probably a low Reynolds number issue: in the speeds covered in this rudder sweep, the rudder chord-Reynolds number, Rnc, is in the range 100,000 to 150,000 and so will be prone to laminar separations and possible long bubble events.

As discussed earlier under Laminar Flow Separation in the MODEL PREPARATION section, whilst FloSim deals with short bubble events, (i.e., laminar separation followed quickly by transition and turbulent reattachment), it does not have an *operational* treatment for long bubble separations where transition and turbulent reattachment occur later, and so catastrophic laminar separations from near the leading edge (i.e., upstream of the transition studs at 0.25c) would have no chance at reattachment. As a temporary "fix" in FloSim for the present calculations, when a very low Rnc is encountered on an appendage, an artificial viscosity is applied to maintain a minimum Rnc of 500,000 for that appendage. This could be the cause of FloSim's over-prediction of rudder side force here. Though FloSim has a long bubble treatment under development this still needs work to be done to get it to an operational standard.

FloSim's cases were run on a laptop computer, often two cases simultaneously; the execution times were short compared to the Star-CCM+ times on a multiple processor computer, for example FloSim's time for the appended configuration in asymmetrical conditions took about 2 hours on the laptop compared with 11.5 hours for Star-CCM+ on a 16 processor machine.

CONCLUSIONS

The principal objectives of this work were to validate the advanced BEM code, FloSim, against measured data and to compare the accuracy of its predictions against those from more advanced CFD methods based on RANS, represented here by the Star-CCM+ code, the original development aim being to "lean towards the 'real world' accuracy of RANS". Essentially, the results presented here show these objectives have been achieved, at least for a relatively straight-forward yacht configuration.

The FloSim development also aimed to maintain the low level computer requirement of the panel method and to avoid the high computer resource needs of the RANS codes. This aim also has been achieved, the FloSim runs here taking from 0.2 hours for a symmetric canoe body to 2 hours for the appended configuration in asymmetric conditions on a laptop computer versus 4.7 hours and 11.5 hours, respectively, for Star-CCM+ RANS code running on a 16 processor computer.

Potentially, the relatively accurate and fast turnaround for case runs will be useful for handicappers trying to make equitable handicap rules for racing boats, and in the design environment for new boats it will allow many design options to be explored in a relatively short time before committing a new design to final RANS evaluation and/or tank testing for performance confirmation.

FURTHER WORK

The success of FloSim's predictions at the higher Froude numbers here is largely due to the use of the Wave-Breaker model. The empirical foam factor parameter, D_f, in this model was reset in this project based on a *single point* measurement at Fn 0.8; clearly, therefore, this must be regarded as an initial step in the model development; more data cases are needed to build on this. Although the present results demonstrate the Wave Breaker model is effective in attenuating the downstream waves following a breaking bow wave, there are indications that the calculated wave phasing may be in error, however, in the absence of measured wave profiles this could not be confirmed in the present project. Further work is needed to treat other configurations with hopefully more measured data that includes wave profiles and possibly velocity measurements for comparison. FloSim is currently monitoring particle energy (potential and kinetic); ultimately, it is envisaged that the initial model in which the D_f parameter is described as a function of Fn, can be changed to a basis of *particle energy* in the flow approaching a bow wave crest. Such a model would then be sensitive to the conditions of bow entry and its dependence on bow bluffness, trim, heave and yaw conditions in a maneuver, and also wave encounters.

ACKNOWLEDGEMENTS

The authors sincerely appreciate the efforts of the Sailing Yacht Research Foundation to encourage open sharing of the valuable data published from the Wide Light Project. We sincerely thank the Foundation sponsors, leadership and Technical Committee.

REFERENCES

Keuning, J.A., Sonnenberg, U.B., "Developments in the Velocity Prediction based on the Delft Systematic Yacht Hull Series" The Modern Yacht Conference March 1998

Teeters, J., Pallard, R., Muselet, C., "Analysis of Hull Shape Effects on Hydrodynamic Drag in OffShore Handicap Racing Rules" SNAME 16th CSYS, Annapolis, MD, 2003.

Prince, M., Claughton, A, " The SYRF Wide Light Project" SNAME 22nd CSYS, Annapolis, MD, 2016.

Maskew, B., "Calculation of the Three-Dimensional Potential Flow around Lifting Non-Planar Wings and Wing-Bodies using a Surface Distribution of Quadrilateral Vortex Rings," Loughborough University of Technology, Report TT 7009, 1970.

Maskew, B., "Program VSAERO Theory Document," NASA CR-4023, 1987.

Maskew, B., "A Nonlinear Numerical Method for Transient Wave/Hull Problems on Arbitrary Vessels", Annual SNAME Meeting, New York, N.Y., November 1991.

Maskew, B., "Numerical Analysis of a Complete Yacht using a Time-Domain Boundary Element Method," 3rd International Seminar on Yacht and Small Craft Design, Castellucio di Pienza, Italy, 1993.

Maskew, B., DeBord, F., " Upwind Sail Performance Prediction for a VPP including "Flying Shape" Analysis" , SNAME 19th CSYS, Annapolis, MD, 2009.

Maskew, B. and Dvorak, F.A., "The Prediction of CLMAX Using a Separated Flow Model," Paper No. 77.33.01 Proc. 33rd AHS Meeting, 1977; see also, J. American Helicopter Soc., April 1978

Longuet-Higgins, M.S. and Cokelet, E.D., "The Deformation of Steep Surface Waves on Water: I. A Numerical Method of Computation", Proc. R. Soc. London, A350, 1976, pp. 1-26.

Duncan, J.H.: "An experimental investigation of breaking waves produced by a towed hydrofoil" Proc. R. Soc. London. A 377. 331-348, 1981

Duncan, J.H.: "The breaking and non-breaking wave resistance of a two-dimensional hydrofoil" J. Fluid Mech., **126**, 507-520 , 1983

Cointe, R. and Tulin, M. P., "A Theory for Steady Breakers", J. Fluid Mech., **276**, 1-20, 1994

Muscari, R. and Di Mascio, A.: " A Model for the Simulation of Steady Spilling Breaking Waves", J. Ship Research, Vol 47, No. 1, March 2003

Wolfson Unit MTIA, Report No. 2546D, University of Southampton, Nov 2015

Yongsheng, L. and Wei, S., "Laminar-Turbulent Transition of a Low Reynolds Number Rigid or Flexible Airfoil", 36th AIAA Fluid Dynamics Conference, June 2006, San Francisco, CA

The Un-restrained Sailing Yacht Model Tests – A New Approach and Technology Appropriate to Modern Sailing Yacht Seakeeping

Etienne Gauvain, Wolfson Unit MTIA, Southampton, UK

ABSTRACT

Over the recent years the Wolfson Unit has seen a greater impetus from yacht designers and their clients to quantify and compare the seakeeping qualities of their sailing yacht design choices. Modern high performance yachts, fitted with a wide range of appendages generating lift and creating large moments, provide a number of complex challenges for designers.

Assessing seakeeping behaviour and performance in a seaway is, indeed, important during the design process since the motions cause unsteady effects on the yacht hydrodynamic characteristics, for instance on the lift generating capabilities of the appendages. Hence it may not be justifiable to assume during the optimisation process that the yacht outperforming other design candidates in calm water would also perform well in waves.

Therefore, the Wolfson Unit developed an innovative experimental model testing approach that would be an improvement over existing methods, simulating the 6 degrees of freedom motions and accelerations.

The unique un-restrained sailing test approach uses a mast mounted air screw device to simulate the aerodynamic propulsion from the sails allowing a scaled model of the yacht to be tested at a range of conditions, sea states and wave directions (from head seas to following waves). These tests can be used as a comparative tool to assess controllability and seakeeping characteristics of multiple configurations (e.g. hull shape, appendages, inertias) by quantifying induced motions and providing an estimate of added resistance in waves. Non-linear attitudes such as surfing can be investigated.

Free-running model testing is a technique frequently used in the development of power vessels, but little adoption is made in the sailing yacht world. Furthermore dynamic and seakeeping studies are at present challenging for computational fluid dynamics based tools, encouraging an experimental based approach.

This paper introduces an un-restrained model testing method on sailing yachts. Discussion will also be made on how this new method can be implemented in design decisions and add value during the design and performance evaluation process for sailing yachts.

NOTATION

BCS	Boat Coordinate System	
c	Wave Celerity	[m/s]
Ceh	Centre of effort height above DWL	[m]
CFD	Computational Fluid Dynamics	
CoG	Centre of Gravity	
D	Aerodynamic Drag	[N]
D_F	Drive Force	[N]
DOF	Degrees of Freedom	
F_A	Aerodynamic Force	[N]
FFT	Fast Fourier Transform	
GCS	Global Coordinate System	
H_F	Heeling Force	[N]
L	Aerodynamic Lift	[N]
LPP	Lines Processing Program	
R_{AW}	Added Resistance in Waves	[N]
RANSE	Reynolds Averaged Navier-Stokes Equations	
RAO	Response Amplitude Operator	
RMS	Root Mean Square	
STN	Station Number	
T	Wave Period	[s]
V_A	Apparent Wind Speed	[m/s]
V_M	Boat Speed at Model Scale	[m/s]
V_S	Boat Speed	[m/s]
$V_{S\,CW}$	Calm Water Boat Speed	[m/s]
V_T	True Wind Speed	[m/s]
VPP	Velocity Prediction Program	
X	Surge	[m]
Y	Sway	[m]
Z	Heave	[m]
α	Airscrew Angle to Centreline	[degrees]
β_A	Apparent Wind Angle	[degrees]
β_T	True Wind Angle	[degrees]
ε	Aerodynamic Force Angle	[degrees]
ζ	Wave Amplitude	[m]
θ	Pitch (+ve bow up)	[degrees]
λ	Wave Length	[m]
φ	Roll (+ve starboard sailing)	[degrees]
ψ	Yaw (+ve starboard sailing)	[degrees]

INTRODUCTION

This paper describes the procedure of the un-restrained sailing yacht testing method developed at the Wolfson Unit. The methods presented are illustrated for an IMOCA60 fitted with and without a foil generating dynamic lift.

The main objective of this paper is to stimulate discussion on such technical methods within the Yacht Engineering and Design community. This paper illustrates a method and is not intended to provide a comprehensive dataset.

BACKGROUND

Balance of Air and Water Forces

A sailing yacht maintains a steady course when the aero and hydrodynamic forces and moments are in equilibrium. The aero and hydrodynamic forces are assumed to act through the aerodynamic centre of effort and the centre of lateral resistance respectively as shown in Figure 1 (Claughton et al., 2006).

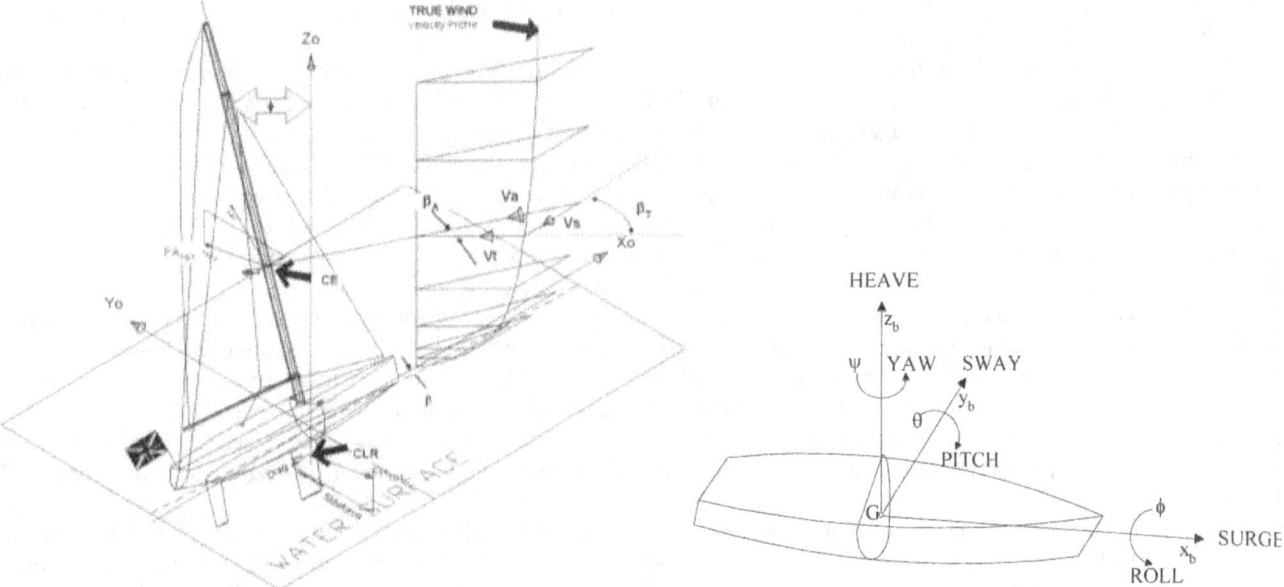

Figure 1 – Forces acting on a sailing yacht   Figure 2 – Axis system for ship motions

Yacht Motions in a Seaway

In a seaway forces and moments are imparted on the hull from the waves. These cause motions in all six degrees of freedom as defined in Figure 2. In turn, these motions cause unsteady effects on the yacht aero and hydrodynamic characteristics, for instance on the lift generating capabilities of the appendages and the sails.

When sailing upwind the yacht experiences an added resistance in waves and a corresponding loss of speed, due to the heave and pitch motions generating damping waves that carry energy away from the yacht.

Extensive research was undertaken by Gerritsma (1971, 1973, 1988, 1993) on the seakeeping performance and steering properties of sailing yachts, using both theoretical and experimental techniques. He explained that the hydrodynamic coupling in sway, roll and yaw, due to the vertical distance between the lateral centre of the keel and rudder and the CoG, cannot be neglected (Gerritsma, 1973).

Korvin-Kroukovsky (1954) highlighted that it is extremely desirable to consider the entire motion in a plane of symmetry and particularly to investigate the effects of surge coupling and of variation in test conditions in regard to surging motion.

Lloyd (1998) explained that, in oblique waves, ship motions in both the vertical and lateral planes occur. For a symmetrical ship, the vertical plane motions are independent of the lateral plane motions. In oblique waves of small amplitude, the lateral plane motions will not affect the vertical plane motions and therefore these may be considered in isolation.

However, due to the asymmetry of a yacht when sailing heeled and yawed, and with the dynamic angle of attack on the appendages, this simplification may not necessarily be appropriate and it may be desirable to consider the entire motion in all planes when investigating sailing yacht motions in a seaway.

Numerical Methods

Various numerical methods have been developed over the years to model the motions of ships in waves. Most of them were based on linear strip theory, then with some non-linear extensions incorporated. But strip-based methods have limitations, especially in the evaluation of the local flow properties near the yacht ends and, according to Lloyd (1998), they are likely to be inaccurate in oblique waves where in practice the large motion amplitudes would be limited by steering the ship.

Sclavounos and Nakos (1993) developed a three-dimensional panel method for determining the motions of ships in a seaway and investigated the seakeeping and added resistance of IACC yachts. Correlations with towing tank experiments were made. 'Non-linearities' were then introduced in this code for trying to predict the performance of yachts in following waves.

Renilson (1981) developed a mathematical model for determining the broaching conditions of ships in following seas. His work was based on the conventional maneuvering equations and validated with a series of model tests techniques, including constrained and free-running (yet un-propelled) model tests. Renilson concluded that the principal factor causing a broach was the large wave-induced yaw moment combined with the small restoring moment available from the rudder operating with reduced effectiveness. Harris et al. (1999) then developed a mathematical model to predict the performance of a yacht operating in following waves.

Although these methods have been developed to calculate with sufficient accuracy the motions of a ship in a given seaway, the dynamic effects specific to modern light displacement yachts, including the effects of motions on the lift generation capabilities of the appendages and on the aerodynamic forces, are neglected.

RANSE based CFD tools are currently being developed. For example Azcueta (2002) conducted simulations of a yacht sailing upwind with up to 4 DOF set free (surge, heave, pitch and roll) and reaching in quartering waves with 3 DOF during the first 20s of the simulations and the surge released thereafter. Also using a RANSE solver, Mazas et al. (2017) conducted upwind and downwind seakeeping studies on an IMOCA60 with the geometry free to heave and pitch (upwind) and also free to sway (downwind), but fixed in the other DOFs, i.e. the yacht sailing at a constant speed.

Yet dynamic and seakeeping studies are, at present, challenging for multi-phase flow simulations. In fact, when conducting seakeeping simulations, one must obtain a statistical description of the forces acting upon the vessel and of the vessel motions. This requires a considerably longer simulation time, and hence considerably more computational expense, than for a steady state simulation. In regular seas the simulation should be run until the behaviour reaches a statistically stationary state, and then averaged over a number of wave cycles. For more complex sea-states, described by a spectrum of wavenumbers, the required integration time to collect adequate statistics may be an order of magnitude longer (or more) than for a regular sea.

Furthermore, the propagation of waves at the free-surface must be accurately captured from the inlet of the CFD domain to the outlet, with minimal/no reflections at the domain boundaries. This requires increased mesh size over a calm-water simulation in order to adequately resolve the water/air interface. This can be achieved relatively efficiently by employing adaptive mesh refinement, which increases the mesh resolution in the vicinity of the wave interface only, however some additional expense is inevitable. If adaptive mesh refinement is not available, the entire mesh volume swept by the waves must be refined to a level that captures the wave propagation correctly, which typically leads to a significant increase in mesh size and computational expense.

Experimental Techniques

Model tests are an additional tool to evaluate a yacht performance, especially for designs lying outside the envelope for which sufficient empirical knowledge exists. Various techniques are available including: semi-captive, existing free-sailing methods and un-restrained tests.

a. Semi-Captive Tests

The Wolfson Unit has been conducting model tests on sailing yachts since its inception in 1967 and has continuously refined its equipment, procedure and analysis techniques allowing tests of yachts fitted with the latest appendages concepts, including vertical lifting boards and dynamic righting moment devices.

In semi-captive mode the model is towed using a purpose designed single or 3-post dynamometry, as shown in Figure 3, which allows the model freedom to heave and pitch, but provides restraint in yaw, sway and roll. The model is ballasted to the specified displacement and trimmed to float parallel to DWL. For each test, a trim moment is applied that equates to the sail aerodynamic forces acting on the yacht.

Measurements are made of the resistance, trim, heave, sideforce, roll moment and yaw moment. The results are analyzed and regressed into a VPP.

Figure 3 - Photos of a model under semi-captive test in a heeled, appended condition

Semi-captive tests can be performed in head seas, as shown in Figure 4, where measurements of vertical accelerations and added resistance in waves can be made. These have been extensively used to evaluate a boat's behaviour in head-seas. Although sailing yachts rarely sail in 180 degrees head seas, the motions and added resistance in oblique waves can be estimated from tests in head seas. To simulate upwind sailing conditions, the wave period is modified to match the encounter periods that would be experienced whilst sailing.

However the model, being towed, is restrained in surge. Free-to-surge tank tests have been performed using spring loaded apparatuses.

Figure 4 – Model of the 100ft ocean racer 'Comanche' under semi-captive head-seas test

b. Existing "Free-Sailing" Methods

The first version of the "Free-sailing" testing method, as described by Allan et al. (1957), consisted of towing a model from the top of a mast, located at the centre of effort of the sail, using a constant tension in a "winch load". The model was free to attain equilibrium in heel and yaw corresponding to the towing speed. Measurements were made of forces parallel and perpendicular to the tank axis in the horizontal plane as well as the torque needed to maintain equilibrium.

Murdey (1978) simplified this system by applying the tow force perpendicular to the mast as shown in Figure 5.

Gommers and van Oossanen (1983) developed a solution to move the mast position relative to the hull, to ensure the model is directionally stable throughout the test. Equilibrium conditions are obtained after 15-20 seconds.

Stensgaard (1985) modified the system at NRC for testing 1/6 to 1/10 scale models.

Murdey (1987) and Larsson (1990) have both highlighted the benefits of free-sailing tests: the main advantage being the potential for minimizing the number of experiments required to produce the performance prediction of a sailboat, the designer can observe the hull in realistic conditions of yaw and heel. However for calm water performance, Murdey (1987) concluded, after 10 years of experience with free-sailing tests, that the semi-captive method with data regressed in a state-of-the-art VPP is the optimum combination.

The method was used for testing in waves at MARIN (Gaillarde et al., 2007 and Eggers, 2012, 2018). However, although the methods described are called "free-sailing", the model is still connected to a carriage in order to apply and measure the "winch load". Therefore they can be difficult to implement in waves over a wide range of conditions and headings, for assessing non-linear attitudes, controllability and for maneuvering tests.

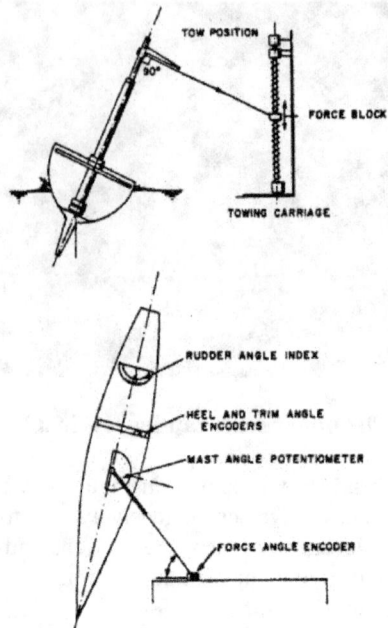

Figure 5 - Free Sailing System at IMD (Ottawa) (copyright Murdey, 1987)

c. Un-Restrained Tests

The Wolfson Unit has been performing free running (i.e. self-propelled) tests of power crafts since the late 1960s and has continuously developed and refined its experimental procedure, technology and measuring equipment. Deakin and Buckland (2010) presented the advances in radio controlled testing at the Wolfson Unit. Since then, technology has improved considerably, allowing tests with more powerful propulsors and lighter equipment at higher speeds, with the use of Li-Po batteries, electric brushless motors and digital control technology.

Free running tests have also been performed on sailing yachts. Following the large number of capsize incidents occurring during the 1979 Fastnet Race, Claughton (1984) investigated the dynamic stability of sailing yachts in large breaking waves. For comparing the characteristics of a series of typical hull forms from the 1960s till the early 80s, Claughton used a combination of model tests techniques including free running radio controlled models. These models were fitted with a motor driven propeller and rudder to allow them to be manoeuvred through the waves as shown in Figure 6. This research identified the design influences on the vulnerability to capsize and also highlighted the benefit of the free running tests in demonstrating how breaking waves could be approached.

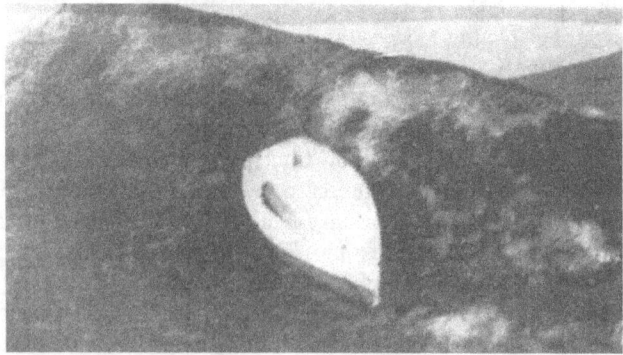

Figure 6 - Free running yacht model in breaking waves

The first incarnation of the present un-restrained tests technique performed on sailing yachts at the Wolfson Unit was carried out in 2008. The unique un-restrained sailing yacht test approach consists of a scale model entirely free to run, with onboard propulsion system and remotely controlled. It uses a mast mounted air screw device to simulate the aerodynamic propulsion from the sails allowing the scaled model of the yacht to be tested at a range of conditions, sea states and wave directions (from head seas to following waves) as shown in Figure 7. The concept of the air screw removes the presence of the "winch load" and allows applying near constant aerodynamic forces during a run or adjusting the thrust for constant boat speed (x-velocity). The un-restrained yacht tests can be used as a comparative tool to evaluate differences between several model configurations and quantify handling, behaviour and response in calm water and in various seastate/heading combinations.

Figure 7 – Sailing yacht under un-restrained tests

UN-RESTRAINED TEST METHOD

PROCEDURE

The un-restrained sailing yacht test follows the procedure described in Figure 8.

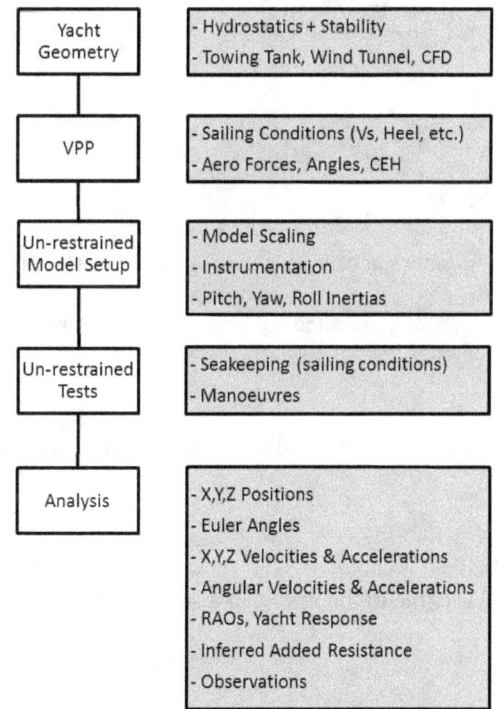

Figure 8 – Un-restrained yacht tests procedure

LPP and VPP

The yacht hydrostatics and stability characteristics were determined using the Wolfson Unit WinLPP program. VPP calculations were performed using WinDesign. The VPP output is then used for determining the model aerodynamic parameters at the selected sailing conditions (boat speed, heel angle at a β_T and V_T).

Hydrodynamic Calm Water Forces

The hydrodynamic forces in calm water can be determined from restrained model tests.

Aero Force Vector Magnitude and Direction

In the boat coordinate system, the following assumption is made to set the airscrew power:

$$F_A = Propeller\ Thrust$$

The aerodynamic force angle from the boat centerline can be defined by:

$$\varepsilon = cos^{-1}\left(\frac{D_F}{F_A}\right)$$

Where:

$$D_F = L \times \sin\beta_A - D \times \cos\beta_A$$

$$H_F = L \times \cos\beta_A + D \times \sin\beta_A$$

$$F_A = \sqrt{D_F{}^2 + H_F{}^2}$$

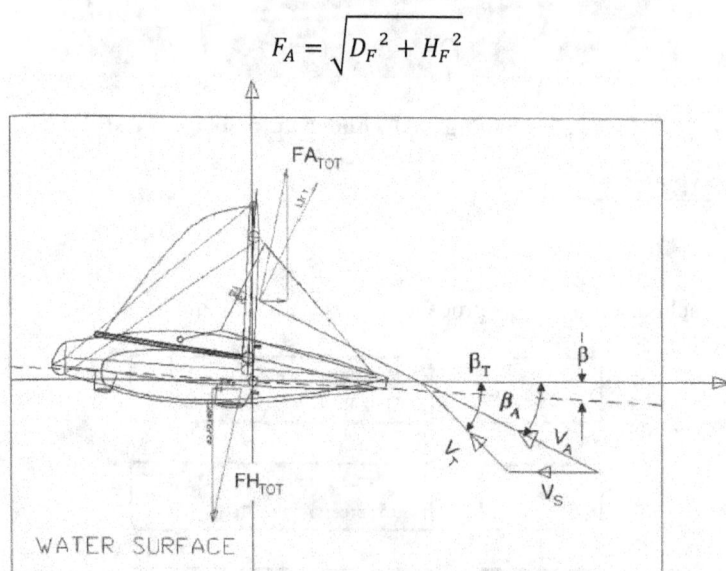

Figure 9 – Components of forces acting on a sailing yacht in the plane of the water surface

Additional vertical force component can be generated by orientating the air screw, e.g. to replicate offwind sails with upward lift component.

Centre of Effort Height

The CEH predicted from the VPP is used for each condition and sail set configuration. This is used to set the height of the air screw hub centre.

Centre of Effort Longitudinal

The mast is located at a fixed longitudinal position and the air screw is fitted to the mast at a known position. The mast position or the position of the air-screw relative to the mast can also be changed to replicate the XCE determined from the VPP.

VCG / LCG / Radii of Gyration

The model is ballasted to the yacht VCG and LCG using inclining experiments in air and/or by calculation. The yaw, pitch and roll inertias are set to match those of the scaled vessel.

Velocity

The model velocity is recorded using the Qualisys motion capture system. Calm water runs are performed to validate the estimated model velocity from the VPP for a given F_A and CEH combination.

Heel Angle

The heel generated by the heeling moment from the aerodynamic forces is calculated prior to the tests. The dynamic heel angle at a specific F_A and CEH combination is measured in calm water.

Gyroscopic Effect and Torsional Moment

The air screw is manufactured in plastic with relatively light blades tip; and does not generate a notable gyroscopic effect or torsional moment on the mast, that would affect the model inertias.

Moment Correction due to Friction Coefficient Delta between Model and Full Scale

A moment is applied by moving ballast on the model. This correction is due to a lower Reynolds number at model scale and the model friction coefficient is therefore higher than the ship one. If the model is towed from above the centre of wetted area a bow down trim is induced by the relatively high viscous resistance of the model.

Scaling Laws

It is appreciated that the method is not perfect due to the mismatch between the scaling laws of the aero and hydrodynamic forces. This needs to be further investigated in the future.

Pitch and Roll Moment due to CE Longitudinal, Vertical & Transversal Position

For simplification, the mast position is set for both the upwind and downwind sailing conditions. An additional sail trim moment can be applied on the model to correct for the XCE determined from the VPP.

By changing the air screw height the model inertia is modified. Therefore moving a ballast weight along the mast (or in the model) will compensate for this change of inertia.

Air Screw Angle and Aerodynamic Forces Corrections due to Model Advanced Speed

The air screw performance characteristics (thrust and rpm curves) were determined from bench tests. Further tests were carried out in the low speed section of the 7x5 wind tunnel at the University of Southampton (Figure 10) over a range of advance wind speeds and propeller angles in order to quantify their effect on F_A and ε. The air screw is then set at the corrected angle to the model centreline and the throttle is adjusted to achieve the required F_A.

Figure 10 – Airscrew tested in the wind tunnel

A selection of the airscrew performance results from the wind tunnel tests is presented in Figure 11, at a given throttle set and over a range of wind speeds.

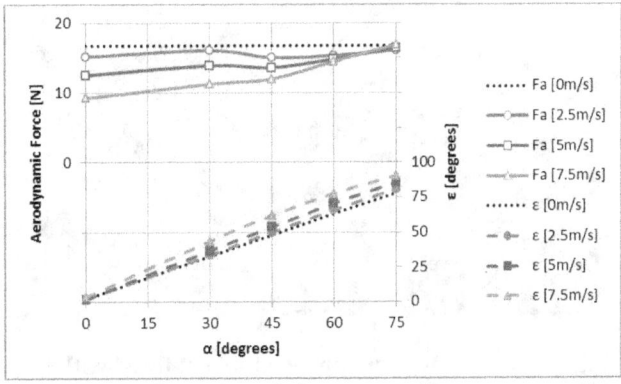

Figure 11 – Selection of airscrew performance results

46

ANALYSIS

Euler Angles, Velocities and Accelerations

Euler Angles, X, Y and Z accelerations and angular velocities about these axes are measured at specific locations on the model. Position, angles, velocities and accelerations of the model are measured with the motion capture system.

Typically the yacht response can be expressed as a response amplitude and an RMS can also be derived.

RAO's

By considering the yacht as a linear system the response can be given in a non-dimensional form as a response amplitude operator (RAO) for a large number of harmonic waves with different wave length. The RAO's for the 6 degrees of motion, as defined in Figure 2, represent a non-dimensional response on a basis of wavelength or frequency. RAO's are determined from Fast Fourier Transfer (FFT) functions in the frequency domain. To obtain RAO's the translations X, Y and Z are divided by the wave amplitude ζ and the rotations ϕ, θ and ψ are divided by the waveslope $(2\pi\zeta/\lambda)$:

$$RAO_{X,Y,Z} = \frac{X,Y,Z}{\zeta}$$

$$RAO_{\phi,\theta,\psi} = \frac{\phi,\theta,\psi}{2\pi\zeta/\lambda}$$

Added Resistance in Waves

The inferred added resistance in waves ΔR_{AW} can be estimated from the change in model speed compared to calm water at a pre-defined condition, such that:

$$R_{AW}\ Factor = \Delta V_S{}^2 \qquad \text{and} \qquad \Delta R_{AW} = R_{CW} \times R_{AW}\ Factor$$

Where: R_{CW} is the calm water resistance of the model in the corresponding sailing condition and V_S in m/s.

Signal Analysis

In order to visualize the yacht response with the incident waves with respect to time the signal traces are analyzed. The wave profile at the yacht x-position in the tank at a time t is derived from the wave probes data measured.

IMOCA60 MODEL TESTS

MODEL

A 1:7 scale model of an IMOCA60, previously used for restrained model tests, has been modified and used for the tests illustrated in this paper. The model was fitted with active movable twin rudders, keel fin and bulb as well as a foil providing dynamic stability (similarly to a DSS configuration), as shown in Figure 12. The tests were carried out with the appended model fitted with the foil fully deployed and removed. The model was also fitted with a deck and a self-propulsion arrangement. The model was ballasted to the specific scaled sailing condition. Yaw, pitch and roll inertias were set to match those of the scaled vessel.

Figure 12 - IMOCA60 appended model fitted with foil

SETUP

The model was fitted with a mast supported by a tripod. The airscrew is connected to an air-cooled 3-phase electric brushless motor, attached to the mast by a fitting that allows adjustments in height and rotation about Y and Z axes. The propulsion system is presented in Figure 13.

Li-Po batteries power the system via a voltage regulator and speed controller. A servo allows the rotation of the mast. Servos are connected to the rudders via tiller arms. Remote control of the steering, throttle and data logging/triggering is achieved via a radio controlled transmitter. Prior to testing, the helm setting was adjusted on the transmitter to produce the required helm angle range and rate.

The implementation of an autopilot and PID controller is not included in this paper. Future development will include the incorporation of such equipment, already adopted during powercraft testing, to adjust in real-time the aerodynamic force and angle to the yacht heading.

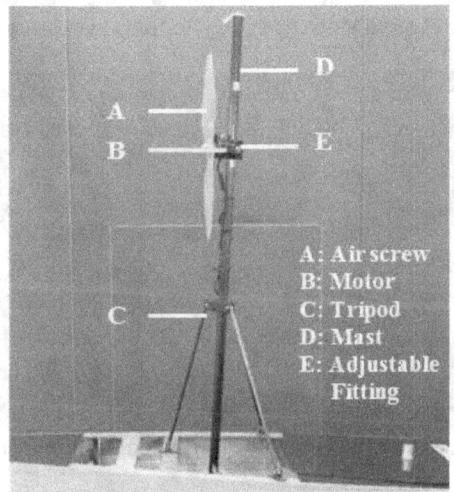

Figure 13 – Detail of the propulsion system

TEST FACILITIES

For the purpose of this paper, the tests were performed in the Boldrewood Towing Tank, University of Southampton, which is 138m long by 6m wide and 3.5m deep. Model seastates up to 0.5m significant height could be achieved. It would be preferable to use maneuvering basins to cover non axial headings and to help steering during turns and general re-positioning of the model.

Figure 14 - Boldrewood Towing Tank, University of Southampton

MEASUREMENTS

On the IMOCA60 model, vertical accelerations at two locations (STN1.9 and STN10), longitudinal acceleration at one location (STN3.7) and transverse acceleration at one location (STN4.6) were measured with accelerometers, namely Acc1, Acc2, Acc3 and Acc4 respectively. The model motions were measured using two methods:
- A heading/roll/pitch gyro.
- A Qualisys motion capture system to track the model and record its velocity in the tank longitudinal axis, as well as its motions. Passive markers were fitted to the model. A 6DOF rigid body was then defined with its origin at the model CoG.

Seven motion capture cameras were located alongside the tank such that a capture volume is obtained (Figure 15). One camera was set to video recording mode to overlay the model trajectory as shown in Figure 16.

Wave probes were positioned at 5 locations along the tank wall over the length covered by the motion capture, in order to simplify and verify the wave profiles mapping during the analysis.

On-board measurements were logged using a data logger at a sampling frequency of 100Hz. The on-board logger and the Qualisys measurement system were triggered simultaneously for data synchronization.

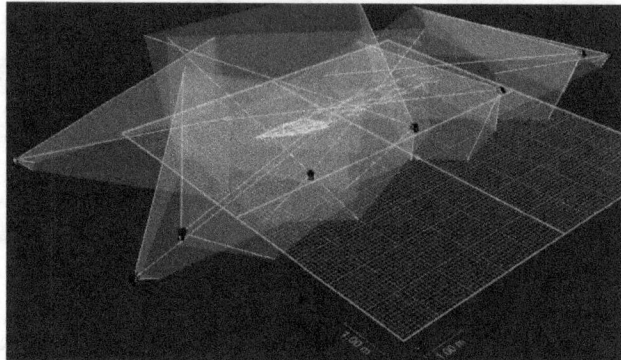

Figure 15 - Motion capture camera view cones with 6DOF rigid body and trajectory of the IMOCA60 model

Figure 16 - Motion capture overlaid trajectory of the IMOCA60 model under un-restrained sailing test

TEST CONDITIONS

Un-restrained seakeeping tests were carried out across a combination of speed and seastate (wave height and period) over a range of headings (180: head seas, 0: following seas) as shown in Table 1. For the purpose of this paper, tests were carried out in regular waves. Once the required seastate had fully developed within the basin the model was driven on a straight line course at the required speed and heading. Data was acquired over the period of steady speed and heading.

Table 1 – Test conditions

Vs cw [m/s]	Heading [degrees]	H [m]	T [sec]	Foil On / Off
3.9	180	0.7	4	On
4.1	180	0.7	4	Off
5.9	0	3	8	Off
7.6	0	3	8	Off

RESULTS

In order to illustrate the methods presented a series of results are shown for the IMOCA60 model fitted with and without the foil generating dynamic righting moment. The data presented hereafter corresponds to the Yacht Rigid Body with its local coordinate system origin located at the centre of gravity.

The wave profiles were calculated and mapped to the wave probes measurements for synchronization. Where:

Wave Celerity:

$$c = \frac{g \times T}{2\pi}$$

Wave Length:

$$\lambda = \frac{g \times T^2}{2\pi}$$

Wave Amplitude:

$$\zeta = \frac{Wave\ Height}{2}$$

Head Seas Motions

The following data correspond to the model with and without the foil, presented at full scale. Note that these tests were performed with an equal drive force applied for both configurations, in order to compare the sailing performances with the same rig. This results in a calm water boat speed difference between the two configurations due to the presence of the foil.

Figure 17 shows the full-scale variation of wave height, heave, pitch, X- and Z- velocities, with the boat X-position (i.e. corresponding to the wave characteristics at the boat CoG position along the tank longitudinal axis), in head seas (regular waves, H = 0.7m, T = 4sec) at a calm water boat speed of 3.9 m/s and 4.1 m/s respectively with and without the foil. Figure 17 presents the results in head-seas for one wave cycle. Table 2 presents a selection of the tests results.

Heave and Pitch

Figure 17 shows that the heave motion is synchronized with the wave height variations. The boat centre of gravity lies below the wave peak and above the wave trough due to the wave length and encounter frequency combination. The foil is generating a vertical force, resulting in a higher heave, yet with a similar response amplitude (0.24m).

The minimum and maximum pitch angles occur at approximately ¾ along the ascending waveslope and ¾ along the descending waveslope respectively.

Roll and Yaw

Figure 17 shows that the roll and yaw motions are synchronized. Their minimum and maximum occur at neutral wave height, along the ascending and descending slopes respectively. Note that the yaw response results from wave induced motions and the steering applied to hold a straight course.

The dynamic righting moment generated by the foil reduces the roll angle by approximately 1.2 degrees over one wave cycle.

X-Velocity and Acceleration

It can be seen from Figure 17 that the minimum surge velocity at the CoG (2.7 m/s) occurs just after the top of the wave peak; the maximum surge velocity (3.1 m/s) occurs just before the bottom of the wave trough. The latter can be explained by the wave to boat lengths ratio and encounter frequency combination in this condition.

Z-Velocity and Acceleration

Table 2 shows that the vertical accelerations RMS at STN1.9 and STN10 are reduced with the foil on.

It can be seen from Figure 17 that the minimum vertical velocity (-0.57 m/s) occurs half way through the descending waveslope, i.e. at a wave height of zero; the maximum vertical velocity (+0.51 m/s) occurs half way through the ascending waveslope, i.e. also at a neutral wave height.

Speed Loss and Inferred Added Resistance in Waves

With a lower speed loss the model fitted with the foil experiences a lower added resistance in waves.

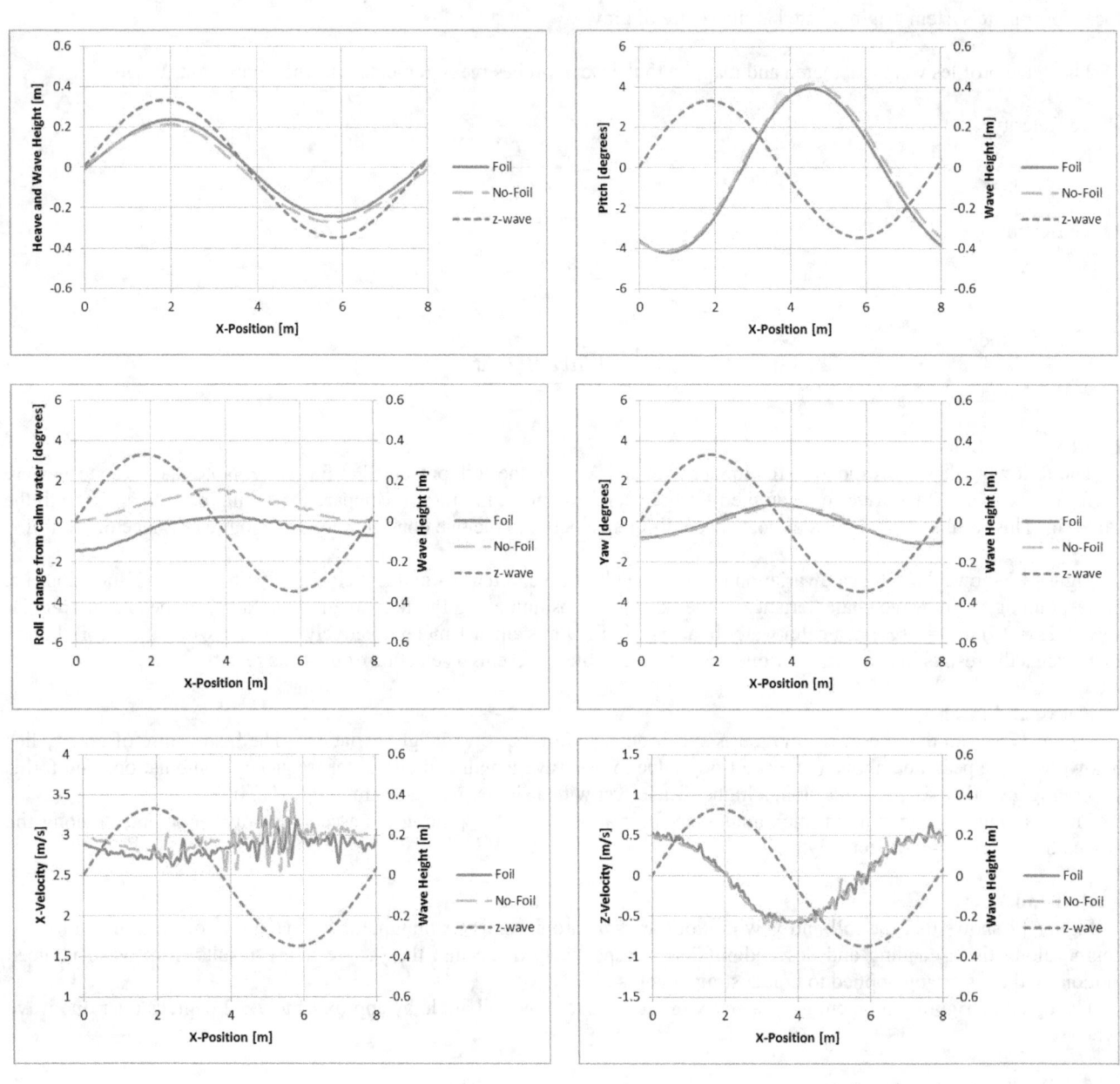

Figure 17 – Variation of motions at CoG with X-Position in head seas H = 0.7m and T = 4sec
Vs Calm Water (Foil) = 3.9 m/s and Vs Calm Water (No-Foil) = 4.1 m/s
Data presented for one wave cycle.

Observations

Although the tests presented here were performed at the same F_A for both configurations, it can be observed that the presence of the foil does not affect significantly the behaviour of the yacht sailing in the head-seas condition tested. This can be explained by the regime, away from heave excitation frequency, not showing significant differences.

In terms of controllability, the un-restrained approach is also a mechanism to investigate different strategies for sailing these waves, that can be witnessed during testing.

Table 2 – Selection of results in head-seas

	Units	Foil	No-Foil
Vs Calm Water	m/s	3.9	4.1
H	m	0.7	0.7
T	sec	4	4
c	m/s	6.2	6.2
λ	m	25	25
Vs	m/s	2.8	2.9
Speed Loss	m/s	1.1	1.2
R_{AW} Factor	–	1.11	1.24
RMS Acc1	m/sec^2	2.498	2.506
RMS Acc2	m/sec^2	1.709	1.746
RMS Acc3	m/sec^2	0.329	0.302
Response Amplitude Heave	m	0.240	0.242
Response Amplitude Pitch	degrees	4.08	4.14

Following Seas Motions

The following data correspond to the model without the foil and is presented at full scale.

Figure 18 shows a typical run of the IMOCA60 model surfing in following seas.

Figure 18 – IMOCA60 model surfing in following seas

Figure 19 and Figure 20 show the full-scale variation of wave height, heave, pitch, X- and Z- velocities, with the boat X-position (i.e. corresponding to the wave characteristics at the boat CoG position along the tank longitudinal axis), in following seas (regular waves, H = 3m, T = 8sec) at calm water boat speeds of 5.9 m/s and 7.6 m/s.

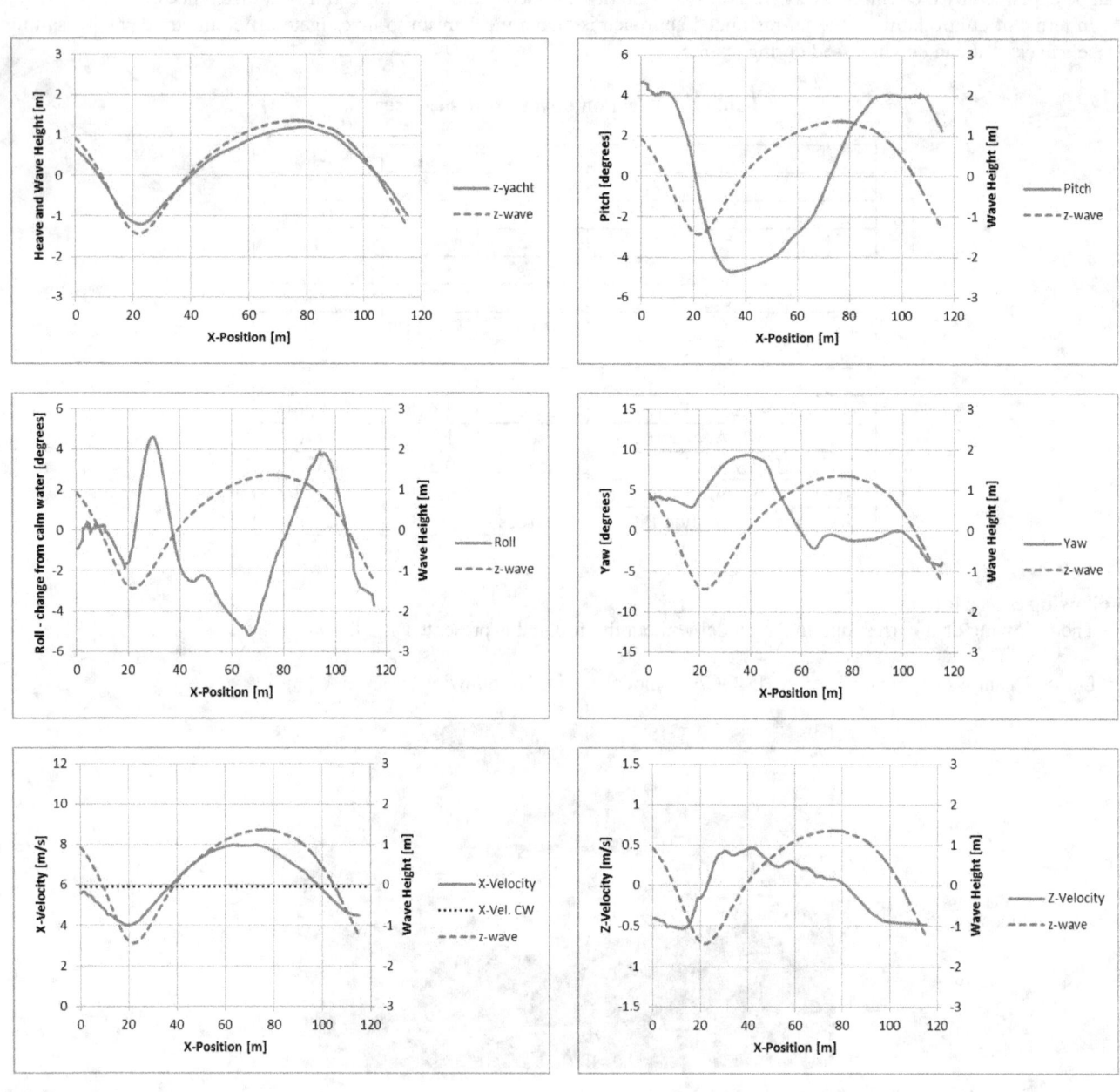

Figure 19 – Variation of motions at CoG with X-Position in following seas H = 3m and T = 8sec
Vs Calm Water = 5.9 m/s

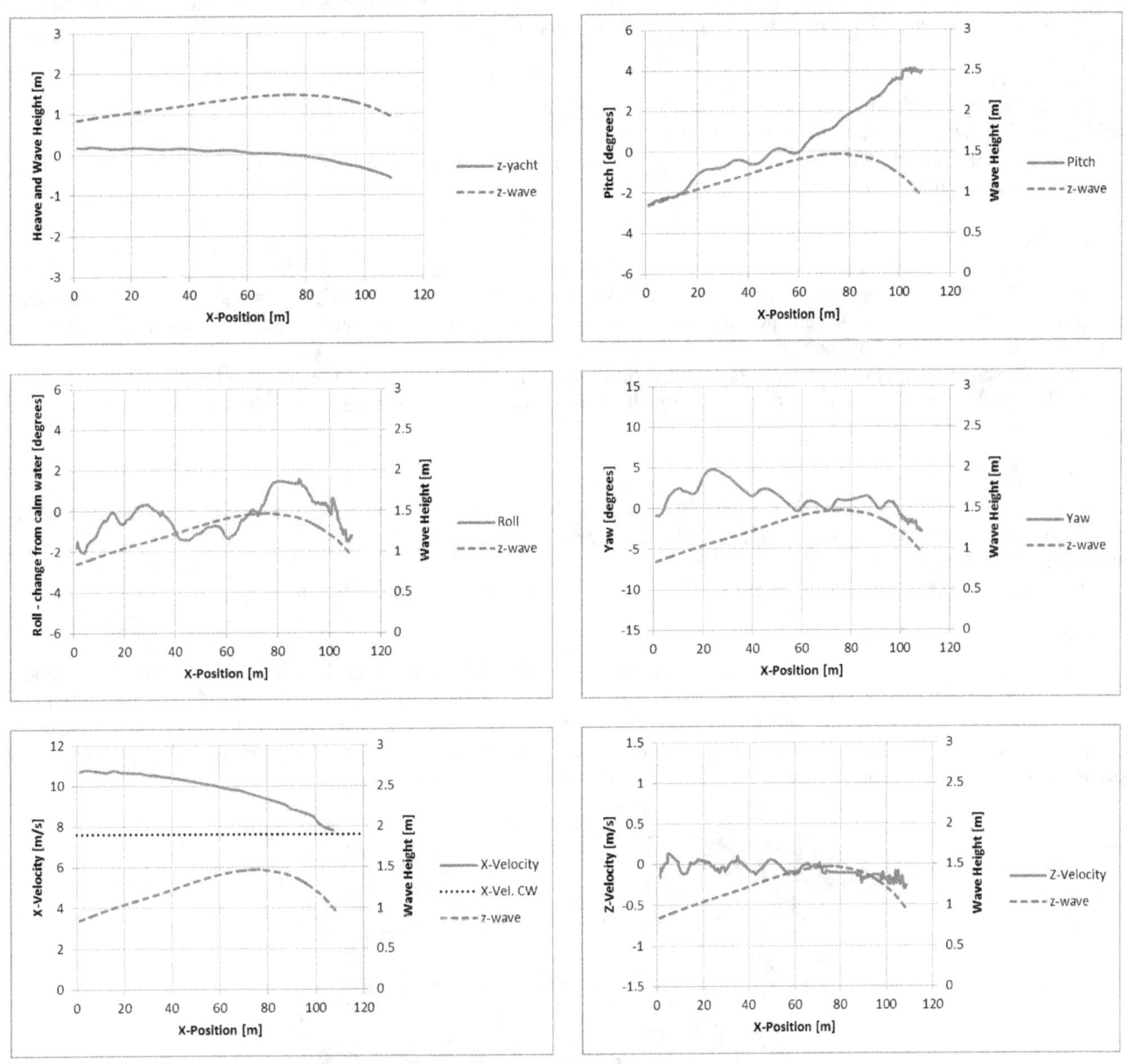

Figure 20 – Variation of motions at CoG with X-Position in following seas H = 3m and T = 8sec
Vs _{Calm Water} = 7.6 m/s

Heave and Pitch

Figure 19 shows that the heave motion is nearly synchronized with the wave motion. The maximum positive pitch motion leads the maximum wave height at the centre of gravity by approximately one quarter of the encounter period. This is in agreement with the vertical plane motions of a ship in regular following seas as described by Lloyd (1998).

Figure 20 shows that, during surfing, the model heave relative to the wave surface is wave profiling. When the wave is slowly overtaking the model, the pitch is increasing with the wave height.

Roll and Yaw

During surfing in following seas, the yacht motions in roll and yaw are relatively small and little steering is required to hold a straight course. This results in low changes in yaw motions as shown in Figure 20.

When being overtaken by the wave, the yacht experiences larger roll and yaw motions, as shown in Figure 19; steering is required to maintain a straight course.

X-Velocity

Fluctuations in X-Velocity at the CoG show the surge motions of the model.

Figure 19 shows that, at a calm water boat speed of 5.9 m/s, the X-Velocity at the CoG is not synchronized with heave and pitch motions. The model decelerates in the trough (Vs_{MIN} = 4 m/s) and accelerates in the peak (Vs_{MAX} = 8 m/s), i.e. being overtaken and carried by the wave.

Figure 20 shows that, at a calm water boat speed of 7.6 m/s, the model is surfing the wave and accelerates with it (Vs_{MAX} = 10.8 m/s). Over the measured distance the boat speed reduces and averages 9.8 m/s which corresponds to a speed delta of 2.2 m/s compared to calm water.

Z-Velocity

Similarly to the head-seas condition, Figure 19 shows that the minimum vertical velocity occurs at a neutral wave height on the descending waveslope (-0.5 m/s) and the maximum vertical velocity occurs at a neutral wave height on the ascending waveslope (0.5 m/s).

Figure 20 shows that, during surfing, the vertical velocity is relatively constant about 0 m/s.

Observations

The performance of a sailing yacht is not only characterized by its speed and behaviour in calm water and in a seaway, but also by its steering properties (Gerritsma, 1973). Controllability, such as the tendency to broach in running conditions, can be witnessed and investigated during the tests. For example Figure 21 shows the model in following seas with the rudders at full lock, i.e. applying a restoring yaw moment, to straighten its trajectory and avoid broaching. This is of particular interest with appendage loading due to dynamic running trim and heel.

Figure 21 – IMOCA60 model in following seas with full rudders applied to avoid broaching

MANEUVERING

Similarly to power craft free running tests, the un-restrained yacht technique can also be used to determine maneuvering characteristics of the vessel in calm water. This could potentially lead, in the future, to the investigation of additional maneuvers specific to sailing yachts (bearing away, luffing up, tacking, gybing), in calm water and in a seaway.

DISCUSSION

Method

The method described has been successfully used for sailing yachts including an IMOCA60 fitted with and without a foil generating dynamic righting moment.

Compared to existing methods such as semi-captive tests or other "free-sailing" tests, the un-restrained test method, with its airscrew arrangement, allows for measurements of all 6 DOF motions and accelerations over a range of sea-states and wave directions, from head-seas to following waves. Although (to the author's knowledge) coupling, per se, between the vertical plane motions and those of the lateral plane is not proven, first and higher order effects between them can be simulated. In fact, one of the main parameters that is measured during the un-restrained tests is the model X- velocities and accelerations, i.e. surge motions. The results presented herein highlight the surge motions of an IMOCA60, i.e. velocity variations in a seaway including the boat speed increase during surfing.

Another aspect of the un-restrained approach is the incorporation of some of the dynamic motions effects on the aerodynamic propulsion: aero force and angle varying with dynamic β_A and V_A from a constant throttle in steady state.

State-of-the-art motion capture system, onboard instrumentation and remote control equipment allow for a wide range of testing possibilities: e.g. controlling the steering, the aerodynamic force magnitude and angle. Future development will include the incorporation of a PID controller, already adopted during powercraft testing, to adjust in real-time the aerodynamic force and angle to the yacht heading.

Implementation

The yacht designer's toolbox is relatively empty when it comes to assessing the seakeeping performance of design concepts and choices. Seakeeping and maneuvering are often relegated to an empirical based approach, which is fine when designs are similar but not necessarily valid with new design concepts. Furthermore, in addition to having an impact on the sailing performances, motions are critical for the structural and mechanical engineering of the yachts. These tests can be used as a comparative tool to assess controllability and seakeeping characteristics of multiple configurations. This can be implemented in design decisions in the early stages of a project, in a timely manner, or later on for retrofitting purposes.

This new method can add value during the design and performance evaluation process for sailing yachts. In fact it is of great interest to the designers to know how the various hull parameters and appendages characteristics influence the motions, and vice versa, how the waves affect the hull resistance and appendages lift generating capabilities. The model tests can help them to extend their knowledge and database to evaluate motions, speed potential and comfort of designs. This is not only useful for racing yacht designs. The risk limitation factor is important, especially for production and superyachts. Finally these tests can be used to validate dynamic computational based tools and refine numerical models.

CONCLUSIONS

Model tests can highlight potential problems with behaviour, handling and control, in calm water and in a seaway. The use of models gives the designer and owner the confidence to pursue with or modify the design. Scale testing is often used in the development of new concepts, e.g. INEOS Team UK and American Magic are currently sailing on scaled trial boats in preparation for the 36th America's Cup. The un-restrained test method can be considered as a scale down from this.

The un-restrained yacht testing approach, presented in this paper and illustrated on an IMOCA60 model, offers a new tool, with improvements over existing semi-restrained model testing methods. It is not seen as a replacement for existing techniques but as an addition, in support of the design and performance evaluation of sailing yachts, especially in waves.

The main features of the un-restrained method are:

- Entirely free to run self-propelled model
- 6 DOF motions and accelerations
- Simulation of non-linear conditions (e.g. surfing)
- Inferred added resistance in waves
- Model configurations comparison
- Seakeeping and controllability assessment
- Range of sailing conditions
- Range of sea-states, wave spectra and directions
- Limited number of assumptions and simplifications
- Possibility to test at constant drive force
- Possibility to test at constant boat speed
- Possibility to test at a varying/quasi-steady drive force
- Manual or pre-set PID control
- Validation data for numerical models and computational tools

ACKNOWLEDGEMENTS

The author is grateful to Guillaume Verdier for the permission to use one of his models. Acknowledgements should also be given to the author's colleagues at the Wolfson Unit MTIA and Boldrewood Towing Tank, for their support and help.

REFERENCES

Allan, J.F., Doust, D.J. and Ware, B.E., "Yacht Testing", Trans. RINA, Vol.99, 1957

Azcueta, R., "RANSE simulations for sailing yachts including dynamic sinkage & trim and unsteady motions in waves", High Performance Yacht Design Conference, 2002

Claughton, A., Wellicome, J., Shenoi, A., "Sailing Yacht Design: Theory", 2nd edition, 2006

Claughton, A. and Handley, P., "An investigation into the stability of sailing yachts in large breaking waves", University of Southampton, Department of Ship Science Report, No.15

Claughton, A., "The dynamic stability of Sailing Yachts in Large Breaking Waves", International Conference on Design Considerations for Small Craft, 1984

Deakin, B., "Capsize and Stability of Sailing Multihulls", Wolfson Unit MTIA Report No.1238, for the UK Marine Safety Agency, August 1995

Deakin, B. and Buckland, D., "Recent Advances in Radio Controlled Model Testing", The Second Chesapeake Power Boat Symposium, Annapolis, MD, March, 2010

Eggers, R.: "Prediction methods for Team AkzoNobel in the Volvo Ocean Race and Wind Assisted Ship Propulsion for merchant vessels", HISWA Symposium, 2018

Eggers, R., Gaillarde, G. and Koning, J.: "A review of unsteady hydrodynamic behavior of sailing yachts and methods to study it", Proceedings of the 4th High Performance Yacht Design Conference, 2012

Gaillarde, G., de Ridder, E-J., van Walree, F. and Konning, J., "Hydrodynamic advice of Sailing Yachts through Seakeeping Study", Proceedings of the 18th Chesapeake Sailing yacht Symposium, 2007

Gerritsma, J., "Course keeping qualities and motions in waves of a sailing yacht", Proceedings of the third AIAA Symposium on the Aero/Hydrodynamics of sailing, California, 1971

Gerritsma, J. and Moeyes, G., "The seakeeping performance and steering properties of sailing yachts", 3rd HISWA Symposium, Amsterdam, 1973

Gerritsma, J. and Keuning, J.A., "Performance of light- and heavy displacement sailing yachts in waves", The Second Tampa Bay Sailing Yacht Symposium, St. Petersburg, Florida, 1988

Gerritsma, J., Keuning, J.A. and Versluis, A., "Sailing yacht performance in calm water and waves", 11th Chesapeake Sailing Yacht Symposium, SNAME, 1993

Gommers, C.M.J., van Oossanen, P., "Design of a Dynamometer for Testing Yacht Models", Proceedings of the 20th American Towing Tank Conference, Davidson Laboratory, Stevens Institute of Technology, Hoboken, N.J., USA,1983

Harris, D., Thomas, G. and Renilson, M.R., "Downwind Performance of Yachts in Waves", 2nd Australian Sailing Science Conference, 1999

Korvin-Kroukovsky, B.V. and Lewis, E.V., "Ship Motions in Regular and Irregular Seas", Memorandum No.106, Experimental Towing Tank Stevens Institute Technical, July 1954

Larsson, L., "Scientific Methods in Yacht Design", Annual Reviews Fluid Mechanics 1990

Lloyd, A.R.J.M., "Seakeeping: ship behaviour in rough weather", edited in 1998

Mazas, L., Andrillon, Y., Letourneur, A., Kerdraon, P. and Verdier, G., "Comparison of hydrodynamic performances of an IMOCA 60' with straight or L-shaped daggerboard", Proceedings of the 4th International Conference on Innovation in High Performance Sailing Yachts Innov'Sail, 2017

Murdey, D.C., et al., "Techniques for Testing Sailing Yachts", ITTC87 discussion on "Advances in Yacht Testing Technique"

Murdey, D.C., "Yacht Research at NRC", National Research Council of Canada, DME/NAE Quarterly Bulletin, No. 1978 (3)

Renilson, M.R., "The broaching of ships in following seas", PhD Thesis, 1981

Sclavounos, P.D., and Nakos, D.E., "Seakeeping and Added Resistance of IACC Yachts by a Three Dimensional Panel Method", 11th Chesapeake Sailing Yacht Symposium, 1993

Stensgaard, G. and Roddan, G.A., "Design and Implementation of a Modified NRC Style Sailboat Dynamometer at B.C. Research", B.C. Research Report No. 6-02-499-1, December 1985

Van Oossanen, P., "The Development of the Twelve Metre Yacht Australia II", Proceedings of the 7th Chesapeake Sailing Yacht Symposium, 1985

Experimental measurement and simplified prediction of T-foil performance for monohull dinghies

Sandy Day, University of Strathclyde, Glasgow Scotland
Margot Cocard, University of Strathclyde, Glasgow Scotland
Moritz Troll, University of Strathclyde, Glasgow Scotland

Exocet Moth - Photo © Katie Hughes

ABSTRACT

The rise in interest in large foiling yachts, such as those in the America's Cup, has spurred a corresponding interest in foiling applications in monohull sailing dinghies, both for boats designed specifically for foiling, and through retro-fit foil kits. The present study considers the assessment of foil systems for such boats, and specifically the prediction of the necessary foil performance of T-foils with flaps.

The main lifting foil for a moth dinghy with flap was tested at full-scale in a towing tank at a range of speeds, trim angles and flap angles, and immersions, to measure vertical and lift and drag. Results are then compared with two simplified models typical of those utilized in Velocity Prediction Programs for preliminary design. The first model uses section lift and drag data obtained using the well-known XFOIL code allied to a simple correction for 3D effects while the second model deploys a numerical lifting line theory in conjunction with section data.

A number of practical conclusions are drawn regarding the set-up and sailing of foiling dinghies with flapped T-foils. Results show that whilst the simple and rapid models typically used in basic VPPs may not accurately represent the relationships between angle of attack / flap angle with lift and drag, the predicted relation between lift and drag is reasonably accurate in most cases.

NOTATION

AR	Aspect Ratio (geometric)
b	Foil span
c	Foil chord
e	Oswald Efficiency Factor
C_D	Drag Coefficient
C_{Di}	Induced Drag Coefficient
C_{Dw}	Wave Drag Coefficient
C_{Dj}	Junction Drag Coefficient
C_{Dws}	Wave and Spray Drag Coefficient
C_L	Lift Coefficient
C_{Lh}	Lift Coefficient at immersion h
DA	Drag Area
D_j	Junction Drag
D_{ws}	Wave and Spray Drag
h	Immersion
LA	Lift Area
Re	Reynolds Number
t	Thickness
TR	Taper Ratio
V	Speed
α	Angle of attack (horizontal foil)
Λ	Sweep Angle
ρ	Density
φ	Flap Angle (horizontal foil)

Other symbols are defined as required in the text.

INTRODUCTION

Foiling dinghies have become popular in the last decade, led by developments in the International Moth. Foiling boats won races in the Moth world championships as long ago as 2000, and the Moth world championship was won for the first time in a foiling boat in 2005. The foil set-up for the Moth is restricted by the class rules, so that foils must exit the hull below the static waterline. This has led to the universal adoption in Moths of a foil system comprising two T-foils mounted on the centerline. This layout is therefore quite different to the foil configurations often seen on larger foiling catamarans, typically employing a pair of L- or J-foils as the main lifting surfaces along with T-foil rudders.

In the Moth system, the main horizontal foil is rigidly attached to the vertical (centreboard), although on some boats it is possible to adjust the rake of the vertical slightly to account for different wind conditions by moving the top of the foil fore and aft. The vertical is typically raked forward at around 7° to reduce the risk of ventilation. The ride height is controlled via a trailing edge flap on the main horizontal foil, which is activated through a mechanical linkage by a wand mounted on the bow of the boat, or, in more boats, on a short bowsprit. Within this mechanical system, the relationship between the angular movement of wand and flap is adjusted using a gearing system, whilst the neutral position of the flap relative to the wand position can be adjusted via a ride height or bias adjuster.

The lifting foil on the rudder is rigidly attached to the vertical, and the rudder lift is normally controlled by moving the top of the rudder assembly fore and aft using a screwed rod turned via a twist grip in the tiller extension, thus raking the rudder vertical and changing the angle of attack of the lifting foil. This is generally adjusted relatively infrequently to account for changes in wind speed, or changes from upwind to downwind sailing.

Modern Moths are capable of speeds in excess of 30 knots, and foil tacking is normal at the higher level of the racing fleet. While the boats vary in design, competitive boats are relatively expensive, with most boats beings 100% carbon fibre, and with increasing use of high modulus carbon in foils. Even so, the boats can be fragile with all-up sailing weight around

35kg. In this context, a number of production one-design foiling boats have been developed in recent years using T-foil systems based on standardized, and, in some cases, lower cost, implementations of the Moth system. These include monohulls such as the *Waszp*, *Skeeta* and *Onefly*, catamarans with twin T-foil systems such as the *Whisper* and *Stunt 9* and trimarans such as the *F101*.

Most of the development of foiling Moths has been achieved via full-scale trials of equipment, although increased interest in foiling boats is encouraging developments in modelling tools. However, whilst a number of studies have been published examining performance prediction for T-foil boats in general and moths in particular (e.g. Findlay and Turnock (2008)), Bögle *et al.* (2010), Mackenzie (2014)), relatively few recent studies have focussed on experimental measurements of flapped T-foils.

Historical studies of fully-submerged hydrofoils (e.g. Ramsen & Vaughan (1955), Wadlin *et al.* (1955) Ripken (1961)) were typically aimed at design of foils for large powered vessels. As such, the foil geometries are typically very different, and generally did not employ flaps, at least in the test programs. Binns (2017) has carried out a series of detailed measurements investigating ventilation around T-foils, both in the towing tank, and from a powered catamaran, following an earlier study (Binns *et al.* 2008) on an unflapped T-foil broadly similar to a Moth foil, although with lower aspect ratio.

Beaver & Zseleczky (2009) carried out a highly detailed study of the fluid flow around a Moth Dinghy, including tank testing of the hull, several main foil, and rudder foil, and aerodynamic testing for the hull and sailor; Andersson *et al.* (2017) executed a comprehensive set of hydro-and aerodynamic studies in the successful design, build, and sailing of the "Foiling Optimist", including tank test studies of hull with and without foils and wind tunnel testing of the sails. These studies formed the starting point for the present work.

The aim of the present work was to develop a test rig suitable for testing flapped T-foils in a towing tank and to use the test rig to develop a data set for validation of computational analysis of foil phenomena. In the longer term it is intended to use the data to validate high-fidelity predictions of complex phenomena associate with the foils, including effects of yaw and heel, and foil tip immersion; however, the present study is confined to upright foils with no yaw over a range of angles of attack and flap angles. Since CFD studies may not always be well-suited to early stage design, and velocity prediction, a second aim is to investigate the accuracy of some relatively simple models which may be deployed in velocity prediction programs for foiling boats.

TANK TESTING

The towing tank study was conducted in two stages. A preliminary stage took place in early 2018 to commission and troubleshoot the test set-up, whilst the main test campaign took place in late 2018, as the first part of a study into foil performance.

T-foil tested

The foil utilized for the present tests was chosen for pragmatic reasons, largely related to availability. The foil is from a rather old "*Bladerider*" design of International Moth (originally designed around 2006) owned by one of the authors. However, it appears to be similar to the foil described as "Vendor 1" tested by Beaver and Zseleczky (2008). The foil is regarded as slow by modern Moth standards, with the horizontal in particular being thicker, less stiff, and more fragile than modern competitive Moth foils. The key dimensions of the foil are shown in Table 1.

Table 1 - Main dimensions of foil

Horizontal	Span	0.988 m
	Chord at root (extrapolated due to bulb)	0.125 m
	Chord at 90% span	0.045 m
	Mean Chord (area / span)	0.095 m
	Thickness / Chord	12.8%
	Camber / Chord	3.1%
Vertical	Span (from bottom of hull)	1.0 m
	Chord at root (bottom of hull)	0.118 m
	Chord at tip	0.1135 m
	Thickness / Chord	14.5%

The foil was mass-produced using carbon skins and a foam core, unlike modern foils which are generally solid carbon fibre (and now increasingly often high modulus carbon fibre). However, for the purposes of the present study, it is sufficiently similar to a contemporary Moth foil to allow the test rig to be validated and useful results to be gained for comparison with numerical predictions. The foil is also more comparable than a state-of-art Moth foil to some of the rather more basic foils utilised in production foiling boats. The foil is shown in Figure 1.

The flap comprises very close to 35% of the chord for the vast majority of the foil; it is attached using a hinge consisting of a bead of a flexible jointing compound on the upper surface. A squashed bulb in the center of the horizontal allows adequate reinforcement of the joint between the horizontal and vertical foils. The vertical foil is installed in the boat with forward rake, intended to reduce the incidence of ventilation, and the angle of attack of the horizontal foil is mounted to reflect that.

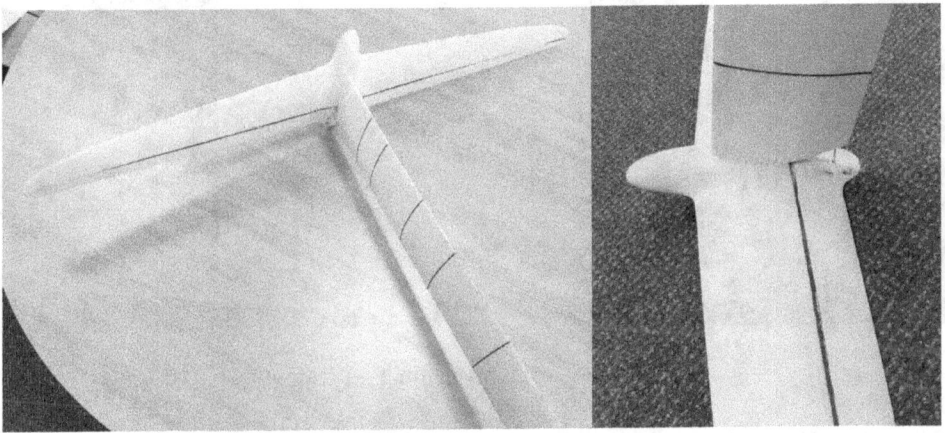

Figure 1 -T-foil tested

The precise geometry of the foil is not publicly available. However, the manufacturers published CAD drawings of the geometry of a series of jigs which could be used at a range of span-wise locations for holding the flap at the neutral angle whilst refurbishing the hinge. By scaling the jig drawings, it could be seen that the sections were close to geometrically similar along the span. The largest section had the highest number of points, and was used to generate an offset table for the horizontal section. This is shown in Figure 2. The vertical section appeared to match very closely with a NACA 66012 section scaled to 14.5% thickness. In the numerical study no attempt was made to model the foil bulb; instead the profile was extrapolated to the centreline.

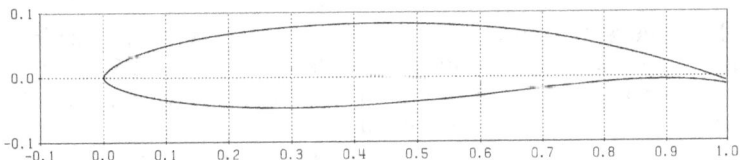

Figure 2 - Horizontal Foil Cross section

Equally some other details which may have some impact on the hydrodynamics were ignored in the numerical modelling – in particular the gap in the lower surface of the horizontal which allows the flap to rotate in the positive (downward) sense, and the cut-out at the rear lower part of the vertical which allows the flap to rotate in the negative (upward) sense.

Test Rig and instrumentation

Test were carried out in the towing tank of the Kelvin Hydrodynamics Laboratory in Glasgow, Scotland. The tank is 76 m long, 4.6 m wide, and 2.5 m deep, with a water depth for these tests of 2.1 m. The towing carriage has a maximum speed of just over 4.6 m/s. For the present tests a bespoke test rig was constructed which was attached to the vertical member of the standard towing post.

The vertical component of the foil was mounted in a frame which could be pivoted in rake (or pitch/trim) relative to the fixed part of the test rig to allow the angle of attack of the horizontal foil to be varied. The vertical position of the foil could be easily adjusted within the frame to vary immersion. The fixed part of the test rig was mounted on two tri-axial load cells. These load cells were more sensitive in the X and Y directions than in the Z direction, and were therefore oriented so that X and Y corresponded to the vertical lift and drag forces. Each mount had a bearing releasing moments in the pitch axis, whilst the upper mount was also fitted with a low friction slide to release the vertical force. The test rig support could thus be regarded as equivalent to a classical simply-supported beam mounted vertically. The flap on the foil was actuated via a screw mechanism acting on the bell-crank at the top of the vertical foil which in turn activates the flap control rod running through the vertical foil. A pointer was installed to indicate the flap angle at the top. The test rig is shown in Figure 3.

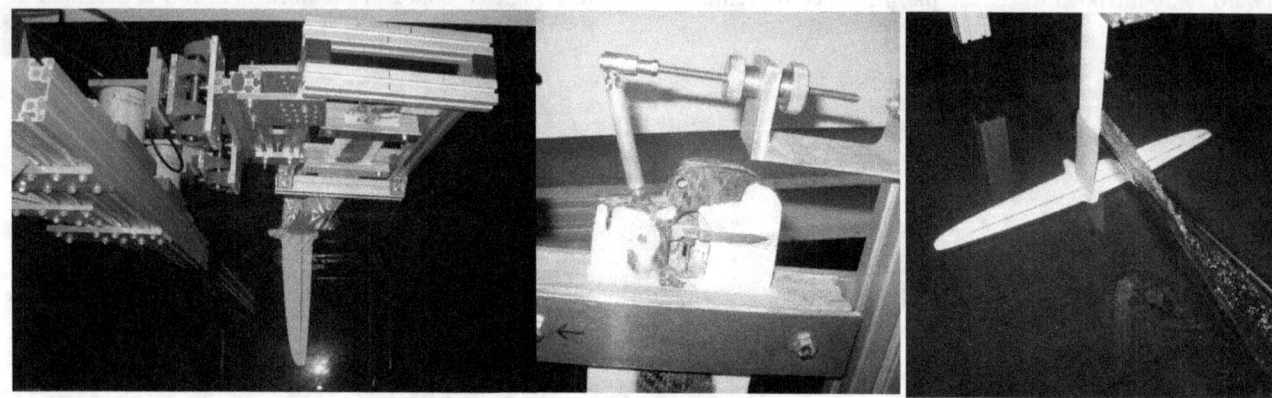

Figure 3 Towing Tank Test Rig

The load cells were calibrated individually in all three directions, and the cross-coupling between load in each direction and the measured load in the other two directions was determined separately for each cell, with particular attention paid to the coupling between lift and drag. The test rig was then assembled with the vertical in place and the rig performance was verified by applying horizontal loads at various vertical locations and vertical loads at various horizontal locations corresponding to typical test conditions.

It should be noted that whilst the rotation of the foil assembly and of the flap can be measured accurately relative to the starting position, it is difficult to be completely precise about the zero position of both the angle of attack of the horizontal and of the flap angle. This is discussed later in comparison with numerical results. The uncertainty in the main foil arises because it is common practice to use shims to adjust the angle of attack of the horizontal relative to the vertical prior to bonding the foils together permanently, so every foil assembly is likely to be slightly different.

A further issue with the flap angle is that it was observed in some cases that the pointer indicating the flap angle was moving under load. This was presumed to be due to the relatively slender push rod (2.5 mm stainless steel) which controls the flap bending inside the vertical. Once this had been realized, the pointer was videoed during the tests so that the value indicated by the pointer could be determined under load after the tests. This was predominantly noticeable in cases at the highest speeds with large angle of attack and large flap angle, which are less relevant to normal sailing conditions for the Moth – for example, in some cases, the pointer moved more than 2° at the highest speeds with angle of attack of 6°. For these reasons, where possible, these angles are not used as independent variables in presenting the results.

This highlights the practical issues involved in precise control of the flap in the boat under sailing conditions for which the speeds (and loads) can be much higher. In recent years, Moths have moved away from thin steel rods towards the use of much larger diameter carbon tubes for the pushrod components running between the wand and the top of the vertical, but the lack of space inside the thin trailing edge of the vertical foil presents a challenge in stiffening this component.

The tank testing procedure was extremely straightforward once the rig was installed; the immersion, flap angle, and angle of attack were set, and the carriage accelerated up to speed, whilst measuring the load in the load cells and the speed of the carriage. In the first version of the rig it had proved difficult to alter the angle of attack; however, after a re-design, this was easily altered, so once the range of speeds was complete for a given angle of attack / flap angle combination, the angle of attack was altered. Once all angles of attack were complete the flap angle was then changed.

The majority of the tests were carried out at a submergence of 457 mm (18"), corresponding to 4.8 × mean foil chord.

This is rather deep for a moth foil in sailing condition, but was chosen for several reasons. The maximum carriage speed in the Kelvin Lab is 4.5 m/s which is not far above take-off speed for a Moth (typically around 3.5 m/s boat speed for this boat), at which point the boat would be flying quite low, and so a relatively deep submergence of the foil is appropriate in that sense. This depth also matches some data from the previous study by Beaver & Zseleczky (2008), allowing a comparison to be made. Finally, it reduces the impact of wave making & free surface effects on the foil, giving a good baseline condition.

A range of speeds from 0.5 m/s to 4.5 m/s (mean chord Re 0.4×10^5 to 3.6×10^5) with a step of 0.25 m/s was explored for the case of angle of attack $\alpha = 0°$ and flap angle $\varphi = 0°$. Whilst the lower speeds are of no practical relevance for foil sailing, they add some insight into the hydrodynamics of the foil. With a flap angle of zero, the angle of attack was varied from 0° to 6° with a step of 1° at speed increments of 0.5 m/s. For the remainder of the tests, angles of attack from 0° to 6° and flap angles of –6° to +6° with steps of 2° were utilized with speeds varying from 2.0 m/s to 4.5 m/s and a step of 0.5 m/s. This speed range corresponds to a mean chord Reynolds Number range 1.6×10^5 to 3.6×10^5. In order to explore the effects of free surface proximity, a second, shorter, series of tests was carried out with a submergence of 100 mm. A total of nearly 400 tests were carried out, including a substantial number of repeats.

One question which would benefit from further study is the impact of turbulence on the flow over the foil, and in particular, in the nature of the laminar-turbulent transition. The level of turbulence experienced by a foil advancing in a towing tank into essentially still water may well be lower than that in the water in which a foiling boat is sailing when wind speeds are high enough to foil, and especially in wavy conditions, although this is dependent on the waiting time between test runs. Furthermore, it is likely that vibration may play a part – on the towing carriage there is always some small mechanical vibration, whilst in the real boat the foils may "sing" due to hydroelastic effects at higher speeds. In the present tests no attempt was made to measure background turbulence in the water or on the foil.

Results

Results are presented in terms of lift and drag forces (N) and lift and drag areas (equivalent to product of lift and drag coefficient and the relevant reference areas), defined as:

$$LA = Lift / \left(\frac{1}{2}\rho V^2\right)$$

$$DA = Drag / \left(\frac{1}{2}\rho V^2\right)$$

The use of lift and drag areas was adopted in order to avoid any requirement to decompose forces between horizontal and vertical foils with different chords, thicknesses and areas. Variation of lift and drag, and the corresponding areas with speed are shown in Figure 4 and Figure 5.

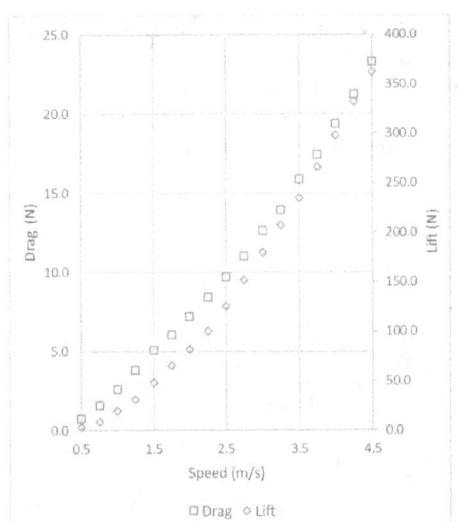

Figure 4 Variation of lift and drag with speed: $\alpha = 0, \varphi = 0$

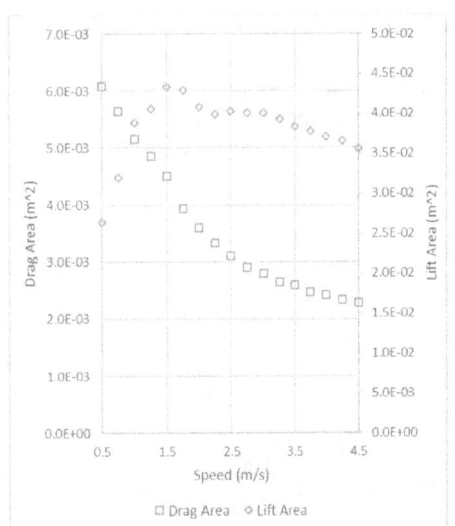

Figure 5 Variation of lift and drag areas with speed: $\alpha = 0, \varphi = 0$

The drag area reduces as speed and Reynolds Number increase, as might be expected, whilst the lift area peaks at 1.5 m/s, corresponding to a mean chord Reynolds number of 1.2×10^5, and then reduces slowly. The variation of lift and drag area with angle of attack with the flap at zero degrees over a range of speeds is shown in Figure 6. As expected the lift slope is slightly less than linear as angle of attack increases, whilst the drag shows slightly more subtle variations; both lift and drag areas reduce with speed and Reynolds Number, as expected from Figure 5.

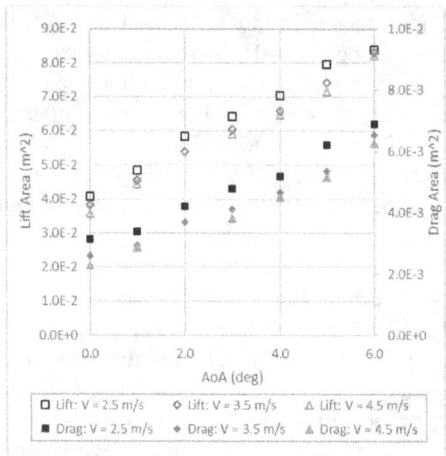

Figure 6 Variation of lift and drag areas with angle of attack: $\varphi = 0$

The variation of foil performance as flap angle and angle of attack were varied is shown for a selection of speeds in Figure 7. Each set of seven points corresponds to a single angle of attack with a range of flap angles from $-6°$ to $+6°$ at $2°$ intervals. In this case, the data is presented as drag area versus lift area squared, in order to eliminate flap angle as an independent variable, since the flap angles varied under load in some cases as discussed in the previous section. This is arguably the most important curve in terms of velocity prediction.

It is interesting to note that the curves for angles of attack from $0°$ to $4°$ almost collapse onto a single line. This has the practical implication of suggesting that the performance of the foil in sailing condition is not especially sensitive to the precise set up of the horizontal relative to the vertical within a relatively broad range of angles of attack. The only real exception to this is for the rather extreme angle of attack of $6°$, which yields higher drag. It is also apparent that the six-degree angle of attack curve shows a small dip in the middle, particularly at the lowest speed. This point corresponds to the condition of zero flap angle. A similar, although less pronounced, feature can be observed in the other curves.

The variation of foil performance between speeds of 2.0 m/s and 4.0 m/s is shown in Figure 8. These speeds correspond to mean-chord-based Reynolds Number range of 1.6×10^5 to 3.2×10^5. As expected from Figure 5, both lift and drag area reduce as the speed increases, but the drop in drag outweighs the drop in lift, and hence the foil can be seen to perform more efficiently at the higher speed. The change is most pronounced at the lower speeds (between 2 m/s to 3 m/s); it can be seen from Figure 7 that the differences between 3.5 m/s to 4.5 m/s are much smaller than those shown in Figure 8.

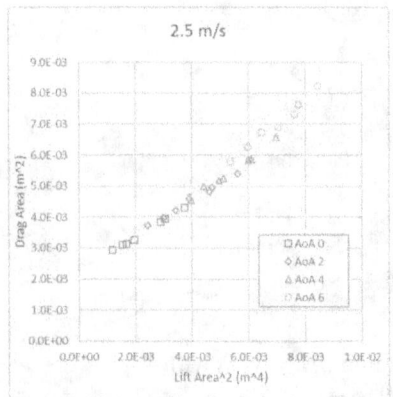
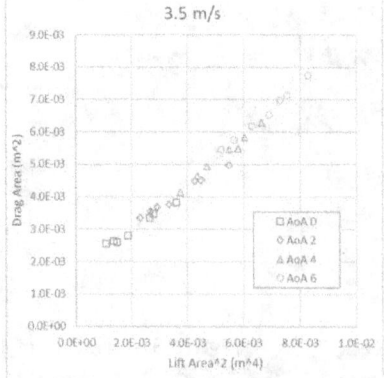
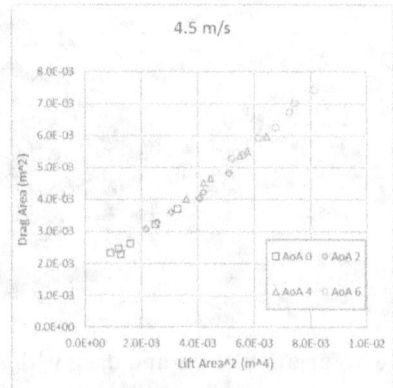

Figure 7 Variation of lift and drag areas with angle of attack and flap angle

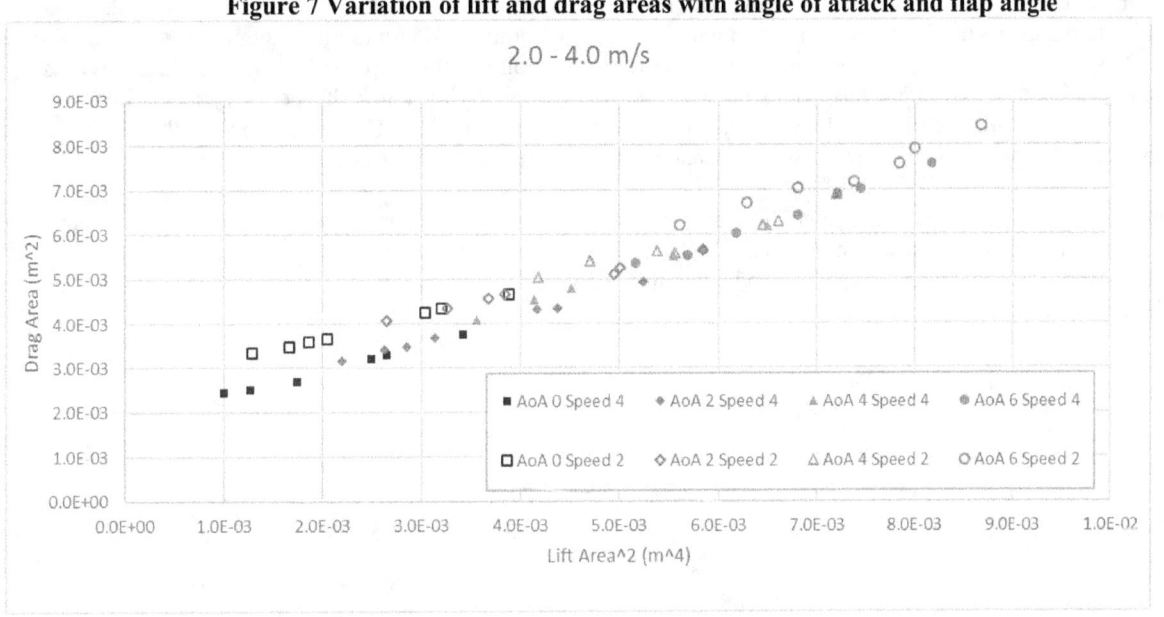

Figure 8 Sensitivity of Lift and Drag Areas to speed

Figure 9 shows the variation of lift and drag as immersion is reduced from 457 mm (4.8 × mean chord) to 100 mm (1.05 × mean chord) for an angle of attack of zero degrees and flap angles of zero and six degrees. It can be seen that the variation is rather small when the flap angle is neutral, but that both lift and drag drop substantially as submergence is reduced in the condition with greater flap angle, when lift is greater. It should be noted that this is not expected to be a realistic condition, as it is highly unlikely that a Moth would be set up to have such a large flap angle at a ride height high enough to yield this low immersion.

As immersion reduces, less of the vertical foil is immersed, leading to a drop in viscous drag; at the same time the wave drag of the horizontal foil may be expected to increase as the foil approaches the surface. Moth sailors typically sail high in flat water for maximum speed, so it may be expected that there is a net gain in performance in realistic sailing conditions as immersion reduces.

The small increase in lift at the highest speed with zero flap angle is less expected, although this may be within uncertainty due to the flap angle. It is worth mentioning that for the shallower immersion for which the waves generated are larger, there may be an effect of depth Froude Number at the higher speeds, further complicating an already complex issue.

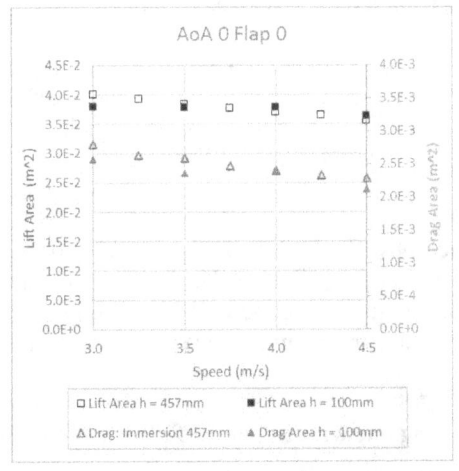
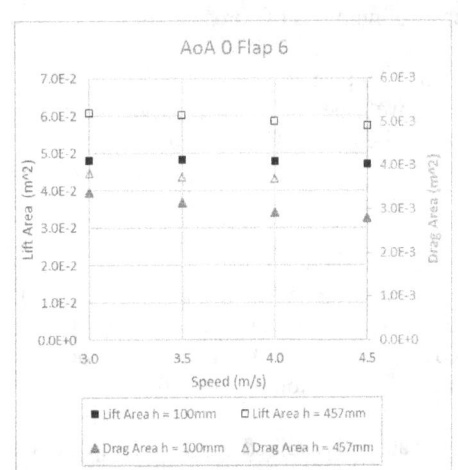

Figure 9 Sensitivity of Lift and Drag Areas to immersion: $\alpha = 0°$, $\varphi = 0°$ & $\alpha = 0°$, $\varphi = 6°$

Since shallow immersion is likely to be related to higher speeds, the fastest tests are of most practical interest here. The comparison for all cases tested at 4.5 m/s is shown in Figure 10. The solid markers show the points at the 100 mm immersion, with flap angles –6°, 0° and 6°. The open markers show the points at 457 mm immersion, with flap angles from –6° to +6° at 2° steps. It can be clearly seen that there is a cross-over point in this plot. At low lift conditions, with small angles of attack and small flap angle (as would be the case for a Moth flying high) the performance is improved by reducing the immersion, with slightly less drag generated for the same amount of lift. Here it is expected that the reduction in drag of the vertical is greater than the increase in resistance of the horizontal due to free surface proximity.

Conversely in higher lift conditions, the shallow immersion gives reduced performance, with more drag at the same level of lift. As mentioned before these conditions are highly unlikely to occur in practice (at least intentionally). Hence this plot illustrates why flying high is generally likely to be fast in a Moth.

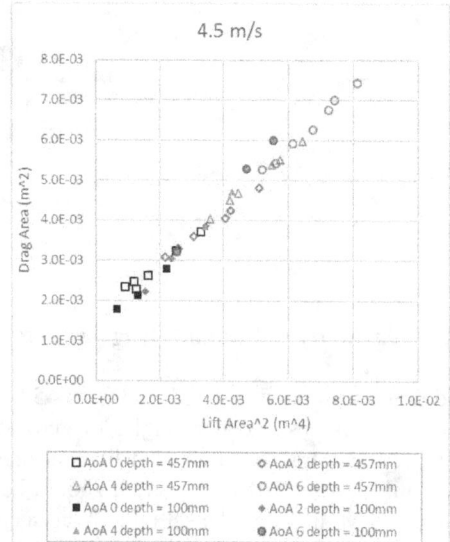

Figure 10 Sensitivity of Lift and Drag Areas to immersion: $V = 4.5$ m/s

SIMPLIFIED MODELLING OF T-FOIL

A high fidelity model of the behaviour of the foils is likely to require the use of a CFD code with free surface capability; this is undoubtedly a challenging task due to the complex geometry of the foil details such as the flap and hinge, the bulb, and the cut-outs, leading to a need for a dense mesh; at the same time, hydrodynamic features such as the thin sheet of water running up the vertical, the spray from the vertical foil, and the wave pattern generated (see Figure 3) will also present challenges in terms of domain size and mesh density. This is likely to lead to very substantial implications for computer resource in terms of numbers of cores and run time required, even for a small number of cases.

Whilst CFD results should in principle lead to the highest possible fidelity of prediction, one of the goals of the present study was to examine how well the hydrodynamic behaviour of the foils could be predicted by relatively simple models, particularly those which can be run very rapidly on a simple PC, and may be deployed in a VPP. To this end two highly simplified approaches have been deployed, both of which are many orders of magnitude faster than a CFD model.

Simplified Model 1

The simplest model deployed here for the T-foil broadly follows the approach described by Andersson *et al.* (2017), which was successfully deployed in the design of the "Foiling Optimist". The section lift and drag characteristics of the horizontal foil section are analyzed over a range of speeds, angles of attack and flap angles, using the well-known publicly-available XFOIL 6.99 code (Drela (2013)), employing a coupled panel method and integral boundary layer approach. The same code was used to determine section drag coefficients for the vertical foil.

The 3D lift coefficient was then obtained using the simple equation based on lifting line with elliptical lift distribution:

(1)
$$C_L^{3D} = C_L^{2D} / \left(1 + \frac{2}{AR}\right)$$

Here C_L^{2D} is the section lift coefficient obtained from XFOIL at a nominal Reynolds Number based on mean chord, and AR is the geometric aspect ratio. The lift is then obtained in the usual manner. The profile drag is obtained from the XFOIL section drag coefficient (again at a nominal Reynolds Number) applied to the area of the foil. Induced drag is obtained from the lifting line result for an elliptical foil: ,

$$(2) \qquad\qquad C_{Di} = C_L^2/\pi AR$$

Following Beaver and Zseleczky (2009), the coefficient of wave drag for the horizontal foil based on planform area was taken from Hoerner (1965) (section 11-26) as:

$$(3) \qquad\qquad C_{Dw} = C_L^2 \frac{h}{c} \frac{C_D}{C_{Lh}^2}$$

where h is the submergence, and Beaver and Zseleczky (2009) give $\frac{C_D}{C_{Lh}^2} = 0.25$.

Wave and spray drag for the vertical foil are estimated using empirical values for drag coefficient found in Hoerner (1965) (section 10-13) as:

$$(4) \qquad\qquad D_{ws} = \frac{1}{2}\rho V^2 C_{Dws} t^2$$

Here C_{Dws} is the combined wave and spray drag coefficient based on strut thickness t, given as 0.54. Finally the drag due to flow around the junction between main foil and vertical is estimated based on a drag coefficient based on thickness using a further equation from Hoerner (1965) (section 8-11):

$$(5) \qquad\qquad D_j = \frac{1}{2}\rho V^2 C_{Dj} t^2 \quad \text{and} \quad C_{Dj} = 17\left(\frac{\bar{t}}{c}\right)^2 - 0.05$$

Here $\bar{t}$ is the mean thickness of the horizontal and vertical foils. This approach may be very easily implemented in a simple environment such as MS Excel once the section lift and drag coefficients from have been obtained from XFOIL.

In the present implementation, for each foil section, XFOIL was run for 3126 cases comprising 6 Reynolds numbers ranging from 1.0×10^5 to 1.0×10^6, 13 angles of attack from $-4°$ to $8°$, and 17 flap angles, from $-8°$ to $8°$. In all cases free transition was utilised with a value of the XFOIL parameter $Ncrit$ set to 4, corresponding to a turbulence level of around 0.6%. An example data set for $Re = 4 \times 10^5$ is shown in Figure 11.

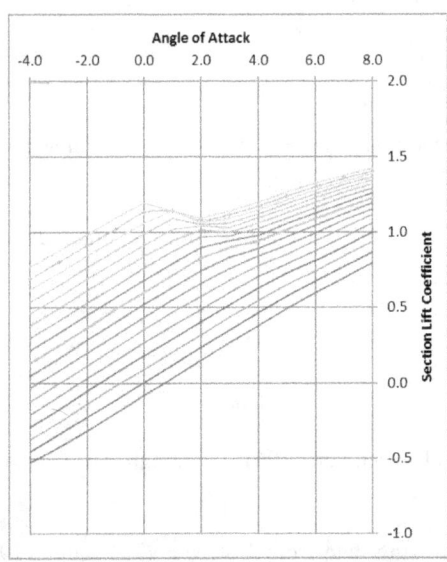 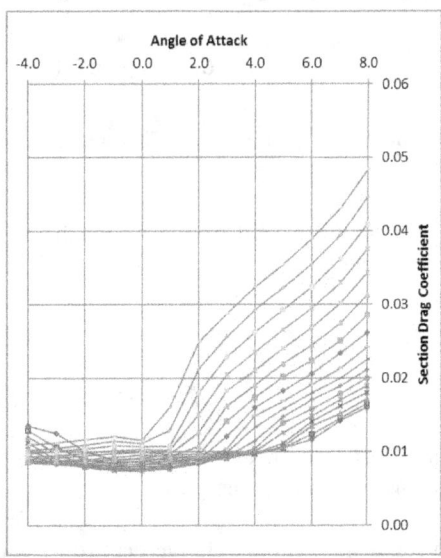

Figure 11 XFOIL Section lift and drag coefficients for Bladerider foil: $Re = 4 \times 10^5$, $\varphi = -8°$ to $8°$

As discussed previously, the level of turbulence in both sailing and tank test conditions is somewhat uncertain, and laminar-turbulent transition may well not occur in the tank test or sailing conditions via the "natural" transition modelled here in XFOIL. Further investigation of turbulence levels and transition models are planned for future studies.

The section lift and drag data generated was stored in an Excel sheet; cubic spline surface interpolation was then used to find the coefficients for a given Reynolds number and angle of attack for each of the flap angles, and then cubic spline interpolation was used again to find the value for the given flap angle.

Simplified Model 2

Whilst a model very similar to that described above had clearly been used with great success in design of the foiling optimist, it was of interest to see to what extent it could be improved without dramatically increasing the requirement for computational resource. An approach was developed using a numerical implementation of a lifting line method in MS Excel, using a total of 21 span-wise stations across the main foil.

The lifting line approach utilised data from XFOIL to estimate a linearized lift slope and zero-lift angle for the foil section at the given flap angle, based on the mean chord Reynolds Number. The lifting line solution was used to calculate local angles of attack at each span-wise station and induced drag coefficient. The local angles of attack were then used in turn in conjunction with the local Reynolds Numbers and the XFOIL section drag coefficients to estimate profile drag. The profile drag of the vertical foil was estimated using the XFOIL data as in the first model described above. A number of corrections were then explored for the various lift and drag values obtained in this manner.

The effect of the free surface was included using an approach described in Daskovsky (2000), deploying an approximation to 3D biplane theory. In this approach, the free surface influences both the lift and the drag of the foil. The lift coefficient is multiplied by the factor:

(6) $$(1 + 2/AR)/(1 + 2K(1 + \sigma)/AR)$$

Here

(7) $$K = \frac{16(h/c)^2 + 1}{16(h/c)^2 + 2} \quad \text{and} \quad \sigma = 1/(1 + 12\,h/b)$$

where h is the submergence, c is the chord (taken here as mean chord) and b is the foil span. With the same approach, the induced drag coefficient is multiplied by the factor $(1 + \sigma)$. This approach is shown to yield similar results to the model proposed by Wadlin (1955), based on a horseshoe vortex near a free surface, although Wadlin's model is rather more complex to implement. This approach is based on infinite Froude Number. For finite Froude numbers wave-making effects should also be considered. Daskovsky suggests the use of the approximate relationship:

(8) $$C_{Dw} = \frac{C_L^2}{2F_c^2} \exp(-(2h/c)^2/F_c^2)$$

Here F_c is the Froude Number based on chord. This was found to give broadly similar results to the approach suggested in Vladimirov (1955).

The effect of the vertical foil on the lift from a fully-submerged foil was modelled according to a recommendation in Gibbs & Cox (1954) for strut interference. The lift and induced drag coefficients are multiplied by the factors:

(9) $$C_{Lstrut} = C_L/(1 + \gamma)$$

$$C_{Di\,strut} = (1 + \gamma)^2 C_{Di}$$

where $\gamma = 0.8t/b$ for a central strut. Further study of this effect is found in Ripken (1961), although the data in this study is presented in a manner which is not easy to use.

The effect of the wave and spray around the vertical foil is calculated utilising data from Coffee and McKann (1953), who suggest that the wave and spray drag be estimated from:

(10) $$D_{ws} = \frac{1}{2}\rho V^2 C_{Dws} ct$$

The published data for moderate chord Froude numbers and a 12% thick section, shows that C_{Dws} is approximately 0.0275. It should be noted that the reference area in this equation is different from that of equation (4). Similar results were obtained by Ramsen and Vaughan (1955).

In the current Moth flap, the hinge is made of flexible sealant, whilst in more modern foils it is often made with Kevlar cloth impregnated with flexible epoxy. As a result, the flow from the lower surface cannot flow through the hinge to the upper surface. The hinge of the flap is flush with the upper surface, and the foil thus has a substantial cut-out on the lower surface designed to allow the flap to deflect down. This clearly impacts the flow on the lower surface. A substantial body of published data exists for drag due to flaps for aircraft (e.g. Wenzinger and Harris (1939)). It should be noted that in the published studies of cases for which the geometry is broadly similar to present foil, the flaps are typically slotted, and fluid may flow through the gap between the main foil and flap which may lead to different behaviour from the Moth foil. Nonetheless the correction suggested by Wenzinger and Harris (1939) was adopted here – this adds a drag coefficient increment of 0.0012 for lift coefficients below 0.6 and then linearly increases to 0.0022 for lift coefficient of 1.0.

Finally, the effect of the bulb was acknowledged in a crude manner by scaling the viscous drag coefficient according to the ratio of the planform area including the bulb to the planform area without the bulb. This added approximately 3% to the area. No attempt was made to correct for the cut-out in the vertical foil, and the bulb.

COMPARISON WITH EXPERIMENT DATA

The comparison of the two models with the experiment data of Figure 4 for zero angle of attack and zero flap angle is shown in Figure 12 below.

Here the solid points are the laboratory data; the curves labelled "2D model" are calculated using model 1 from section 0 while the curves labelled "LL" are calculated using model 2 of section 0. Since there was a small degree of uncertainty related to the precise value of the zero of the angle of attack of the main foil, curves are presented for both 0° and -0.5°.

Both models over-predict lift, with relatively small differences between them; model 2 is slightly better than model 1. Model 1 predicts drag reasonably well at an assumed angle of attack of 0°, whilst model 2 predicts drag better at assumed angle of attack of -0.5°. For reliable VPP predictions, however, the relationship between angle of attack/ flap angle and drag or lift is arguably less important than the relationship between drag and lift. Furthermore, the model must also perform well at different flap angles and different immersions.

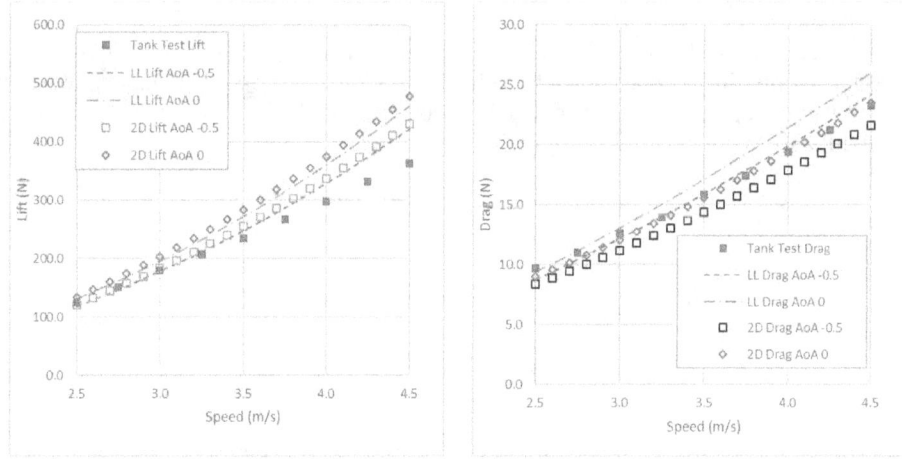

Figure 12 Comparison of approximate models with test data for varying speed: α = 0, φ = 0

Figure 13 shows data for a range of flap angles and angles of attack at a speed of 4.0 m/s. Each curve corresponds to a fixed angle of attack and consists of points representing seven flap angles from –6° to 6° in 2° steps. In these plots the data is again plotted as Drag Area versus Lift Area squared to reduce the effect of the uncertainty of flap angle and angle of attack in the measured data.

It can be seen from these curve that the general trend of the relationship between drag and lift predicted by the very simple model 1 matches fairly well with experiment data, even if the values of lift and drag at a given flap angle are relatively inaccurate. The slightly more sophisticated model 2 generally performs better than model 1 in virtually all cases, although the agreement is noticeably less good for higher angles of attack, which may affect prediction of take-off speed.

This data suggests that where the simple model is implemented in a VPP model the hydrodynamic drag of the foil will be broadly correct for a fixed lift value for the boat when flying (related to the weight), but that the angle of attack and/or flap angle required to achieve that lift value will be incorrect. However, use of this approach would lead to errors if the VPP predicted large positive or negative flap angles where values of lift may be predicted which are not achievable in practice; this might affect take-off speed prediction.

If the VPP uses a model representing the mechanical linkage between the wand and the flap, relating flying height to flap angle, then use of data from these models could lead to the VPP predicting the boat flying at a height which is different to that achieved in practice, with the corollary that the foil immersion will then be incorrect. This will then have a secondary effect on the drag. However, whilst sailing a modern Moth, it is common to adjust both the zero position of the flap relative to the zero position of the wand (using a "ride height" or "bias" adjuster) and the ratio of flap angle movement to wand angle movement (using a "gearing" adjuster) and, so this model is likely to be somewhat arbitrary in any case.

The effect of submergence is shown in Figure 14, in which the data is presented for the tested immersion of 100 mm, or $1.05 \times$ mean chord at an angle of attack of 0° and flap angles of 0° and 6°. With flap angle of 0° there is relatively little change from the deeper submersion case, as seen in Figure 9.

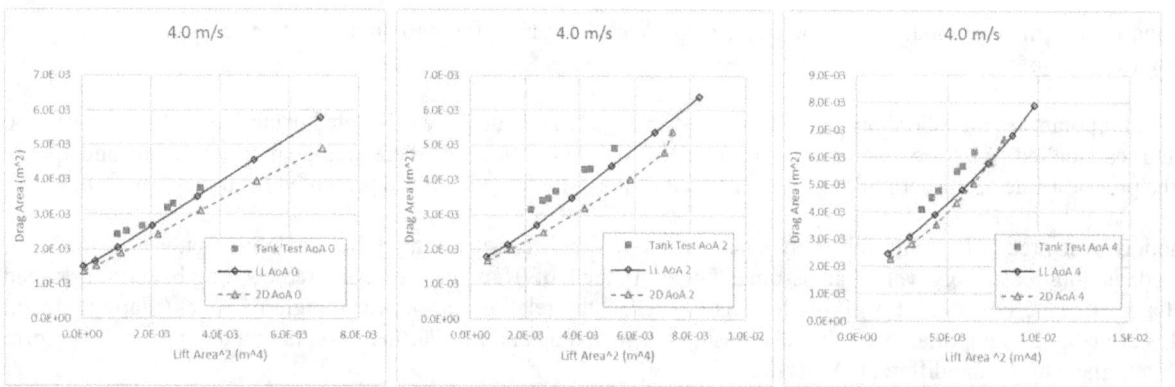

Figure 13 Comparison of approximate models with Test Data: V = 4.0 m/s

The lifting line model performs fairly well, especially at the higher speeds, predicting drag quite accurately, and slightly under-predicting lift, while the simpler model over predicts both lift and drag. At the higher flap angle neither approach works well in this respect; the simpler approach predicts drag quite accurately while dramatically over estimating lift, while model 2 underestimates both lift and drag for large flap angles, as seen in Figure 12.

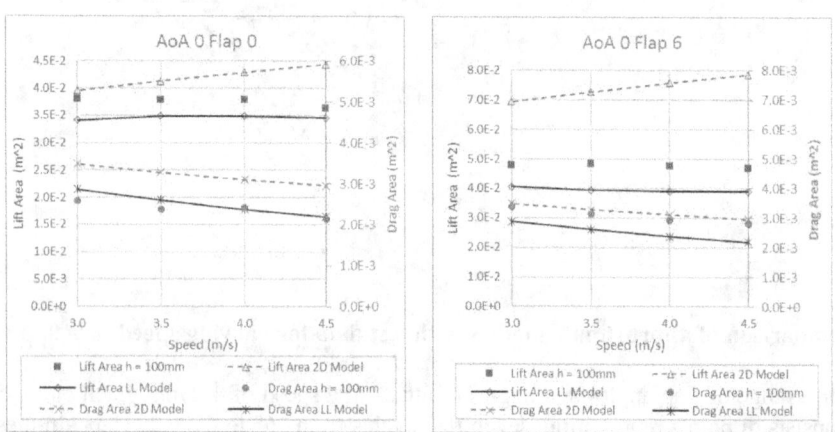

Figure 14 Comparison of approximate models with Test Data: Lift & Drag v. Speed: h = 100 mm

As mentioned previously, however, for performance prediction the most important issue is the relationship between drag and lift. This is shown for the two models at the highest speed in Figure 15. In a manner which may be compared to the discussion of the results of Figure 13, it can be seen the relationship between drag and lift is predicted reasonably well in both models, even though the relationship between angle of attack and drag and lift is not predicted so well for large flap angles. This is especially true in the cases of most practical relevance to small immersion; i.e. those with low angles of attack and small flap angles.

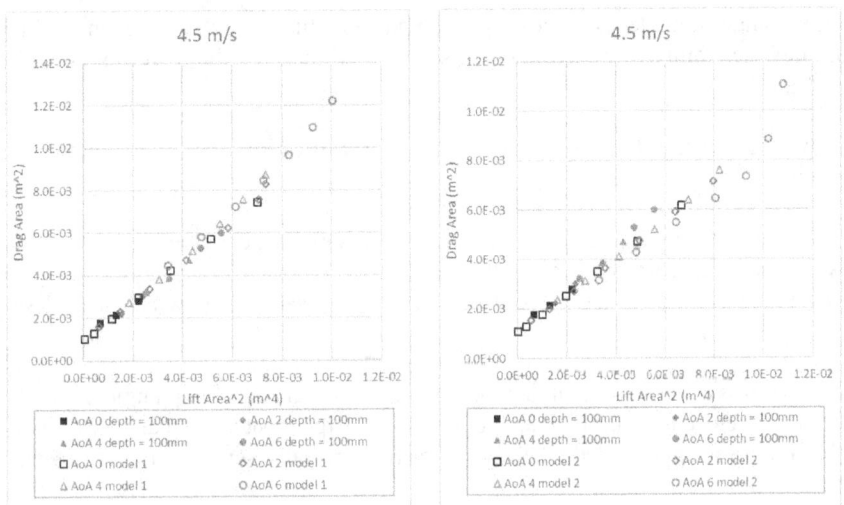

Figure 15 Comparison of approximate models with Test Data: Lift v Drag, h = 100 mm

It should be noted that many combinations of models and corrections could be assembled from the range of semi-empirical models and simplified theoretical models deployed here or in other studies. For example, model 1 could usefully utilize approximate corrections for lift and drag related to planform efficiency (Oswald Factor) such as those suggested by Niţă & Scholz (2012) as

(11)
$$e = 1/\left(1 + f\left(TR - \Delta TR\right)AR\right)$$

with:

$$\Delta TR = -0.357 + 0.45\exp\left(0.0375\Lambda\right)$$

and

$$f\left(TR\right) = 0.0524TR^4 - 0.1500TR^3 + 0.1659TR^2 - 0.0706TR + 0.0119$$

This yields a value of $e = 0.97$ for the current foil. This will lead to improved predictions in some cases.

In general, however, it must be said that further systematic test data is highly desirable for identifying the best simple corrections, especially in terms of free surface effects, in order to identify with confidence the most reliable approach. It should also be noted that it is extremely likely that such simple models will never be able to capture all of the complex flow physics, and that more sophisticated models will be required for high-fidelity predictions.

Nonetheless it can be seen that both of the models examined are broadly successful in predicting the performance of a flapped t-foil; the lifting line approach required a little more effort, but can still easily be set up in an Excel spreadsheet. The simple model can very easily be built into a VPP; whilst the calculation of the lifting line approach may be a little slow for implementation within a VPP, the foil performance may easily be tabulated prior to running the VPP in a manner similar to that used for the XFOIL data in the present calculation. The VPP can then interpolate data for foil performance whilst solving the equations of equilibrium.

CONCLUSIONS

A test rig for evaluating the performance of a flapped T-foil has been designed, built and tested, using a foil from an International Moth dinghy.

Tests were carried out over a range of angles of attack, and flap angles which might be encountered in sailing a Moth; speeds were restricted by the tank capability to a maximum of 4.5 m/s, which is at the lower end of foiling speeds for a Moth.

As well as providing a benchmark data set for analysis of flapped T-foils, a number of practical conclusions may be drawn from the test which can inform future work.

At high angles of attack and large flap angles, when high lift coefficients may be developed, it was found that the control rod fixing the flap angle was bending. From the view the model test this lead to some uncertainty in the flap angle (although it was measured by video during the test). From the view of the sailor, this will lead to less precise control. However, a design solution is not obvious.

When plotting drag against lift for a variety of combinations of angles of attack and flap angle, it is seen that the curves at moderate angles of attack collapse close to a single line, with only very large positive flap angles yielding higher drag for a given lift. This suggests that the performance of the boat will not be especially sensitive to the precise angle fixing the main foil relative to the vertical foil.

The effect of foil immersion has been examined for a small number of cases. In low lift conditions, (small angle of attack and flap angle) which would be expected for Moths flying high, it can be seen that drag is reduced at a given lift compared to deeper immersion. It is presumed that this effect results from the loss of drag on the vertical foil outweighing the gain in drag and loss of lift on the horizontal foil.

This result is expected from observation of Moth sailors who generally fly as high as possible in a given set of conditions. However, for high lift conditions (large angle of attack and flap angle), the reverse is true, and the foil has more drag at the lower immersion. This condition is not of great practical relevance, since Moths flying high will not have large flap angles under normal conditions.

Two simple models have been examined for predicting the performance of the T-foil. Both are sufficiently simple to be implemented on a spreadsheet. The first model uses 2D section data in conjunction with simple empirical equations and some approximate results from lifting line theory; the second model is based on a solution of lifting line theory deploying 2D section data.

It is found that both models predict the relationships between drag and lift fairly well over the range of conditions tested, although the drag is somewhat underestimated in high-lift conditions. The model based on the lifting line generally out-performs the simple model by a small margin. However, both could be regarded as suitable for implementation in a VPP at a level of fidelity appropriate for early-stage preliminary design studies. However, more test data is needed for further validation, especially with regard to some of the more complex flow phenomena, and it is very likely that more sophisticated models will be required to capture all features of the flow physics

Acknowledgements

The authors would like to acknowledge the technical staff of the Kelvin Hydrodynamics Lab: Steven Black and Bill Wright for building the test rig and Dr Saishuai Dai and Grant Dunning for support in testing. We also acknowledge Thomas King and Dr Weichao Shi for carrying out the preliminary study used to troubleshoot the test rig design.

References

Andersson, A., Barreng, A., Bohnsack, E., Larsson, L., Lundin, L. Sahlberg R., Werner E., Finnsgård C., Persson A., Brown, M. and McVeagh, J., "The Foiling Optimist", Proceedings 4th International Conference on Innovation in High Performance Sailing Yachts (Innov'sail), Lorient, France 2017

Beaver, B., Zseleczky, J., "Full Scale Measurements on a Hydrofoil International Moth", Chesapeake Sailing Yacht Symposium, Annapolis, USA, 2009

Binns, JR Brandner, PA and Plouhinec, J, "The effect of heel angle and free-surface proximity on the performance and strut wake of a moth sailing dingy rudder t-foil" Proceedings 3rd High Performance Yacht Design Conference Auckland, 2008

Binns, JR, Ashworth Briggs, A, Fleming, A, Duffy, J, Haase, M and Kermarec, M, Unlocking hydrofoil hydrodynamics with experimental results, Proceedings 4th International Conference on Innovation in High Performance Sailing Yachts (Innov'sail), Lorient, France 2017

Bögle, C., Hochkirch, K., Hansen, H., Tampier-Brockhaus, G, "Evaluation of the Performance of a Hydro-Foiled Moth by Stability and Force Balance Criteria", 31. Symposium Yachtbau und Yachtenwurf, Hamburg, Germany November 2010

Coffee, C. W. and McKann, R. E. "Hydrodynamic Drag of 12- and 21- percent Thick Surface Piercing Struts" NACA Technical Note 3092. 1953

Daskovsky, M, "The hydrofoil in surface proximity, theory and experiment", Ocean Engineering, Volume 27, Issue 10, Pages 1129-1159, October 2000

Drela M., "XFOIL", http://web mit.edu/drela/Public/web/xfoil/, 2013

Hoerner, S., "Fluid Dynamic Drag", Hoerner Fluid Dynamics, Albuquerque, USA, 1965

Findlay, M. W., Turnock, S. R. "Investigating sailing styles and boat set-up on the performance of a hydrofoiling Moth dinghy", Proc. 20th International HISWA Symposium on Yacht Design and Construction, Amsterdam, Netherlands, 2008

Gibbs and Cox inc. "Hydrofoil Handbook Vol II: Hydrodynamic Characteristics of Components" ONR Hydrofoil Research Project, 1954

Mackenzie, J.R. "Can a Flapless Hydrofoil Provide a Realistic Alternative To a Standard Moth Foil With a Flap?" HISWA Symposium Amsterdam, Netherlands, 2014

Niţă, M. and Scholz, D. "Estimating the Oswald Factor from Basic Aircraft Geometrical Parameters" Deutscher Luft- und Raumfahrtkongress 2012

Ramsen, J. A. and Vaughan, V. L. "Hydrodynamic Tares and Interference Effects for a 12-Percent Thick Surface-Piercing Strut and an Aspect-Ratio-0.25 Lifting Surface. NACA Technical Note 3420, 1955

Ripken, J. F. "Interference effects of a strut on the lift and drag of a hydrofoil" Bureau of Ships project No SF 013 02 01, Task 1702, ONR Contract Nonr-710(39), 1961

Vladimirov, A. N. "Approximate Hydrodynamic Design of a Finite Span Hydrofoil" NACA Technical Memorandum 1341, 1955

Wadlin, K.L., Shuford, C.L., Jr., McGehee, J.R., "A theoretical and experimental investigation of the lift and drag characteristics of hydrofoils at subcritical and supercritical speeds". NACA Report No. 1232. 1955

Wenzinger, C.J. and Harris, T. A. "Wind-tunnel investigation of an NACA 23012 airfoil with various arrangements of slotted flaps" NACA Report 664, 1939

Impact of Composite Layup on Hydrodynamic Performances of a Surface Piercing Hydrofoil

V. TEMTCHING, Ecole Navale,Brest-France and SEAIR, Foil Resource Center, Lorient-France
B. AUGIER, IFREMER, Laboratoire du comportement de structures en Mer, Brest-France
T. DALMAS, IFREMER, Laboratoire du comportement de structures en Mer, Brest-France
N. DUMERGUE, IFREMER, Laboratoire du comportement de structures en Mer, Brest-France
B. PAILLARD, Alternative Current Energy, Bordeaux, France

ABSTRACT

Composite materials are good candidates for hydrofoils manufacturing, ensuring a good balance between strength and weight. In the high performances sailing yacht domain, hydrofoils are thin structures, highly loaded that experience significant displacements. This study investigates experimentally and numerically the influence of the laminate layup on the hydrodynamic performances of a surface piercing hydrofoil. Four hydrofoils with a constant chord, geometrically identical with different composite layups are mechanically characterized and tested in a hydrodynamic flume. The foils are designed to have a significant tip displacement of 5 to 10% of the span. Experimental results highlight a bending-twisting effect that leads to significant change in the hydrodynamic performances of the structures. Two different FSI numerical approach: from a potential code coupled with beam theory to the full coupling of a shell structural code and a VOF hydro model with free surface are presented and the first one is compared to the experiments with great results. The two approaches are two complementary bricks in the design process to compute the effect of passive deformation on hydrodynamic performances of the foils and therefore the yacht stability.

Key words: Bending-twisting coupling, Composite materials, Fluid Structure Interactions, Hydrofoils.

NOTATION

α, AoA	Angle of Attack [°]
DSBT	Differential Stiffness Bend-Twist
E_f	Young modulus of the fiber
E_l	UD 's Young modulus in longitudinal direction $[MPa]$
E_m	Young modulus of the resin
e_p	UD 's thickness $[mm]$
E_t	UD Young modulus in transverse direction direction $[Mpa]$
FFT	Fast Fourier Transform
FSI	Fluid structure Interactions
G_{lt}	Shear modulus of the UD in direction lt $[Mpa]$
G_m	Shear modulus of the resin $[Mpa]$
LT	Laminate Theory
ν_f	Poisson coefficient of the fiber
ν_m	Poisson coefficient of the resin
ν_{lt}	Poisson coefficient of the UD in direction lt

PAC	Passive Adaptive Composites
POM	Polyoxymethylene material
UD	Unidirectional ply made of resin and fiber
U	Velocity [m/s]
VOF	Volume of Fluid
$V_f[\%]$	Volume of fiber in the composite
VAM	Variational Asymptotic Method
VLM	Vortex Lattice Method
Y	normal displacement of the foil $[mm]$
Z	foil dimension in spanwise $[mm]$

INTRODUCTION

For the top high performances foiling yacht design, the design process is a complex combination of free surface hydro-dynamic simulation coupled with highly loaded composite structure analysis. The hydrodynamic loading leads to significant deformations of the structure that must be consider in the design. These deformations can have an impact on the hydro-dynamic performance and the global yacht equilibrium (Balze et al., 2017). Within the available literature, a part of the described problem has been studied as the performance of a surface piercing straight foil (Young and Brizzolara, 2013), the coupled effect of fluid structure interaction FSI on cavitation (Ducoin et al., 2012) or the Velocity Program Prediction due to the apparent wind seen by the foiler (Hagemeister and Flay, 2017).

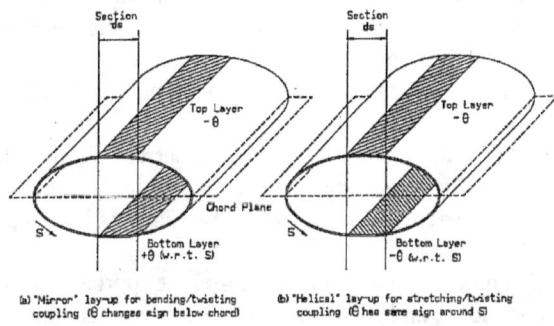

Figure 1: type of asymmetric lay-ups required to produce (a) bending-twist and (b) tension-twist coupling. (Veers et al., 1998)

The structural deformations can have a devastating effect on structure and are highlighted by the fluttering effect on the keel. The bend-twist coupling in the structure can be a good approach to control the lifting force through FSI with a passive adaptive method. Composite structures are extensively used for the appendages in the sailing yacht domain, putting more complexity in the behavior prediction.

Bend twist coupling in composites has been extensively used for wind turbine blade applications. Works on wind turbine have described that the phenomenom depends on the plies orientations in the layup Fig 1 (Veers et al., 1998) or more recently (Capellaro, 2012), where bend-twist coupling is evaluated with the plies orientation for glass and carbon fibers.

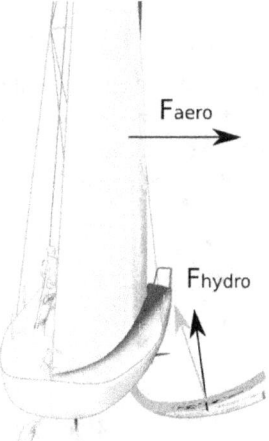

Figure 2: Effect of deformation of the hydrofoil on the hydrodynamic resultant on the 747 flying Mini.

Regarding hydrofoil application,(Giovannetti, 2017) has recently described the use of different techniques for bending-twist coupling as Passive Adaptive Composites (PAC) that tailors the response of a structure by changing the orientation of the composite plies (Veers et al., 1998) and Differential Stiffness Bend-Twist (DSBT) that utilizes the internal stiffness of a structure to change the aero-hydrodynamic response to fluid load (Raither et al., 2013). They have shown that the PAC can be used to control the lift response to hydrodynamic load in the case of a composite structure (Giovannetti et al., 2018) including by decreasing the tip load. (Young et al., 2018) also studied the effect of bend-twist coupling on the hydro-elastic response of a composite hydrofoil in a cavitation tunnel.

Indeed, bend twist coupling is a target effect for highly loaded structures to reduce the loads by reducing the incidence at the tip. (Gozcu et al., 2015),(Rohde et al., 2015) investigated the effect of this coupling on natural frequencies and mode shape to improve the design and the control of composite structures.

The present study investigates the impact of composite layup on hydrodynamic performances of a hydrofoil piercing the free surface in the case of large displacements up to 10% of the span, at the tip. The application test case, the so called 747 flying mini 6.50 (Figure 2) indeed experiences such important deformation. As a first approach, the problem is simplified by a foil piercing the free surface at $45°$, four composite hydrofoils made up of carbon or glass fibers are built for the experiments.

The first chapter focuses on the mechanical characterization of the composite hydrofoils through a comparison of different experimental techniques. These results are also compared with theoretical models based on Laminate theory (LT)on one hand and with numerical approaches on another hand: the in house "FS6R", based on a Variational Asymptotic Method (VAM) applied to a non linear composite beam theory (Hodges, 2006) and the commercial software ABAQUS.

The second chapter is dedicated to the hydrodynamic experiments carried out in a flume for the four composite hydrofoils, with FSI and free surface interaction. The foils displacements and hydrodynamic loads are measured and compared to a reference case corresponding to a rigid body computation. Vibration analysis is also performed to quantify the resonance frequencies in water.

Last part described the FSI numerical simulations based on two different approaches. The first tool, "FS6R", is a coupled approach between the potential flow code AVL, and an internal code based on beam theory by finite elements that integrates cross section properties calculation. "FS6R" aims to compute FSI on a hydrofoil during the pre-design process. The second tool, much more time CPU consuming, is a coupled approach between the composite shell structure code ASTER and the viscous VOF flow solver interFOAM. Numerical and experimental comparison are presented based on the potential simplified approach when the the second one is a work in progress.

The principal objective of this paper is first to find out is to quantify and measure the effect of composite lay-up on bend-twist coupling and the hydrodynamic performances and to validate the different numerical approaches.

Hydrofoils characterization

This chapter describes briefly the manufacturing process and the mechanical characterization of the hydrofoils in air.

Hydrofoils Manufacturing

Four hydrofoils are manufactured with the help of SEAir, the industrial partner of this research work. The foils are straight structures of $1.35m$ span and $0.114m$ constant chord. The section is a NACA 0013 sandwich structure made of an AIREX web and laminated skin illustrated in Fig 3. In the cut-section, the green AIREX is clearly visible, the black part is the glue Spabond 345 (shows up as orange in CAD figure) and the white part is the glass fiber.

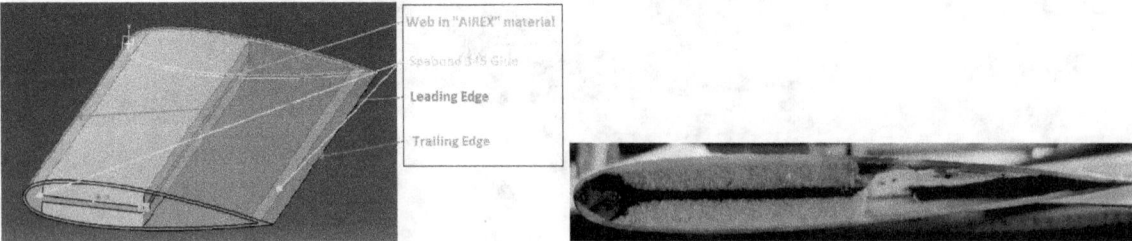

Figure 3: Mechanical structure of the composite foils and cut section of the P_2 foil. The orange glue in the sketch is visible in black in the section.

The four different layups and associated materials are given in Table 1. The manufacturing of the foils is realized by a vacuum lamination method and the overall process goes through several steps including: stratification in the mold, glueing of the web on the skin, intrados-extrados glueing, demolding, foil base manufacturing and finishing.

The foils base (Fig 4) is directly manufactured on the foils by molding using a 3D printed piece to position the hydrofoil in the mold. Laminate plates of $200 \times 200 \ mm^2$ with the layup of Table 1 are manufactured simultaneously for the tensile specimens.

Foils	Material	layup
P_1	Epoxy-Glass2	$[(\pm45)_2/0_{0.5}]_{sym}$
P_2	Epoxy-Glass1	$[(90/-45/0_{0.5}]_{sym}$
P_3	Epoxy-Glass1	$[(90/45/0_{0.5}]_{sym}$
P_4	Epoxy-Carbon	$[(90/0)_{sym}]$

Material	Carbon	Glass1	Glass2
$E_l[GPa]$	120	54	45
$E_t[GPa]$	10	10.4	10
$G_{lt}[GPa]$	4	3.9	5
ν_{lt}	0.362	0.25	0.25
$e[mm]$	0.3	0.2	0.2
$V_f[\%]]$	55	60	60

Table 1: Laminates Layups and Mechanical properties of the unidirectional plies, the matrix is made up of Epoxy. The layup P_1 uses the Glass2 when layup P_2 and P_3 uses the Glass1.

The results expected with these layup configurations are:

- P1: Important bending and no bend twist coupling

- P2: Bend twist coupling with a negative twist and higher bending when negative incidences are investigated

- P3: Bend twist coupling with a positive twist and smaller bending when negative incidences are investigated

- P2: Small bending and no bend twist coupling

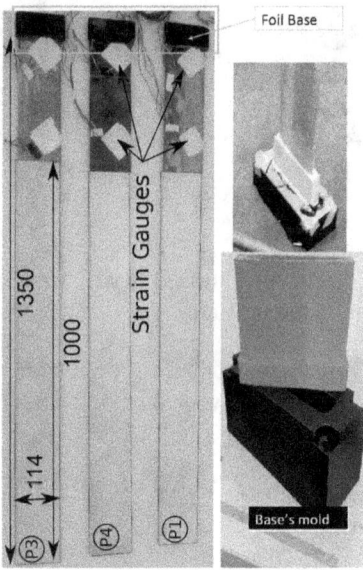

Figure 4: Hydrofoils equipped by strain gauges and foil base in black (Dimensions are in mm). The white coat represents the wetted surface, the foil base 's mold is in black and the connection piece are shown on the right.

Mechanical characterization

The straight foils are made of an extruded constant chord and a constant composite layup along the span. The structure will be then considered as a beam, with an equivalent EI modulus constant along the span. Three different methods are used here to experimentally determine the EI modulus of the 4 composite structures.
Three different methods are used here to experimentally determine the EI modulus of the 4 composite structures:

- A cantilever bending test with known masses

- A natural frequency vibration test

- A tensile test on representative specimens

Experimental setup

For the cantilever bending test and the vibration test, the clamping system used during the hydrodynamic test is mounted on a rigid structure to reproduced the same boundary conditions. Displacements of the structure are measured by a laser telemeter, also used during the hydro test, at the tip of the foils.

Bending tests: Four different calibrated masses $M_1 = 518g$, $M_2 = 1018g$, $M_3 = 2018g$ and $M_4 = 3018g$ are applied to the considered center of hydrodynamic load and the laser measures the vertical displacement Y of the foil at the tip at distance X from the root (See Figure 5). The bending stiffness EI of the structure is calculated with the elastic straight beam relation (1).

$$EI = \frac{FL^2(3X - L)}{6Y} \tag{1}$$

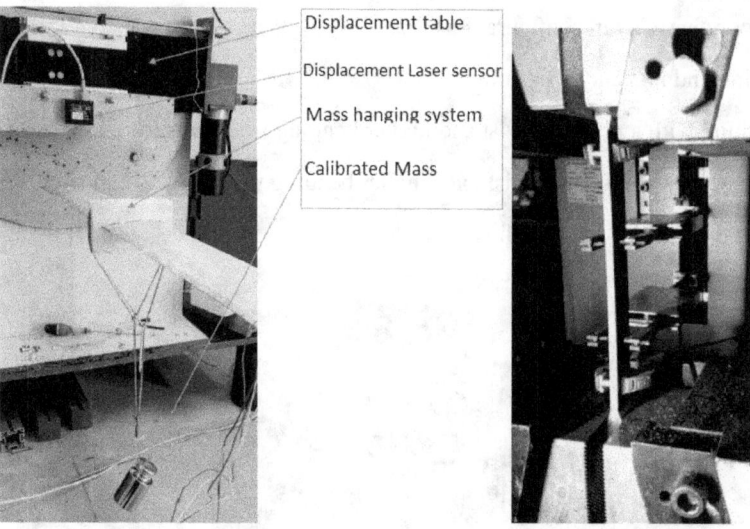

Figure 5: Left: Bending test on a cantilever foil with calibrated mass. The vibration tests are made with the same set up. Right: Tensile tests setup used for the laminate characterisation showing a specimen fixed in the machine's jaws.

Vibration tests: the structure is manually displaced and release; the laser measures the vertical displacement over time at the tip. For each case, two tests are performed and the average is presented in the results. The bending stiffness EI of the structure from vibration analysis is then calculated by (2).

$$EI = \left(\frac{(2\pi f_i L^2)^2}{\lambda_i^4} \right) m \tag{2}$$

f_i: the natural frequency of the mode i
λ_i: a number associated to the proper mode i, depending of the boundary conditions
m: the structural weight per unit of length
L: the length of the structure.
The natural frequency is obtained by a FFT on the temporal signal recorded by the laser as illustrated in Fig 6. In this case the the foil first bending mode is $3.9Hz$ giving a bending stiffness of $EI = 157.87N.m^2$.

Tensile tests: For each layup, three specimens complying with the ISO 527 standard are used and a tensile test is carried out on each. A specimen is fixed in the jaws of the machine (Fig 5) and the displacement speed is fixed.
As output of the machine, the temporal evolution of the force and deformation are recorded.The Young's modulus E is evaluated as the slope of the elastic stress versus the elastic strain. The specimens of each layup gives good agreement results and E is the average value. The bending stiffness EI of the structure is obtained by the product of the measured E and the inertial momentum I calculated on the hydrofoil section, taking into account the skin thickness of the laminates.

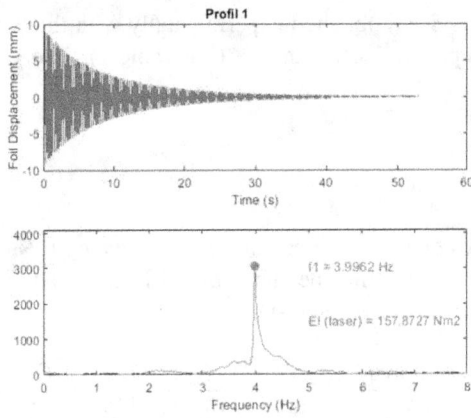

Figure 6: Vibration response of P1.

Calculation of composite structural properties

Three different approaches are used to calculate the mechanical properties of the composite foils:

- An in house tool based on laminates theory

- An in house tool "FS6R" based on a Variational Asymptotic Method (VAM)

- The commercial code ABAQUS

Laminates theory describes a ply in its membrane plane and calculates its properties E_l, E_t, G_{lt}, ν_{lt}, e_p as a function of the fibers and resin properties E_f, E_m, G_m, ν_f, ν_m.
(Gay, 1991) describes the theory and shows the functions calculating the plies properties in the membrane plane and the equivalent properties of a stratified structure in all the directions.
EI is obtained by the product of E from the laminates theories by the inertial momentum I calculated with the real skin thickness.
The commercial software ABAQUS uses its meshed beam cross-sections function which allows the description of a beam cross-section including multiple materials and complex geometry.
The mechanical part of "FS6R" is presented in the next chapter.

Comparison of the hydrofoil stiffness

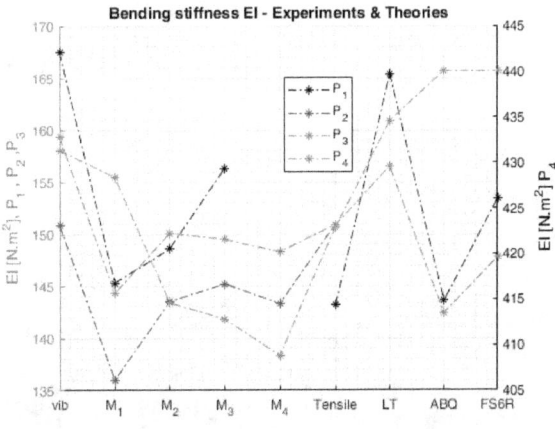

Figure 7: Comparison of the EI measured vibration test, bending test with the mass M1, M2, M3, M4 and tensile and the EI calculated with laminate theory LT, Abaqus ABQ and the in house code FS6R.

Figure 7 compares the bending stiffness EI obtained experimentally with the numerical results from the laminates theory, ABAQUS and FS6R for all the layups P_1, P_2, P_3 and P_4 (see Table 1). The y-axis gives the EI values and the x-axis represents the different methods:
vib: the mean value of the vibration tests .
M_i: The bending test using the mass M_i as load.
LT: Laminate theory using the properties defined in Table 1.

The different experimental techniques give EI modulus close in a range of 13% when comparing a test to another. The high differences observed with the vibration tests may be due to the stiffness of the clamping support. Table 2 gives the mean value of EI computed as an average of the experimental tests.

Foil Layup	P_1	P_2	P_3	P_4
EI $[N.m^2]$	155.6	153.8	145	420.5

Table 2: Bending stiffness of the different foils

A comparison of the analytic and numerical results to these reference values show low discrepancies up to 10% with LT (reached with P_1), 3% with ABQ (reached with P_4) and 5% for FS6R (reached on P2, P3). The results show a good confidence in the different approaches. With these simple extruded structures, one should prefer the very simple FS6R tool when ABAQUS is time consuming for no gain.

FS6R numerical model

FS6R is a code dedicated to the preliminary stages of foils design to model and to analyze the structure through fluid structure interactions simulations. It stands on the lifting line method and beam theory by finite elements and its algorithm is presented in Figure 9.

After defining the foil geometry, the materials and the configurations to simulate, the fluid flow is solved by the open source tool AVL which performs a VLM in-viscid 3D calculations on the whole surface and provides the hydrodynamic forces. A viscous correction is then realized with XFOIL which performs 2D viscous simulations and the structural analysis is performs by an in-house code standing on beam theory by finite elements. This code uses the Variational Asymptotic Method describes in (Hodges, 2006) to calculate the properties of a given section associated to a material such as: the shear center, torsional center, young modulus, inertia, shearing stiffness GJ, etc.

As an output of FS6R calculation, we get the efforts applied on the structure and the distorted shape. More details on FS6R can be found in (V. Temtching and D.R, 2018) where a validation on a 3D trapezoidal foil in POM material is shown.

This part presents the implementation of bend twist coupling in structural analysis of the code.

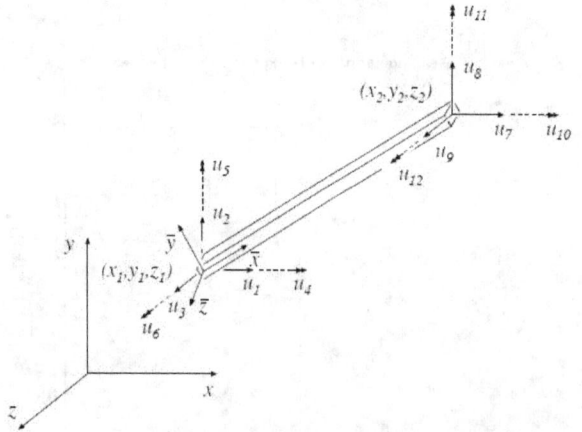

Figure 8: DOF of a 1 element beam. (Hodges, 2006)

Figure 8 shows a 1-element beam with the DOF on the two nodes. We have 3 DOF in translation and 3 DOF in rotation for each node which are the solution of the system (3).

$$K \times U = F \tag{3}$$

K: the stiffness Matrix in the global reference, the matrix is symmetric.
F: Forces applied to the structure in the global reference.
U: DOF of the structure.

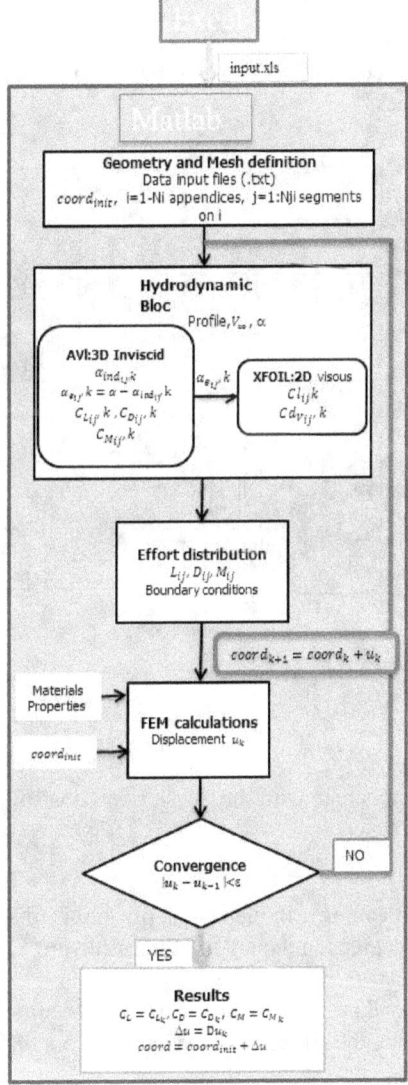

Figure 9: FS6R organizational chart.

To consider bend twist coupling in FS6R, coupling terms usually neglected for quasi-isotropic or orthotropic materials, and set as zero in the stiffness matrix, are replaced by bend twist coupling terms c_{ij}, described in (Capellaro, 2012) as presented in Fig 10.

c_{ij} depend on the coupling percentage $\alpha[\%]$ described in (Capellaro, 2012) for glass and carbon fibers, that varies with the plies orientation, the torsional stiffness and the bending stiffness (see Fig 11). The proportion of plies in the laminate responsible of this coupling is also consider.

The coupling terms are the calculated by (4)

$$C_{ij} = \sum_{\theta=0}^{90} \alpha_\theta \times A_\theta \times \sqrt{k_{ii} \times k_{jj}} \tag{4}$$

82

$$K = \begin{bmatrix} k_{11} & k_{12} & k_{13} & 0 & k_{15} & k_{16} & k_{17} & k_{18} & k_{19} & 0 & k_{111} & k_{112} \\ & k_{22} & k_{23} & c_{24} & k_{25} & k_{26} & k_{27} & k_{28} & k_{29} & c_{210} & k_{211} & k_{212} \\ & & k_{33} & c_{34} & k_{35} & k_{36} & k_{37} & k_{38} & k_{39} & c_{310} & k_{311} & k_{312} \\ & & & k_{44} & c_{45} & c_{46} & 0 & c_{48} & c_{49} & -k_{44} & c_{411} & c_{412} \\ & & & & k_{55} & k_{56} & k_{57} & k_{58} & k_{59} & c_{510} & k_{511} & k_{512} \\ & & & & & k_{66} & k_{67} & k_{68} & k_{69} & c_{610} & k_{611} & k_{612} \\ & & & & & & k_{77} & k_{78} & k_{79} & 0 & k_{711} & k_{712} \\ & & & & & & & k_{88} & k_{89} & c_{810} & k_{811} & k_{812} \\ & & & & & & & & k_{99} & c_{910} & k_{911} & k_{912} \\ & & & \text{Sym} & & & & & & k_{1010} & c_{1011} & c_{1012} \\ & & & & & & & & & & k_{1111} & k_{1112} \\ & & & & & & & & & & & k_{1212} \end{bmatrix}$$

Figure 10: Global stiffness matrix of the structure with the added bend-twist coupling terms c_{ij} in red and torsional stiffness in yellow. $k_{1010} = k_{44}$ is the torsional stiffness.

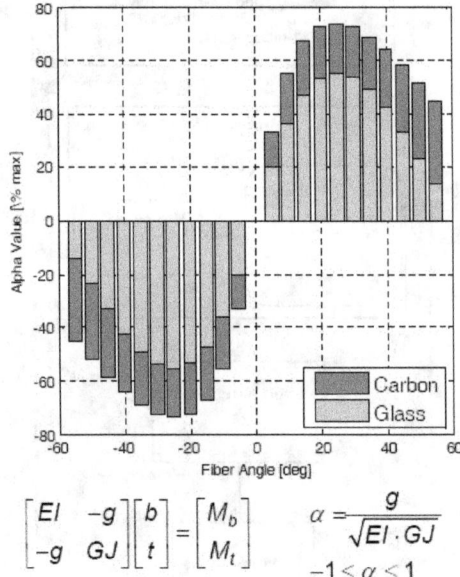

$$\begin{bmatrix} EI & -g \\ -g & GJ \end{bmatrix} \begin{bmatrix} b \\ t \end{bmatrix} = \begin{bmatrix} M_b \\ M_t \end{bmatrix} \qquad \alpha = \frac{g}{\sqrt{EI \cdot GJ}}$$
$$-1 \leq \alpha \leq 1$$

Figure 11: Limit of bend twist coupling percentage with the plies orientation for glass and carbon fibers. (Capellaro, 2012)

θ: is the ply orientation [°].
α_θ: The maximum percentage of bend twist coupling induced by a ply, oriented at θ degrees.
A_θ: the percentage of the ply oriented at θ degrees in the layup of the structure.

These new terms of the stiffness matrix will induce a twist angle when the structure is loaded by a bending for or a bending moment and therefore a bending motion when the structure is loaded by a torsional moment.

Flume experimental setup

The described foils are tested in a flume in order to measure the impact of the composite layup on their hydrodynamic performances.

Hydrodynamic tests are carried out at IFREMER Lorient 's flume in a working section of $2.5m$ in the flow direction and $1.5m$ depth with a maximum velocity of $1m/s$ (see Fig 12). The foil is clamped to a 6-DOF balance measuring the efforts and moments express in the references frame of the foil and displacements are measured with a laser telemeter through an underwater window (See Fig 12).

The lateral displacements of the foils are measured at 3 different heights. The laser sweeps 10 consecutive times the foil in chord wise at the different height in order to average the distance. This method is preferred to a single point measurement to distinguish the bending from the twist. A calibration process is used to correct the diffraction effect due to the different interfaces that the laser encounters.

As shown on Fig 12 and 13, the foils are piercing the free surface with an angle of $45°$ and mounted cantilevered on the

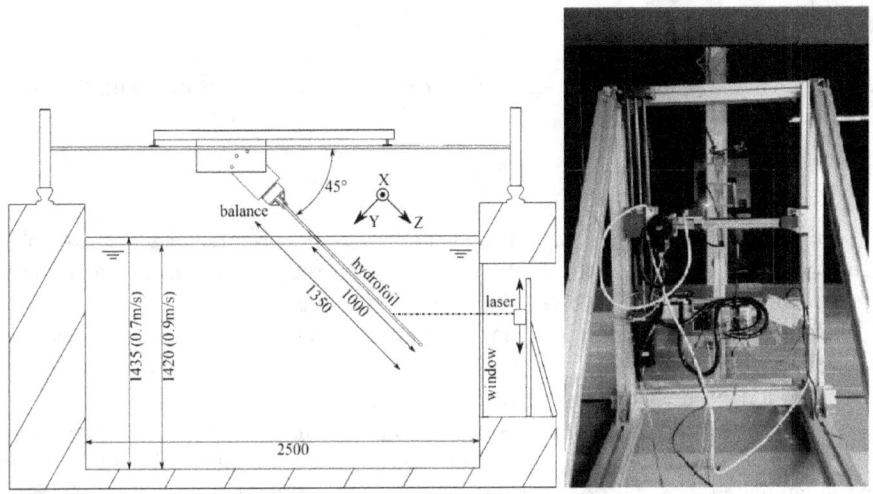

Figure 12: Left: Experimental setup of the hydrofoil test mounted at a 45° angle in a flume. The water level depends of the flow speed. Dimensions are in mm. Right:Calibration set up for the Laser telemeter system. Three defined geometries made up of 3D printed material (shows up in orange) are placed at the level of the path of the laser

balance.

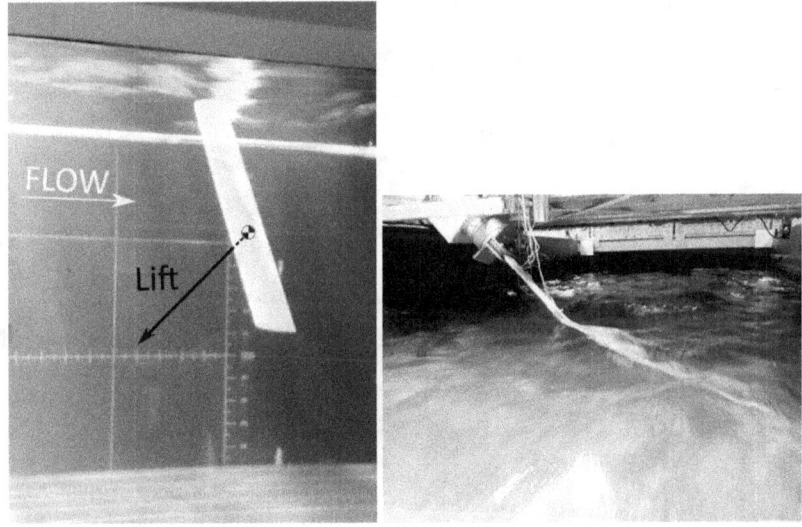

Figure 13: Hydrofoil tested in IFREMER Lorient flume at 0.9ms and AoA= −9°. Picture on the right shows the balance, the clamping system and the strain gauges wires.

The clamping conditions are the same used for the mechanical characterization tests; the flow is aligned with X-axis. Most of the tests are done with a negative AoA to enhance the laser telemeter measurement and to prevent the tip to get to close to the side of the flume (Fig 10). The 45° tilt angle is chosen to represents the interaction of the MINI 747 foil with the free surface and to analyze its effects on the hydrodynamic performances.. It also has the advantage of maximizing the immersed surface with a low confinement effect. The foils are equipped by 2 full bridges strain gauges installed at respectively at $350mm$ and $250mm$ from the embedding. They are placed on the dry part of the foil so they do not impact the flow. The use of two full bridges is chosen to get the hydrodynamic load resultant position on the span. Two speeds $0.7m/s$, $0.9m/s$ and several angles of attack ranging from $[-9°, +3°]$ are investigated. As most of flume, the free surface height varies with the velocity: $1.435m$ for $0.7m/s$ and $1.42m$ for $0.9m/s$.

Hydrodynamic test results

The foil displacements recorded with the laser and the hydrodynamic loads measured with the 6-DOF balance are compared with FS6R simulations.

Foil displacements

Foil displacements are measured at three different span positions with the laser telemeter. The displacements for $-5°$ and $-7°$ at $0.9m/s$ are represented in Fig 14 and in Fig 15, they respectively compare experiments to a FS6R simulations without and with bending twist coupling terms in the stiffness matrix for all the layups.

Without the coupling, the plies orientation sign of $\pm45°$ in the layups is not taken into account and leads to the same simulation results for P_2 and P_3 (See Fig 14). In FS6R computations: $P2$, $P3$ are very close to $P1$. They see the same load and the displacement at the top is around 3% higher due to the different stiffness.

When bend-twist coupling is considered, displacements of P_3 decreases and P_2 increases experimentally and numerically. We observed at the tip displacements up to $77mm$ (5.8% of span) for P_3 glass fiber foil and only $68mm$ (5.01%) for the P_2 with the same type of laminate (P_2 and P_3 only differs by the plies orientation of $-45°$ for P_2 and $+45°$ for P_3) when looking at $-7°$ configuration.

As expected in the design process and in agreement with its highest bending stiffness (depicts in Figure 7), the carbon foil P_4 experiences the lowest displacement with $40mm$ at the tip (3% of its span length).

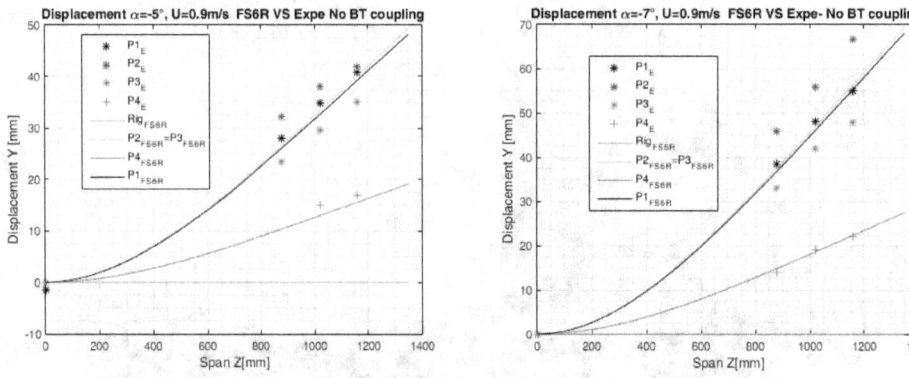

Figure 14: comparison of the measure Z displacement of the foils with a no bending twisting coupling calculation.

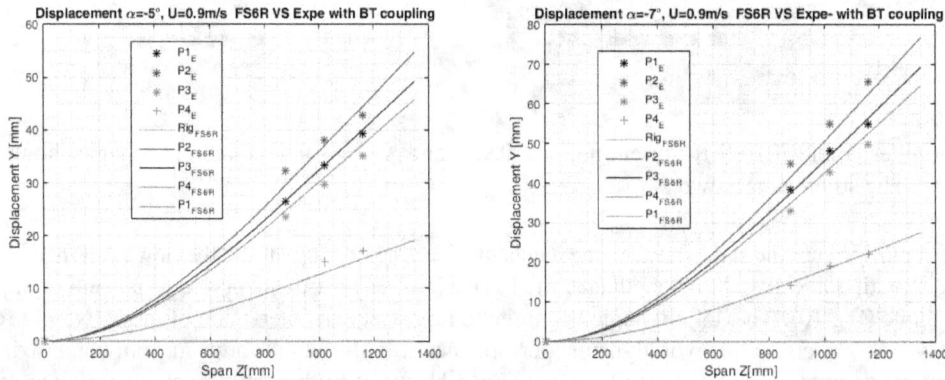

Figure 15: comparison of the measure Z displacement of the foils with a bending twisting coupling calculation.

These measurements highlight different behaviors that are linked both to the composite layup and the materials of the tested foil. First, the carbon foil, with the same number of plies is significantly more rigid than the glass fiber foils. The second behavior is directly linked to the subject of this paper, meaning the bending-twisting coupling. As illustrated in the Fig 13, the coupling coefficient g in the mechanical matrix, described in chap 3, differentiates significantly the simulated displacement of P2 and P3 and leads to a good fit of the experimental results. The -45° orientation of the plies in the

P2 composite structure leads to a negative twist of the structure, with the direct consequence of loading the tip (negative incidence investigated). The opposite phenomenon is expected on P3 where the structure contains $+45°$ plies.

Effects on hydrodynamic

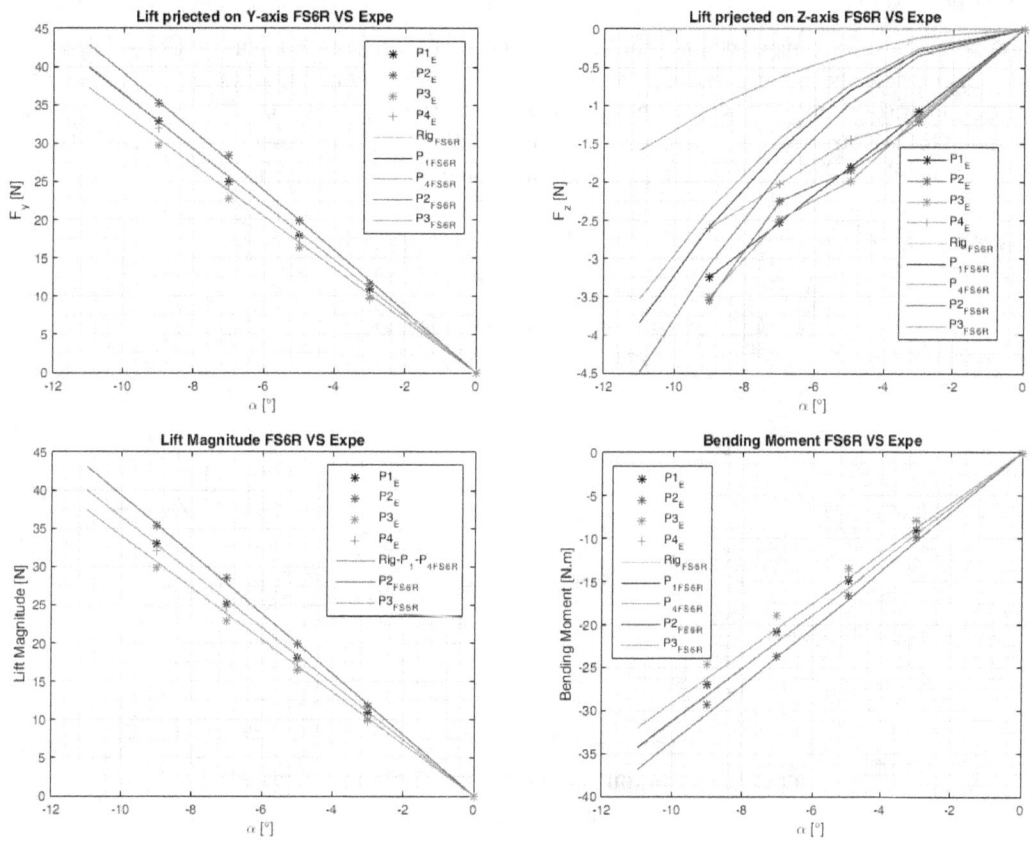

Figure 16: Experimental forces compared to FS6R calculations.

The efforts recorded during the experiments and the computations from FS6R tool for all the incidences at velocity $0.9m/s$ are presented in figure 16. Lift magnitude and its projections on Y-axis (F_y) and Z-axis (F_z) are shown for the 4 foils and for a rigid case (only fluid calculation, no FSI) simulated with FS6R. Many observations can be made with this curves.

Experimental and FS6R results have the same trend and the values fits perfectly excepts for Z projection of the lift force. This value is very small and does not affect the lift force nor the moment leading to a good agreement of both approaches. The maximum discrepancy observed is around 5% on the lift force for P_3 with experiments smaller than FS6R (same observation on the displacements).

The lift force clearly exhibits the different behaviors of the layups: P_1 and P_4 overlap with the rigid case as expected, highlighting no bend twist effect. $P3$ is smaller than the rigid case and P_2 is higher. This behavior is exactly what we expected, because negative incidence is investigated, P_2 with the negative twist is the most loaded when P_3 with the positive twist become the less loaded.

We also observe that when P_1 and P_4 lift magnitude are the same, their projections F_y and F_z are different due to their different bending deflection.

Experimental lift projection on Z-axis does not have a well defined trend which may be due to the balance precision when having small values as for the case.

F_z from FS6R computations has the same trend as the displacements, the force is transferred to the span wise direction due to the foil deflection, modifying the hydrodynamic behavior.

We observe that F_y match very well in both approaches and experimental values are slightly higher. The bending momentum M_x being an image of F_y, we have the same agreement.

We are thus able to estimate the impact of bend twist coupling on lift force distribution.

Since potential flow theory gives good trends for Drag coefficient with underestimated values, Drag force is not presented in this part.

Strain gauges measurements

The measures from the strain gauges don't allow at this stage to recover the resultant force position due to some measurement inconsistencies described in the chapter introduction. The possibility of replacing the 6-DOF balance on the foil by inexpensive and easy to use of strain gauges as full bridges will however be explored. The cross correlation of the balance M_x and the gauges signal is close to one. The dynamic and the resolution of such equipment are perfectly adapted. Transverse deformations and calibration process are the main direction for an upgrade of the system.

Natural frequencies in water

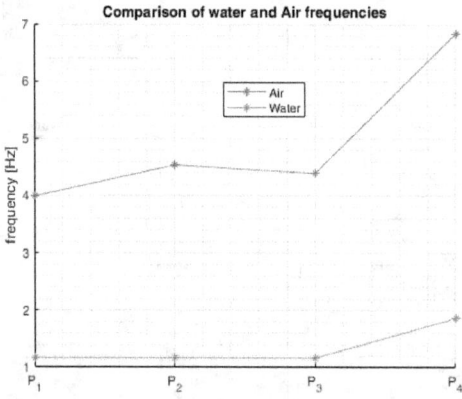

Figure 17: Hydrofoil natural frequencies in air and water.

Figure 17 shows the natural frequencies measured in water in condition identical to the mechanical characterization in air. The natural frequency is, indeed, impacted by the added mass due to the acceleration of the water around the foil which leads to a reduction of the frequency. The added mass can be computed from the natural frequency formula(from Eq. 2) by adding the added mass m_a in the mass term as depicted by (5).

$$f_i = \frac{\lambda_i^2}{2\pi L^2} \times \sqrt{\frac{EI}{m + m_a}} \tag{5}$$

Table 3 shows the added masses calculated for each case.

Foil Layup	P_1	P_2	P_3	P_4
m_a $[kg/m]$	10.03	10.43	9.73	10.84

Table 3: calculated added mass in water

The first bending mode frequency is very low and could be subjected to external solicitations at resonance frequency (wave, ...).

Work in progress: ASTER-InterFOAM coupling

The fluid structure interaction is simulated with OpenFoam for the fluid aspects and code Aster for the composites structure modeling. The fluid simulation is carried out with the volume of fluid phase-fraction based interface capturing approach named interFoam, with a steady mesh. The composites structured is solved with a shell composites layup model. The coupling is done by :

- Launching OpenFoam runs on the deformed foil computed by code Aster

- Launching code Aster runs with loads computed by OpenFoam and projected on the structural mesh (wall shear stress tangential to the shape and pressure normal to the shape)

Going back and forth between fluid and structural solver typically converges in 4 to 6 iterations. Fig 18, shows a first FSI

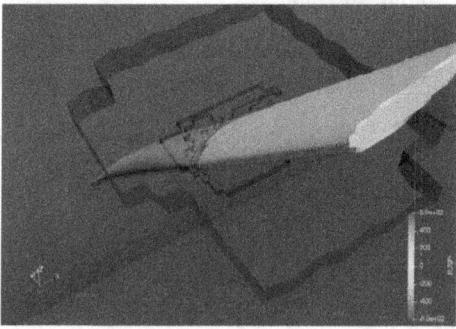

Figure 18: Deformed Hydrofoil computed with InterFoam coupled to code Aster, presented in the flow field. The bottom is the cantilevered part and the tip is at the top in the picture.

simulation computed with the coupling of code ASTER and InterFoam solver, depicting encouraging results.
This approach will allow to take into account, the free surface interaction, to evaluate the drag and to consider more complex structures.

Conclusion

This work presented an experimental and numerical study of the impact of composite layup on the hydrodynamic performances of a surface piercing hydrofoil.
The structural and mechanical characterization of 4 foils from an identical mold but having different layup and material has shown small discrepancies between the different approaches. The comparison, based on the EI modulus of the straight constant section foil, has demonstrated the ability of the Laminate Theory, as well as the FS6R structural module to well estimate the bending stiffness of the composite structure. Discrepancies may be significantly higher with more complex composite layups.
The foils are tested in hydrodynamics flume and show different behaviors due to the layup and the materials. The carbon foil P_4 is significantly stiffer, as expected during the design process. No bend twist coupling is observed for $P4$ and $P1$, showing the same lift force that only differs in the projections. The bending motion does not modify the force magnitude but its distribution on the axis changing the balance between the lift and the side force component.
The experiments investigated with negative incidences clearly highlight the impact of bending-twisting coupling. The foil P_2, with oriented plies at $-45°$ in the laminate, experiences a significantly higher displacement and hydrodynamic loads due the negative twist loading the tip.
The comparison of experimental displacements with FS6R computations is very good (discrepancies are less than 10% with experiments values higher in most of the cases excepts P_3), showing the ability of the code to compute the bending-twisting coupling.
The first step of a FSI coupling between the structural code ASTER and the VOF hydrodynamic solver interFOAM is eventually presented. The coupling, significantly more CPU demanding, brings the final bricks of the design process for foil with FSI, where FS6R has proven to estimates greatly the tendencies and ASTER-interFOAM would provide accurate results on the performances.

REFERENCES

Balze, R., Bigi, N., Roncin, K., Leroux, J., Nême, A., Keryvin, V., Connan, A., Devaux, H., and Gléhen, D. (2017). Racing. *Innovsail International Conference, Lorient, France*, pages 51–58.

Capellaro, M. (2012). Design challenges for bend twist coupled blades for wind turbines and application to standard blades. In *Proceedings of Sandia Wind Turbine Blade Workshop*.

Ducoin, A., Astolfi, J. A., and Sigrist, J.-F. (2012). An experimental analysis of fluid structure interaction on a flexible hydrofoil in various flow regimes including cavitating flow. *European Journal of Mechanics-B/Fluids*, 36:63–74.

Gay, D. (1991). *Materiaux composites; 3rd ed.* Trait des nouvelles technologies. Herms, Paris.

Giovannetti, L. M. (2017). Fluid structure interaction testing, modeling and development of passive adaptive composite foils. *PhD Thesis, -.*

Giovannetti, L. M., Banks, J., Ledri, M., Turnock, S., and Boyd, S. (2018). Toward the development of a hydrofoil tailored to passively reduce its lift response to fluid load. *Ocean Engineering*, 167:1–10.

Gozcu, O., Farsadi, T., Tola, C., and Kayran, A. (2015). Assessment of the effect of hybrid grfp-cfrp usage in wind turbine blades on the reduction of fatigue damage equivalent loads in the wind turbine system. In *Proceedings of the 9th International Workshop on Water Waves and Floating Bodies (10.2514/6.2015-0999)*.

Hagemeister, N. and Flay, R. (2017). Velocity prediction of wing-sailed hydrofoiling catamarans. *Innovsail International Conference, Lorient, France*, pages 11–18.

Hodges, D. H. (2006). *Nonlinear Composite Beam Theory, .* 78-1-56347-697-6. American Institute of Aeronautics and Astronautics, 1.

Raither, W., Bergamini, A., and Ermanni, P. (2013). Profile beams with adaptive bending–twist coupling by adjustable shear centre location. *Journal of Intelligent Material Systems and Structures*, 24(3):334–346.

Rohde, S. E., Ifju, P. G., Sankar, B. V., and Jenkins, D. A. (2015). Experimental testing of bend-twist coupled composite shafts. *Experimental Mechanics*, 55(9):1613–1625.

V. Temtching, B. AUGIER, O. F. and D.R, J.-A. A. (2018). An experimental and numerical study of fsi applied to sail yacht flexible hydrofoil with large deformations. In *Proceedings of 9th International Symposium on Fluid-Structure Interactions Flow-Sound Interactions Flow-Induced Vibration and Noise (Toronto Canada)*.

Veers, P., Lobitz, D., and Bir, G. (1998). Aeroelastic tailoring in wind-turbine blade applications. *Technical report Sandia National Labs., Albuquerque, NM (US)*.

Young, Y. L. and Brizzolara, S. (2013). Numerical and physical investigation of a surface-piercing hydrofoil. *Third International Symposium on Marine Propulsors (SMP13), Launceston, Tasmania, May*, pages 5–8.

Young, Y. L., Garg, N., Brandner, P. A., Pearce, B. W., Butler, D., Clarke, D., and Phillips, A. W. (2018). Load-dependent bend-twist coupling effects on the steady-state hydroelastic response of composite hydrofoils. *Composite Structures*, 189:398–418.

Application of System-based Modelling and Simplified-FSI to a Foiling Open 60 Monohull

Boris Horel[1], Ecole Centrale Nantes, LHEEA res. dept. (ECN and CNRS), France
Mathieu Durand, SIRLI Innovations, LUNA ROSSA Challenge, France

ABSTRACT

The increasing number of foiling yachts in offshore and inshore races has driven engineers and researchers to significantly improve the current modelling methods to face new design challenges such as flight analysis and control (Heppel, 2015). Following the publication of the AC75 Class Rules for the 36th America's Cup (RNZYS, 2018) and since the brand new Open 60 Class yachts are all equipped with hydrofoils, the presented study will propose a system based modelling coupled with a simplified FSI (fluid structure interaction) method that leads to better understand the dynamic behavior of monohulls with deformable hydrofoils.

The aim of the presented paper is to establish an innovative approach to assess appendage behavior in a dynamic VPP (velocity prediction program). For that purpose, dynamic computations are based on a 6DOF mathematical model derived from the general non linear maneuvering equations (Horel, 2016). The force model is expressed as the superposition of 7 major force components expressed at the center of gravity of the yacht: gravity, hydrostatic, maneuvering, damping, propulsion (wind), control (rudders, daggerboard, foils ...) and wave (Froude Krylov and diffraction phenomenon).

As test cases, course keeping simulations are performed on an Open 60 yacht with control loops to simulate the wing trimmer, helmsman and foil trimmer when finding the optimal foil settings is needed. In first hand, IMOCA's polar diagrams are used as reference.

In calm water and in waves, the influence of foil's shapes (foil with shaft pointing downward and tip pointing upward, foil with shaft pointing upward and tip pointing downward) and stiffness (non deformable, realistic, flexible) on the global behavior of the yacht is presented.

NOTATION

A_v Transverse section area (for air resistance) (m²)
A_{w0} Waterplane area at zero speed (m²)
B_{wl} Beam of waterline (m)
B Center of buoyancy
c Chord of the section (m)
C_p Prismatic coefficient
G Center of gravity of the yacht
g Gravity acceleration (m.s^{-2})
I_x Roll moment of inertia (kg.m²)
I_y Pitch moment of inertia (kg.m²)
I_z Yaw moment of inertia (kg.m²)
I_{xy} Roll and pitch product of inertia (kg.m²)
I_{xz} Roll and yaw product of inertia (kg.m²)
I_{yz} Pitch and yaw product of inertia (kg.m²)
LCB Longitudinal position of the center of buoyancy to f$_{pp}$ (m)
LCF Longitudinal position of the center of flotation to f$_{pp}$ (m)
L_{wl} Length of waterline (m)

---

[1] boris.horel@ec nantes.fr

p	Roll angular rate (rad.s^{-1})
q	Pitch angular rate (rad.s^{-1})
r	Yaw angular rate (rad.s^{-1})
Re	Reynolds number of the yacht
S_w	Wetted surface (m^2)
S_{w0}	Static wetted surface (m^2)
T	Draft (m)
TWA	True wind angle (m.s^{-1})
TWS	True wind speed (m.s^{-1})
V	Yacht speed (m.s^{-1})
u	Surge velocity (m.s^{-1})
v	Sway velocity (m.s^{-1})
w	Heave velocity (m.s^{-1})
ρ	Density of water (kg.m^{-3})
ρ_a	Density of air (kg.m^{-3})
η	Incident wave amplitude (m)
V_c	Volume of displacement of canoe body (m^3)

INTRODUCTION

Recent studies have shown that system based modelling can be applied to foiling yacht in order to evaluate their maneuvering abilities for a panel of standard maneuvers (Horel, 2017) but calculations with complex foil geometries were still under development. Since that, improvements have been done to existing models in order to take into account a large panel of foil geometries and simple structural deformations such as hydrofoil's flexion.

In the foil's FSI prediction, regarding the main FSI methods (direct integration, finite elements and modal analysis) and based on experimental validation, a simplified model of dynamic shear force and bending moment is implemented. The accuracy of such a model was investigated by performing experimental test in rough waves on a segmented containership's hull in the 50m×30m×5m hydrodynamic and ocean engineering tank of Ecole Centrale de Nantes.

In this context, the main objective of this paper is to evaluate through simplified modelling whether basic structural deformations of hydrofoils can affect the global yacht's behavior for several sea conditions.

In calm water, numerical results are compared with polar diagrams on conventional IMOCA yacht. In waves, only numerical results are presented and commented regarding the relevant effects of the hydrofoils on the yacht motions.

The application of structural deformations in dynamical simulations is a recent field of research. Also, Open 60 monohulls are complex yachts and the presented work is a first step to better understand the physical phenomena that drive the yacht's performances.

DYNAMICAL VPP

Velocity prediction programs (VPP) aim to provide performance data for sailing yachts which can be expressed as polar diagrams where the target speed is given as a function of the true wind angle (TWA) for several true wind speeds (TWS). These tools are solving a quasi static equilibrium at each time step and the degree of accuracy depends on the accuracy of each force model. Analytical method are commonly used but recent studies have been done on coupling VPP with Computational Fluid Dynamics (CFD) (Roux et al., 2008) (Viola et al., 2015) in order to improve the accuracy of the predictions and to reduce the computing time.

When dealing with dynamical motions, static VPPs are no longer suitable for predicting the performances of the yacht. Then, in order to solve dynamical problems a so called dynamical VPP (DVPP) is used. This DVPP is based on a 6 degrees of freedom (DOF) model derives from the Newton's second law. In this study, the hull of the yacht is assumed to be symmetrical in $(x_bO_by_b)$ and $(y_bO_bz_b)$ planes. However, the canting keel and sails make the moments of inertia of the yacht very asymmetric. Knowing the total force components (X, Y, Z, K, M and N), the accelerations of the yacht can be predicted by solving the following general differential equations:

$$X = m[\dot{u} + qw - rv - x_G(q^2 + r^2) + y_G(pq - \dot{r}) + z_G(pr + \dot{q})] \tag{1}$$
$$Y = m[\dot{v} + ru - pw - y_G(r^2 + p^2) + z_G(qr - \dot{p}) + x_G(qp + \dot{r})] \tag{2}$$
$$Z = m[\dot{w} + pv - qu - z_G(p^2 + q^2) + x_G(rp - \dot{q}) + y_G(rq + \dot{p})] \tag{3}$$
$$K = I_x\dot{p} - I_{xy}(\dot{q} - rp) - I_{zx}(\dot{r} + pq) + (I_z - I_y)qr + I_{yz}(r^2 - q^2) + m[y_G(\dot{w} + pv - qu) - z_G(\dot{v} + ru - pw)] \tag{4}$$

$$M = I_y\dot{q} - I_{yz}(\dot{r} - pq) - I_{xy}(\dot{p} + qr) + (I_x - I_z)rp + I_{zx}(p^2 - r^2) + m[z_G(\dot{u} + qw - rv) - x_G(\dot{w} + pv - qu)] \quad (5)$$
$$N = I_z\dot{r} - I_{zx}(\dot{p} - qr) - I_{yz}(\dot{q} + rp) + (I_y - I_x)pq + I_{xy}(q^2 - p^2) + m[x_G(\dot{v} + ru - pw) - y_G(\dot{u} + qw - rv)] \quad (6)$$

with:

$$\boldsymbol{O_bG} = \begin{bmatrix} x_G \\ y_G \\ z_G \end{bmatrix} \quad (7)$$

$$p = -\sin\theta\,\dot{\psi} + \dot{\varphi} \quad (8)$$
$$q = \sin\varphi\cos\theta\,\dot{\psi} + \cos\varphi\,\dot{\theta} \quad (9)$$
$$r = \cos\varphi\cos\theta\,\dot{\psi} - \sin\varphi\,\dot{\theta} \quad (10)$$

Previous equations are expressed in the yacht's reference frame as defined in Figure 1.

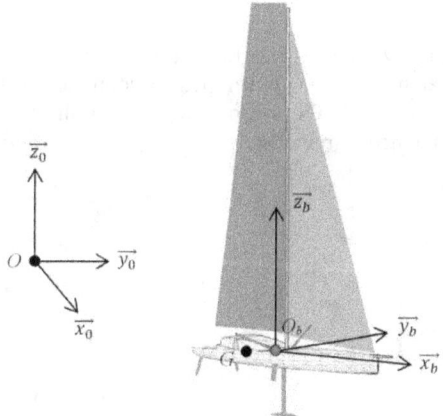

Figure 1 - Reference frames: earth (b_0) and yacht (b_b)

Equations 1 to 6 are solved by applying a 4th order Runge Kutta integration scheme. This leads to calculate the yacht's accelerations and then to predict the yachts motions in surge, sway, heave, roll φ, pitch θ and yaw ψ. The accuracy of the predicted motions is governed by the accuracy of the implemented force models.

FORCE MODELLING

In system based modelling, the complex behavior of the yacht is calculated from taking into account the local phenomena and the global effect of their interactions when mathematical models are available.

This simplified modelling expresses the total loads F on the yacht as the superposition of force components acting on the hull, on the appendages and on the sails. Then, since O_b and G are combined, 7 major force components have been identified and expressed at the center of gravity G of the yacht: gravity (F_{Grav}), hydrostatic (F_{HS}), maneuvering (F_{Man}), damping and radiation (F_{Damp}), propulsion due to the wind (F_{Prop}), control such as rudders, daggerboards, hydrofoils or keel (F_{Ctrl}) and waves including Froude Krylov and diffraction phenomenon (F_w).

$$F = F_{Grav} + F_{HS} + F_{Man} + F_{Damp} + F_{Prop} + F_{Ctrl} + F_w \quad (11)$$

with:

$$F = \begin{bmatrix} X \\ Y \\ Z \\ K \\ M \\ N \end{bmatrix}_{(G,x_b,y_b,z_b)} \tag{12}$$

Gravity forces

An Open 60 is actually composed with at least a bare hull, a canting keel, daggerboards or hydrofoils, rudders and a mast and a boom that have significant influence on the total weight of the yacht. Since each of these $N_{element}$ elements has its own center of gravity, the total yacht's weight distribution can be written as follows:

$$F_{Grav} = \sum_{i=1}^{N_{element}} \begin{bmatrix} F_{Grav_i}^{G_i} \\ GG_i \wedge F_{Grav_i}^{G_i} \end{bmatrix} \tag{13}$$

Hydrostatic forces

In order to take into account the strong geometrical non linearities when the yacht experiences large amplitude motions in calm water and in waves, a nonlinear method was developed to compute the hydrostatic loads on the bare hull and underwater appendages. This method is based on the use of a surface mesh from a *STL* file and the integration of the hydrostatic pressure p_{HS} on the total instantaneous immersed surface S_w composed with N_f elementary surfaces.

$$F_{HS} = \sum_{i=1}^{N_f} \begin{bmatrix} F_{HS_i} \\ GB_i \wedge F_{HS_i} \end{bmatrix} \tag{14}$$

with:

$$F_{HS_i} = -p_{HS}(z_i)dS_w \tag{15}$$
$$p_{HS}(z_i) = -\rho g z_i \tag{16}$$

In equation 16, z_i is the vertical position of the center B_i of the elementary surface compared to O, the origin of the earth's reference frame.

Maneuvering forces

Maneuverability is the study of stable and transient states of the yacht motion in calm water and in waves at low encounter frequency. It consists in evaluating the abilities of a yacht to keep a desired heading or to change her heading after the action of the control appendages. In this context, 3DOF mathematical models in surge, sway and yaw are used to study complex maneuvers and several maneuvering scenario.

In surge, the resistance of the yacht can be estimated following the ITTC recommendations 7.5 02 05 01 (ITTC, 2017) and using the regressions based on the Delft Systematic Yacht Hull Series (DSYHS) (Keuning et al., 1998). The Resistance R_T is expressed in the non dimensional form using the resistance coefficient C_T.

$$C_T = \frac{R_T}{\frac{1}{2}\rho S_{w0}V^2} \tag{17}$$

This total resistance coefficient is expressed as the superposition of 4 identified factors:
- Residuary resistance coefficient of the bare hull, C_R
- Frictional resistance coefficient, C_F
- Wind resistance coefficient, C_{AA}
- Appendage resistance coefficient, C_{APP}

$$C_T = C_R + \frac{S_w}{S_{w0}}C_F + C_{AA} + C_{APP} \tag{18}$$

Empirical formula from ITTC 57 is used to calculate the values of C_F. ITTC mentioned that the frictional coefficient is associated with a form factor k but for high speed marine vehicles with transom stern as Open 60, it is recommended to assumed that $(1 + k) = 1.0$.

$$C_F = \frac{0.075}{(\log_{10} Re \ 2)^2} \tag{19}$$

Wind resistance coefficient formulation is given by using the aerodynamic drag coefficient C_D in hydrodynamic form:

$$C_{AA} = \frac{\rho_a A_v}{\rho S_w} C_D \tag{20}$$

In first approximation, as mentioned in ITTC 78, C_D is assumed to be equal to 0.8 as default value and the transverse section area A_v is calculated from the values of the air draft and the mean width of the hull.

As mentioned in previous studies (Raymond, 2009) (Huetz et al., 2011), the most consistent formulation for residuary resistance coefficient of the bare hull C_R is the DSYHS formulation.

$$C_R = \frac{2g\nabla_c}{S_w V^2} C_{R\,DSYHS} \tag{21}$$

The expression of $C_{R\,DSYHS}$ can be found in previous work on mathematical model for the tacking maneuver of a sailing yacht (de Ridder et al., 2004):

$$C_{R\,DSYHS} = a_0 + \left(a_1 \frac{LCB}{L_{wl}} + a_2 C_p + a_3 \frac{\nabla_c^{2/3}}{A_{w0}} + a_4 \frac{B_{wl}}{L_{wl}} + a_5 \frac{\nabla_c^{2/3}}{S_{w0}} + a_6 \frac{LCB}{LCF} + a_7 \left(\frac{LCB}{L_{wl}}\right)^2 + a_8 C_p^{\ 2} \right) \cdot \frac{\nabla_c^{1/3}}{L_{wl}} \tag{22}$$

In expression 22, values of the coefficients a_1 to a_8 are determined from tables for Froude numbers from $Fn = 0.10$ to $Fn = 0.60$. Also, additional resistance due to the heeling and the sway velocity of the yacht can be added to expression 22.

In previous study on maneuvering models on conventional ship, identification of hydrodynamic derivatives in sway and yaw from several experimental data sets leads to establish empirical expressions as functions of the ship features: draft, width, length between perpendicular, block coefficient, etc. (Tjoswold, 2012) (Clarke, 1983). But for sailing yacht, previous mentioned work from de Ridder et al. (2004) also propose polynomial regressions for the calculations of forces in sway and moments in yaw due to heel angle, sway, roll and yaw velocities and their coupled effects.

Damping forces

Damping forces on the bare hull are known to be due to 2 phenomena: frictional damping and radiated waves. The former can be modeled using Ikeda's formulation as explained in ITTC recommended procedures (2011) while the radiation forces from radiated waves are evaluated in time domain by using Cummins formulation based on linear theory. Added mass and damping coefficients at zero forward speed are calculated using the boundary element methods code NEMOH (Babarit et al., 2015). Then, these coefficients are used for the yacht with forward speed by using a formulation based on a first order development of the slip boundary condition on the hull as described by Delhommeau et al. (1987).

Control forces

The modelling of control forces includes the maneuvering forces from the rudder and the forces on the appendages such as the keel, the daggerboards and the foils. These forces tend to modify the dynamic equilibrium of the yacht while sailing on a desired heading. In this section, particular focus will be paid to the modelling of the forces acting on the hydrofoils.

In our work, a foil is defined using the following parameters:
·   General shape (position of the nodes),
·   Foil sections,
·   Chord of the sections,
·   Linear mass and added mass,
·   Local stiffness.

As can be noticed in Figure 2 for a non conventional modern Dali's foil 2018, complex geometries are taken into account by discretizing the foil with N_{elem} elements. Then, a local reference frame is associated to each element.

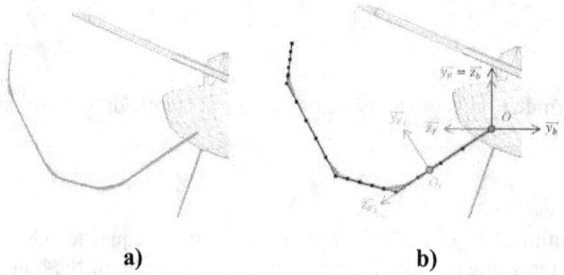

Figure 2 - a) Initial foil geometry, b) simplified foil geometry with oriented finite elements

The global hydrodynamic forces F_{Hyd} on the foil are evaluated from the computation of the lift and drag on each element assuming that the element is encountering a 2D flow. The effective angles due to the yacht motion are used in order to evaluate the angle of the incident flow at each time step. Knowing the position of each element in the ship reference frame, the moment can be expressed at the center of gravity G of the yacht. Also, since the yacht accelerations are known, the added mass forces F_{AM} on the foil can be calculated. In first approximation, a simple linear added mass coefficient m_{yy} is used.

$$m_{yy} = \frac{\pi}{4}\rho c^2 \tag{23}$$

The global force on the foil F_{foil} is expressed from the local forces on each element.

$$F_{foil} = \sum_{i=1}^{Nelem} \left(F_{Grav_i} + F_{Hyd_i} + F_{AM_i} + F_{Inertia_i} \right) \tag{24}$$

The effect of the interactions between the foil and the free surface are taken into account by Faltinsen's formulation (2005) where a submergence Froude number is calculated in the modelling of the lift coefficient reduction near the free surface.

When sailing in waves, the velocity of the incident flow is modified according to the orbital velocity of the water particles.

Wave forces

Wave loads on the bare hull are calculated from potential theory. The total wave potential is known from the superposition of a diffracted wave potential ϕ_d and an incident wave potential ϕ_i. The former is evaluated by using the previous mentioned boundary element methods code NEMOH which gives the complex form $F_d(\omega)$ of the diffraction forces. In time domain, the diffraction forces F_{Diff} can be expressed from wave frequency ω and encounter frequency ω_e as follows:

$$F_{Diff} = \text{Re}\{\eta F_d(\omega_e)e^{\ i(\omega t)}\} \tag{25}$$

The incident wave forces are computed under the so called Froude Krylov assumption where the yacht has no effect on the velocity field around her. Then, from Bernouilli Lagrange pressure relation, diffraction forces can be estimated by integrating the pressure expressed at the center B_i of each elementary surface dS_w on the total wetted surface S_w.

$$F_{FK} = \sum_{i=1}^{N_f} \begin{bmatrix} F_{FK_i} \\ GB_i \wedge F_{FK_i} \end{bmatrix} \tag{26}$$

with:

$$F_{FK_i} = -p_{FK}(z_i)dS_w \tag{27}$$
$$p_{FK}(z_i) = -\rho \left[\frac{\partial \phi_i}{\partial t} + \frac{1}{2}(\nabla \phi_i)^2 \right] \tag{28}$$

The total wave forces F_w are expressed as the superposition of the diffraction and Froude Krylov components.

$$F_w = F_{Diff} + F_{FK} \tag{29}$$

Propulsion forces

The effect of the wind is modelled through a simplified quasi steady model whose lift $C_{L_{max}}$ and drag C_{D_0} coefficients of each individual sail, mainsail, jib and spinnaker are identified from the IMS measurement procedure (ORC, 2016).

As mentioned in the procedure, the reduction of heeling force by the crew trimming and changing sails is modelled by the *Flat* and *Reef* parameters. The former is used to simulate the reduction of the lift coefficient, while the latter is used to simulate the reduction of sail area.

Also, in order to reflect the fact that as sails are depowered, the height of the center of effort is reduced, a *Twist* function depending on the amount of flat is introduced (Jackson, 2001).

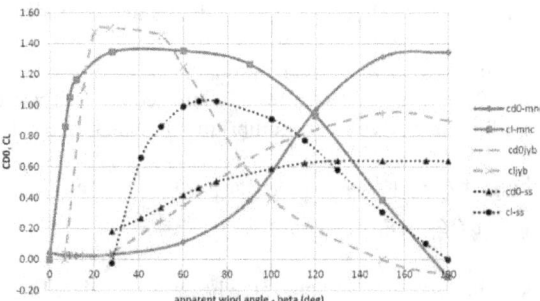

Figure 3 - Basic sail force coefficients (ORC, 2016)

The magnitude and direction of the apparent wind encountered by the sails are modified according to the 6DOF motions of the yacht. Thus, effective angles β_{eff} are used in the calculation of the total lift C_L and drag C_D coefficients (Fossati et al., 2011).

$$C_L = Flat.Reef^2.C_{L_{max}}(\beta_{eff}) \tag{30}$$
$$C_D = Reef^2.\left[C_{D_0}(\beta_{eff}) + \kappa_0 C_L^2 (1 + c_1 Twist^2)\right] \tag{31}$$

In first approximation, the dynamics of the sails is taken into account by adding a simple added mass coefficient $m_{yy_{Sail}}$ calculating from strip theory (Newman, 1977).

$$m_{yy_{Sail}} = \frac{\pi}{4}\rho_a c^2 dz \tag{32}$$

Also, a log wind profile can be used in order to evaluate the true wind speed at the center of effort height above the water (Flay et al., 1995).

STRUCTURAL MODELLING

In order to investigate the effect of simple local deformation on the global behavior of the yacht, a simplified structural model based on beam theory was implemented to quantify the flexion of the hydrofoil.

Direct integration

As previously mentioned, the complex geometry of the foil is discretized with a finite number of elements. It is assumed that the element in contact with the hull is clamped. Then, the dynamical transverse loads F_{Beam_i} supported by the elementary beam, that are, loads that act perpendicular to the longitudinal axis of the beam are expressed as the superposition of gravity, inertial, added mass and hydrodynamic loads.

$$F_{Beam_i} = F_{Grav_i} + F_{Hyd_i} + F_{AM_i} + F_{Inertia_i} \tag{33}$$

Inertial loads are calculated as the mass of each element times the linear acceleration. This force is expressed at the center of gravity of the element and is taken into account in the calculation of the bending moment.

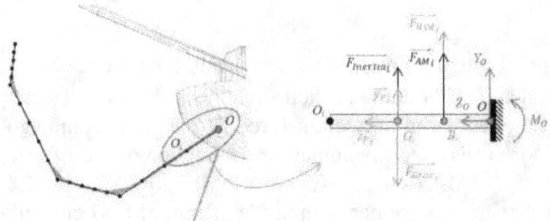

Figure 4 - 2D configuration of the loads on the hydrofoil

Since the hydrofoil's tip is free, the problem to solve is similar to a cantilever problem. As shown in Figure 5, the unknown reaction $\boldsymbol{F_0} = [0, Y_0, Z_0, M_0, 0, 0]$ is considered in the intersection O between the foil and the hull. When considering all the elements and according to the dynamic equilibrium, the reaction can be computing as follows:

$$\boldsymbol{F_0} = -\sum_{i=1}^{Nelem} \begin{bmatrix} \boldsymbol{F_{Grav_i}} + \boldsymbol{F_{Hyd_i}} + \boldsymbol{F_{AM_i}} + \boldsymbol{F_{Inertia_i}} \\ \boldsymbol{OB_i} \wedge \left(\boldsymbol{F_{Hyd_i}} + \boldsymbol{F_{AM_i}}\right) + \boldsymbol{OG_i} \wedge \left(\boldsymbol{F_{Grav_i}} + \boldsymbol{F_{Inertia_i}}\right) \end{bmatrix} \tag{34}$$

For each element, the bending moment M_f in any section of the beam is calculated from a 2D point of view. A coordinate of the sections x is introduced and the beam is sectioned as shown in Figure 5.

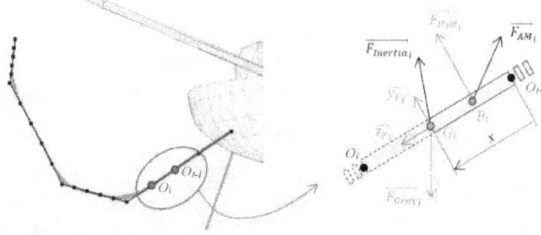

Figure 5 - 2D configuration of the loads on 1 element

For a given stiffness EI of the beam, the 2nd order derivative y'' of the deflection of the foil along $\boldsymbol{y_{F_i}}$ in G_i is calculated from the formulation of the bending moment on the segment $O_{i-1}G_i$.

$$y'' = -\frac{M_{f_i}}{EI} \tag{35}$$

with, for $x_{G_i} > x_{B_i}$:

$$M_{f_i}(x) = -\left[M_0 + (\boldsymbol{OO_{i\ 1}} \cdot \boldsymbol{z_{F_i}}) \cdot (\boldsymbol{F_0} \cdot \boldsymbol{y_{F_i}}) + \sum_{j=1}^{i}(\boldsymbol{G_jO_{i\ 1}} \cdot \boldsymbol{z_{F_i}}) \cdot \left(\left(\boldsymbol{F_{Grav_j}} + \boldsymbol{F_{Inertia_j}}\right) \cdot \boldsymbol{y_{F_i}}\right) + \sum_{j=1}^{i}(\boldsymbol{B_jO_{i\ 1}} \cdot \boldsymbol{z_{F_i}}) \cdot \left(\left(\boldsymbol{F_{Hyd_j}} + \right.\right.\right.$$

$$\left. \boldsymbol{F_{AM_j}}\right) \cdot \boldsymbol{y_{F_i}}\right) + \left(\boldsymbol{F_0} + \sum_{j=1}^{i}\left(\boldsymbol{F_{Grav_j}} + \boldsymbol{F_{Inertia_j}}\right) + \sum_{j=1}^{i}\left(\boldsymbol{F_{Hyd_j}} + \boldsymbol{F_{AM_j}}\right)\right) \cdot \boldsymbol{y_{F_i}} \cdot x\right] \tag{36}$$

When an element is above the water, the forces are equal to zero, but its deflection is a function of the forces acting on the other elements.

Experimental investigation

The formula used in beam theory to calculate the shear force and bending moment were tested during an experimental campaign carried out in the hydrodynamic and ocean engineering tank of Ecole Centrale de Nantes to study the variations of internal forces on a notional hull form in rough waves (Horel et al., 2019). Figure 6 shows a comparison between the measured value of the vertical bending moment and the analytical reconstruction.

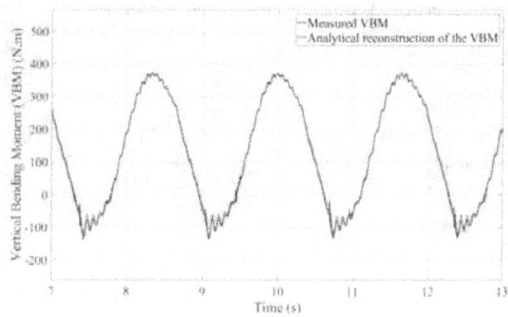

Figure 6 - Comparison between measured bending moment and analytical reconstruction

It can be noticed that analytical formulation reach to approximate with a relatively good accuracy the time evolution of the bending moment in the amidship section.

APPLICATIONS

A conventional and a foiler Open 60 are used to evaluate the ability of the presented models to predict the yacht behavior in calm water and in waves. According to IMOCA Class Rules (2018), their main features are given in Table 1.

Feature	Conventional / Foiler
Length overall, LOA (m)	18.28
Width, B (m)	5.7
Water draft, T (m)	4.5
Total mass of the yacht, m (t)	7.7
Keel's bulb mass, m_{bulb} (t)	3.1
Ballast and crew mass, m_{crew} (t)	0.5
Longitudinal position of the center of gravity of the hull from stern, x_{CoG} (m)	5
Vertical position of the center of gravity of the hull from baseline, z_{CoG} (m)	1.5
Roll radius of gyration, k_{xx} (m)	2.035
Pitch radius of gyration, k_{yy} (m)	4.57
Yaw radius of gyration, k_{zz} (m)	4.57
Mainsail area, A_{MS} (m²)	149.2
Jib area, A_{Jib} (m²)	150
Spinnaker area, A_{Spi} (m²)	400

Table 1 - Open 60 main features

The features of the non conventional modern Dali's hydrofoil 2018 of the foiler Open 60 are given in Table 2. The coordinates X, Y and Z of the origin of the elements (El^t) are given in the foil reference frame (O,x_F,y_F,z_F) as described in Figure 2.

Feature	EI^t 1	EI^t 2	EI^t 3	EI^t 4
X (m)	0	0	0	0
Y (m)	0	-1.35	-1.056	0.641
Z (m)	0	2.088	3.241	4.155
Span, l (m)	2.486	1.19	1.927	0.75
Chord, c (m)	0.53	0.53	0.27	0.14
Linear mass (kg.m^{-1})	67.416	67.416	17.496	4.704
Linear added mass (kg.m^{-1})	226.133	226.133	58.687	15.779
Stiffness, EI (N.m^2)	3×10^6	3×10^6	0.202×10^6	0.015×10^6

Table 2 - Foil main features

Lift and drag coefficients of the foil derived from an asymmetrical H105 profile whose section shape is given in Figure 7 below. The H105 hydrofoil section was designed to avoid laminar separation and ventilation when operating at low speeds and moderate angles of attack, while still having low velocities at small angles of attack to avoid cavitation at high speeds (Speer, 2001).

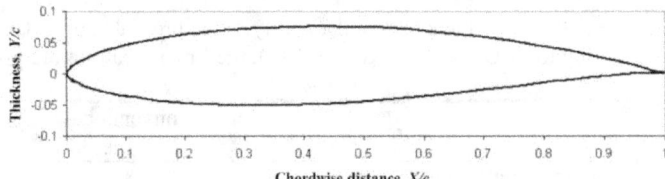

Figure 7 - H105 hydrofoil section shape

Calm water
Polar diagrams for non-deformable (rigid) foils

First, in order to evaluate whether the models get to predict with a reasonable accuracy the performance of the yacht, numerical results from ECN simulations are compared with reference speed polar diagram of the PRB Open 60. Even if the features of the simulated and real yachts are not exactly the same, this comparison reveals that the patterns of the speed polar diagram are similar.

Figure 8 shows the predicted performances of the conventional and foiler Open 60 for non deformable appendages and true wind speed (*TWS*) of 20 knots. The cant of the keel is set to 30 degrees and no tilt is set.

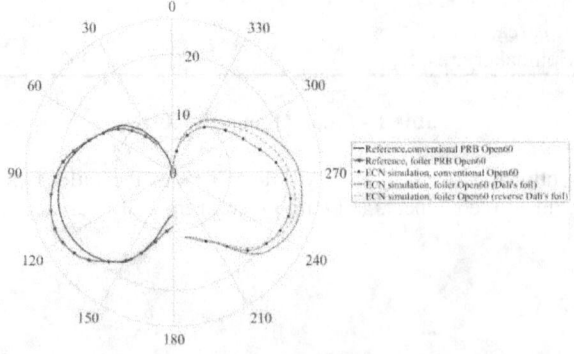

Figure 8 - Speed polar diagram (values in knots) (*TWS*=20knots)

From Figure 8 it can be noticed that the unconventional modern Dali's foils 2018 are helping the yacht to sail faster for a large range of true wind angle (*TWA*), except for downwind conditions. However, these performances are highly depending on the chosen appendages features. Also, for a reverse foil, with the tip pointing downward, the gain in speed compared to a conventional daggerboard Open 60 seems to be significant for a range of *TWA* from 10 to 140 degrees. But with the chosen foil feature, the gain is smaller than the gain of the modern Dali's foils 2018.

From the Polar diagram of total lateral and vertical forces on the foil (Figure 9), it can be notices that the reverse foil acts more like a daggerboard by mainly creating anti drift force while the modern Dali's foil 2018 is creating much more vertical lift force that helps the yacht to increase her performances.

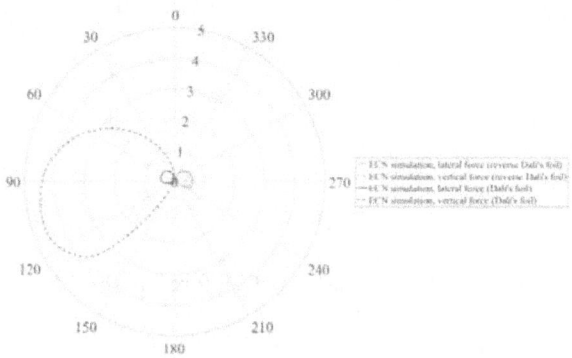

Figure 9 - Polar diagram of forces on the foil (values in tons) (*TWS*=20knots)

Constant acceleration with deformable modern Dali's foil 2018

This case is performed for 3 different stiffness (non deformable, realistic and flexible) in order to evaluate the ability of the structural model to predict the time evolution of the deflection of the foil. The yacht is fixed in sway, heave, roll, pitch and yaw. The acceleration in surge is remained constant and equal to 0.5m.s^{-2} up to a forward speed of 10m.s^{-1}. Above that, the speed is kept constant. The rake angle of the foil is set to 0 degree.

The lateral and vertical deflection of the tip can be respectively evaluated from Figure 10 and Figure 11.

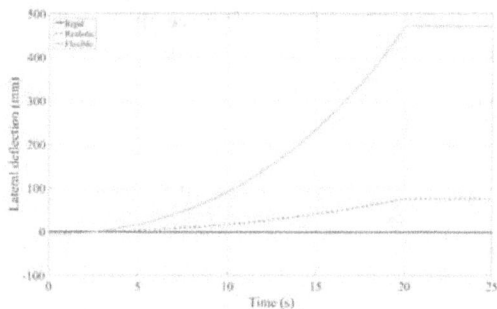

Figure 10 - Lateral deflection of the tip

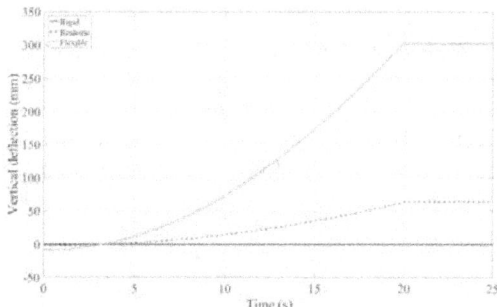

Figure 11 - Vertical deflection of the tip

With realistic stiffness whose values are given in Table 2, the lateral deflection of the tip at 10m.s^{-1} reaches a value of 75mm. When the stiffness is reduced, the deflection is increasing.

In Figure 12 and Figure 13, it can be noticed that the stiffness and the deflection of the foil seem to have a low influence on the drag and lift forces.

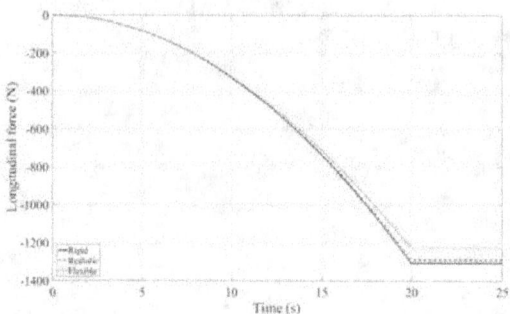

Figure 12 - Longitudinal drag force on the foil

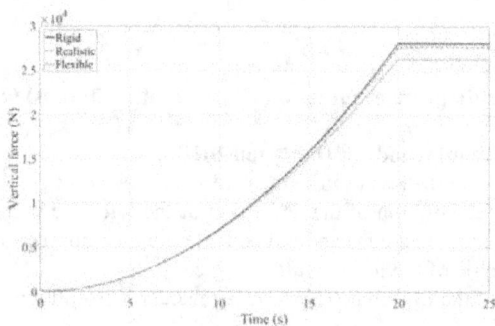

Figure 13 - Vertical lift force on the foil

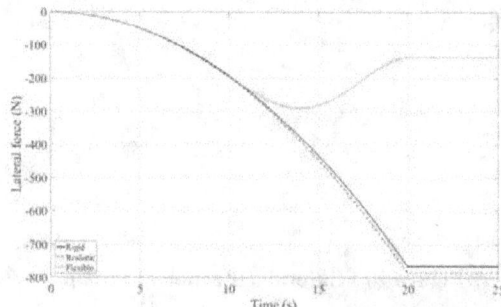

Figure 14 - Lateral drift force on the foil

However, as can be observed in Figure 14, the lateral force on the flexible foil seems to be highly influenced by the deflection of the foil. This deflection changes the loading distribution on the foil and modifies the resulting force.
The global deflection and the loading distribution on the foil are presented in Figure 15.

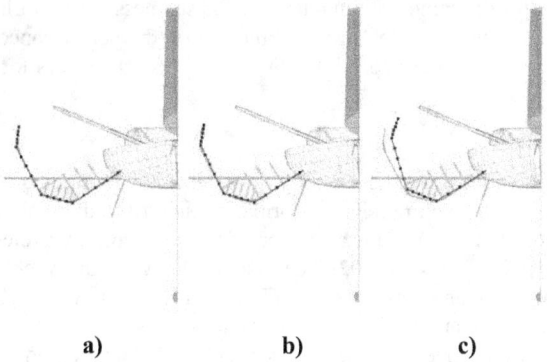

a)   b)   c)

Figure 15 - Global deflection of the hydrofoil, a) non-deformable, b) realistic, c) flexible

Speed polar diagram for flexible foil

The global effect of the flexion of the foil on the yacht's performance is evaluated from the speed polar diagram given in Figure 16.

It can be noticed that for true wind angles from 50 degrees to 130 degrees, the presented method get to capture the variations of the performances between the non deformable foil and the flexible foil. In the particular case of this study, the maximum gap in speed is equal to 1 knot and is experienced for a true wind angle of 110 degrees. This shows that taking into account the effects of the foil's deformation is an important step in the design stage.

This loss in speed can be partly explained from the fact that the flexible foil generate less lift than the non deformable one, creating less righting moment and then does not allow to trim the sails to their best settings.

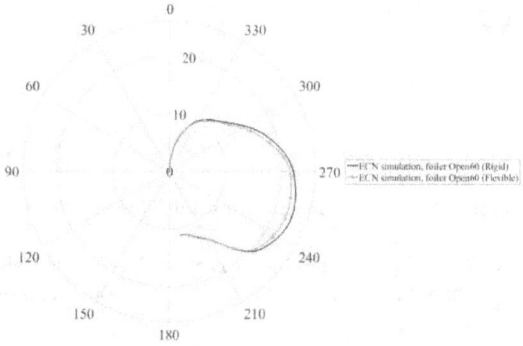

Figure 16 - Speed polar diagram (values in knots) (*TWS*=20knots)

The differences in the vertical force can be seen in Figure 17.

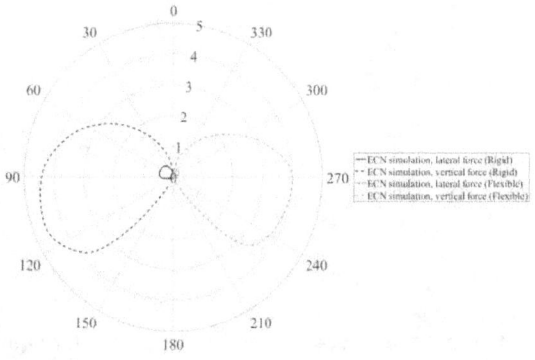

Figure 17 - Polar diagram of forces on the foil (values in tons) (*TWS*=20knots)

These previous results show that for the same optimum targets, the stiffness has an influence on the reached speed.

In the next section the influence of the deflection of a flexible foil on the performance of the yacht sailing in quartering head waves is studied. In order to reach a similar speed, the target *TWA* is 70 degrees for the conventional Open 60 and 50 degrees for the foiler.

Regular sea state

Simulations in regular waves are the most penalizing conditions, since the motions of the yacht are excited with a unique encounter frequency. However, they are given precious information about the dynamic behavior of the yacht. In this test case, the true wind speed is 20 knots and the wave is a regular Airy wave with a wavelength that is twice the yacht length and a wave height to wavelength ratio equals to 0.02. The heading between the yacht's longitudinal axis and the wave direction is 70 degrees. Since the tests are performed in 6DOF, only the rudder deflection is controlled using a PD controller. Sails deflections and rake angle are maintained to constant values.

In these conditions, for a similar target speed between the conventional Open 60 and the foiler, the dynamic behavior can be evaluated from phase diagrams and amplitude spectra.

Figure 18 is a phase diagram in pitch motion that shows the dynamic evolution of the pitch angular rate according to the pitch angle. It can be noticed that the foiler seems to sail with the bow up (negative pitch angle) and experiences less pitch motion than a conventional Open 60. Also, by being influenced by the wave particle orbital velocities, the flexible foil seems to increase the amplitude of the pitch motion.

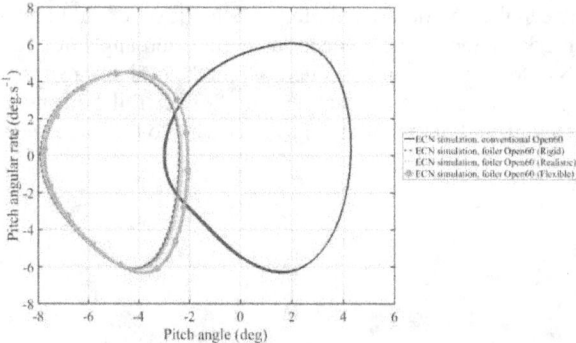

Figure 18 - Phase diagram in pitch (*TWS*=20knots)

The attitudes of the conventional Open 60 and the foiler Open 60 with flexible foils can also be evaluated and qualitatively compared from Figure 19, where the yachts have the same position compared to the wave crest.

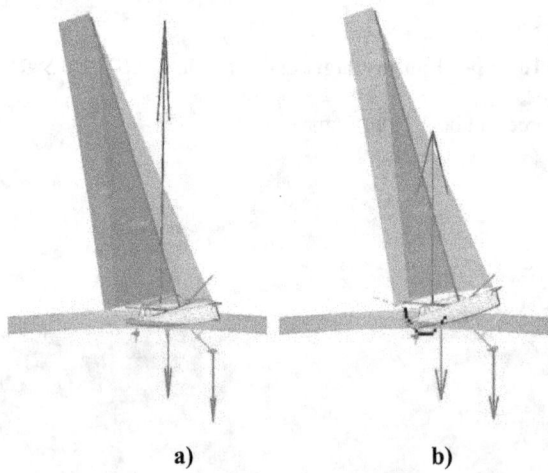

a)   b)

Figure 19 - Attitude of the yacht in waves, a) conventional, b) foiler Open 60 (*TWS*=20knots)

In Figure 20, it can be noticed that the foiler seems experiencing higher amplitude variation than the conventional.

Moreover, the foil seems to act like a low pass filter on the yacht speed variations since the second harmonic of the speed variation is smaller. In our study, the foiler's performances seem to be affected by the wave encounter frequency. Since the true wind angles are different between the foiler and the conventional Open 60, this test case helps to evaluate the motions and to capture the variations of the performance.

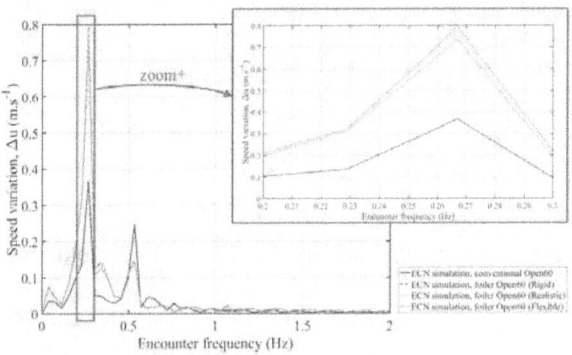

Figure 20 - Speed amplitude spectrum (*TWS*=20knots)

Irregular sea state

The simulation of irregular wave conditions is performed in order to evaluate the influence of the foil structural deformation on the global behavior of the yacht in a more realistic sea state than regular waves. The wave characteristics are chosen according to conditions encountered off the shore of Auckland, New Zealand (Pickrill et al., 1979). The wave spectrum is a Bretschneider spectrum with a significant wave height of 1 meter and a wave peak period of 8 seconds. The same wave phases are chosen for each yacht setup. The wave amplitude spectrum is shown in Figure 21.

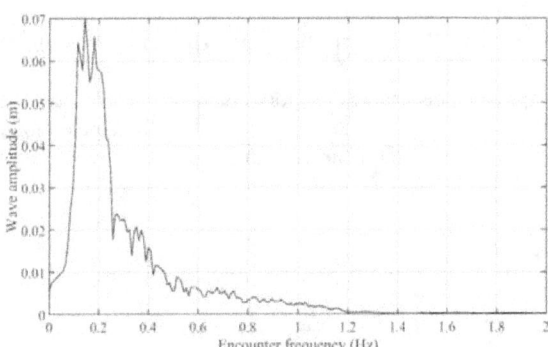

Figure 21 - Wave amplitude spectrum

According to this wave spectrum, the motions of the yacht are evaluated in frequency domain. The amplitude spectrum of the pitch and heave motions are given in Figure 22 and Figure 23. As for regular waves, the flexible foil tends to increase the amplitude of the motions. Also, form the pitch amplitude spectrum, it can be noticed that for all the stiffness, the pitch motion of the foiler Open 60 seems to be also governed by the low frequency content of the wave spectrum.

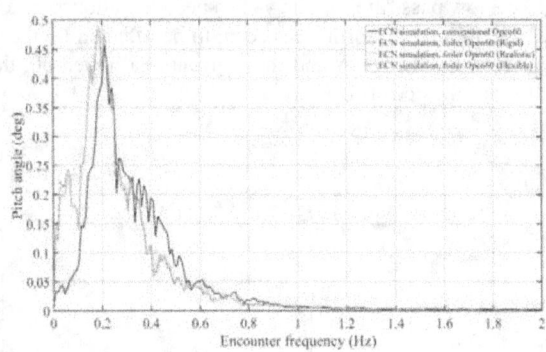

Figure 22 - Pitch amplitude spectrum (*TWS*=20knots)

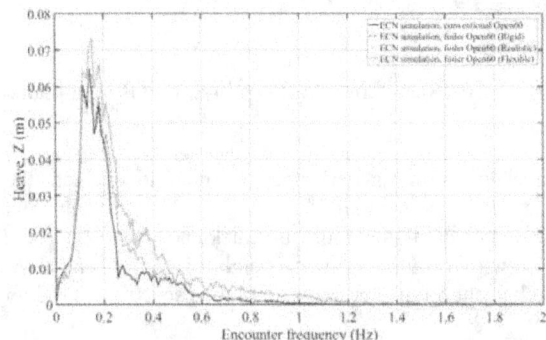

Figure 23 - Heave amplitude spectrum (*TWS*=20knots)

Unlike for regular waves, the amplitude spectrum of the forward speed of the yacht sailing in irregular waves shown in Figure 24, show that the amplitude of the speed variations are similar for conventional and foiler Open 60. The peak frequency is sharper and the maximum value of the peak is higher for foilers. However, the use of a flexible foil tends to decrease the peak amplitude.

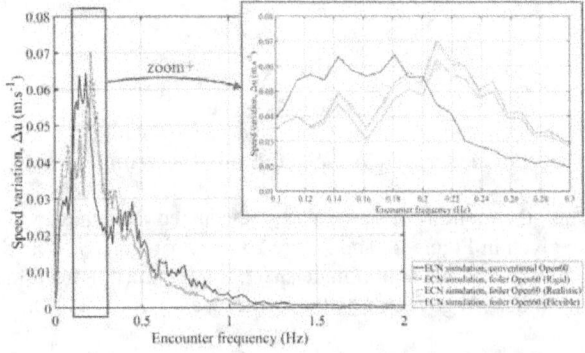

Figure 24 - Speed amplitude spectrum (*TWS*=20knots)

The attitude of the yacht with flexible foil in irregular waves and the qualitative aspects of the foil flexion deformation are also shown in Figure 25.

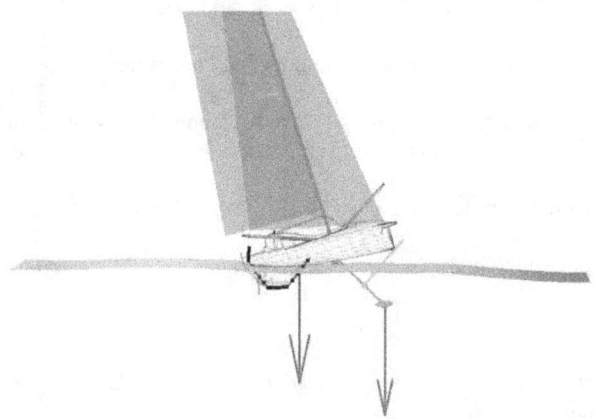

Figure 25 - Attitude of the a foiler Open 60 in irregular waves (*TWS*=20knots)

CONCLUSION

The objective of the paper was to evaluate through simplified modelling the influence of basic structural deformations of hydrofoils on the global yacht's behavior for several sea conditions.

The presented method has the ability to capture the variations of the performance of an Open 60 according to the stiffness of the appendages. The gain and loss were discussed for different test cases in calm water and in waves. Depending on the sea conditions, the effects of the simple deformation such as flexion of the foil were highlighted.

The application of system based modelling coupled with dynamical structural deformations has proven that it can help to understand the global behavior of a yacht with deformable foils. The presented work is a first step to better understand the physical phenomena that drive the yacht's performances.

Then, based on this work, further studies will be carried out in order to evaluate the effect of other modes of deformation, such as torsion of the foil that could be responsible for higher variations in the predicted yacht performances.

Also, validation data from sea trial testing or model tests are required in order to validate such a simplified method.

ACKNOWLEDGMENTS

Authors want to warmly acknowledge PRB sailing team for providing the speed polar diagrams that made possible comparisons presented in this paper.

Authors also want to acknowledge Antoine Connan and his colleagues for their last year student project at ECN on trying to couple the FSI solver *Cast3m* with a DVPP.

REFERENCES

Heppel, P., "Flight Dynamics of Sailing Foilers," Proceedings of the 5th High Performance Yacht Design Conference, Auckland, NZ, 10 12 March, 2015.

RNZYS, Circolo della Vela Sicilia, "AC75 CLASS RULE v1.0," Technical document, 2018.

Horel, B., "Modélisation physique du comportement du navire par mer de l'arrière," PhD Thesis, 2016.

Horel, B., "System based modelling of a foiling catamaran," Proceedings of the 4th International Conference on Innovation in High Performance Sailing Yachts, Lorient, France, 2017.

Roux, Y., Durand, M., Leroyer, A., Queutey, P. & al., "Strongly coupled VPP and CFD RANSE code for sailing yacht performance prediction," High Performance Yacht Design Conference, Auckland (New Zealand), 2008.

Viola, I. M., Biancolini, M. E., Sacher, M., Cella, U., "A CFD based wing sail optimization method coupled to a VPP,"

High Performance Yacht Design Conference, Auckland (New Zealand), 2015.

ITTC, "High Speed Marine Vehicles Resistance Test," Recommended Procedures and Guidelines, 7.5 02 05 01, 2017.

Keuning, J.A., Sonnenberg, U.B., "Approximation of the hydrodynamic forces on a sailing yacht based on the Delft Systematic Yacht Hull Series," The International HISWA Symposium on Yacht Design and Yacht Construction, 1998.

Raymond, J., "Estimation des performances des voiliers au planing," PhD Thesis, 2009.

Huetz, L., Alessandrini, B., "Systematic Study of the Hydrodynamic Forces on a Sailing Yacht Hull Using Parametric Design and CFD," Proceedings of the OMAE 2011, 30th International Conference on Offshore Mechanics and Arctic Engineering, Rotterdam, The Netherlands, June 19 24, 2011.

Tjoswold, S., "Verifying and Validation of a Manoeuvring Model for NTNU's Research Vessel R/V Gunnerus," Norwegian University of Science and Technology, 2012.

Clarke, D., Gedling, P., Hine, G., "The application of manoeuvring criteria in hull design using linear theory," Trans RINA 125:45 68, 1983.

De Ridder, E.J., Vermeulen, K.J., Keuning, J.A., "A Mathematical Model for the Tacking Maneuver of a Sailing Yacht," The International HISWA Symposium on Yacht Design and Yacht Construction, 2004.

ITTC, "Numerical Estimation of Roll Damping," Recommended Procedures and Guidelines, 7.5 02 07 04.5, 2011.

Babarit, A., Delhommeau, G., "Theoretical and numerical aspects of the open source BEM solver NEMOH," In Proc. of the 11th European Wave and Tidal Energy Conference (EWTEC2015), Nantes, France.

Delhommeau, G., Kobus, J.M., "Méthode approchée de calcul du comportement sur houle avec vitesse d'avance," Bulletin de l'Association Technique Maritime et Aéronautique, pp. 467 490, 1987.

Faltinsen, O.M., "Hydrodynamics of High Speed Marine Vehicles," Norwegian University of Science and Technology, pp. 197 199, 2005.

ORC, "ORC VPP Documentation," Offshore Racing Congress, 2016.

Jackson, P., "An Improved Upwind Sail Model for VPP's," The 15th Chesapeake Sailing Yacht Symposium, Annapolis, Maryland, USA, 2001.

Fossati, F., Muggiasca, S., "Experimental Investigation of Sail Aerodynamic Behavior in Dynamic Conditions," Journal of Sailboat Technology, Article 2011 03, 2011.

Newman, J.N., "Marine Hydrodynamics," The MIT press, 1977.

Flay, R.G.J., Vuletich, I.J., "Development of a wind tunnel test facility for yacht aerodynamic studies," Journal of Wind Engineering, 231 258, 1995.

Horel, B., Bouscasse, B., Merrien, A., de Hauteclocque, G., "Experimental assessment of vertical shear force and bending moment in severe sea conditions," In Proc. of the 38th International Conference on Ocean, Offshore & Arctic Engineering, Glasgow, Scotland, UK, 2019.

International Monohull Open Class Association 60 feet, "Class Rules 2018 V2.0," Technical document, 2018.

Speer, T.E., "The BASILISCUS Project Return of the Cruising Hydrofoil Sailboat," The 15th Chesapeake Sailing Yacht Symposium, Annapolis, Maryland, USA, 2001.

Pickrill, R.A., Mitchell, J.S., "Ocean wave characteristics around new Zealand," Journal of Marine and Freshwater Research, 13:4, 501 520, 1979.

Maneuver Simulation and Optimization for AC50 Class

Dr. Heikki Hansen, Oracle Team USA / DNV GL SE, Potsdam, Germany
Dr. Karsten Hochkirch, Oracle Team USA / DNV GL SE, Potsdam, Germany
Ian Burns, Oracle Team USA, San Francisco, USA
Scott Ferguson, Oracle Team USA, San Francisco, USA

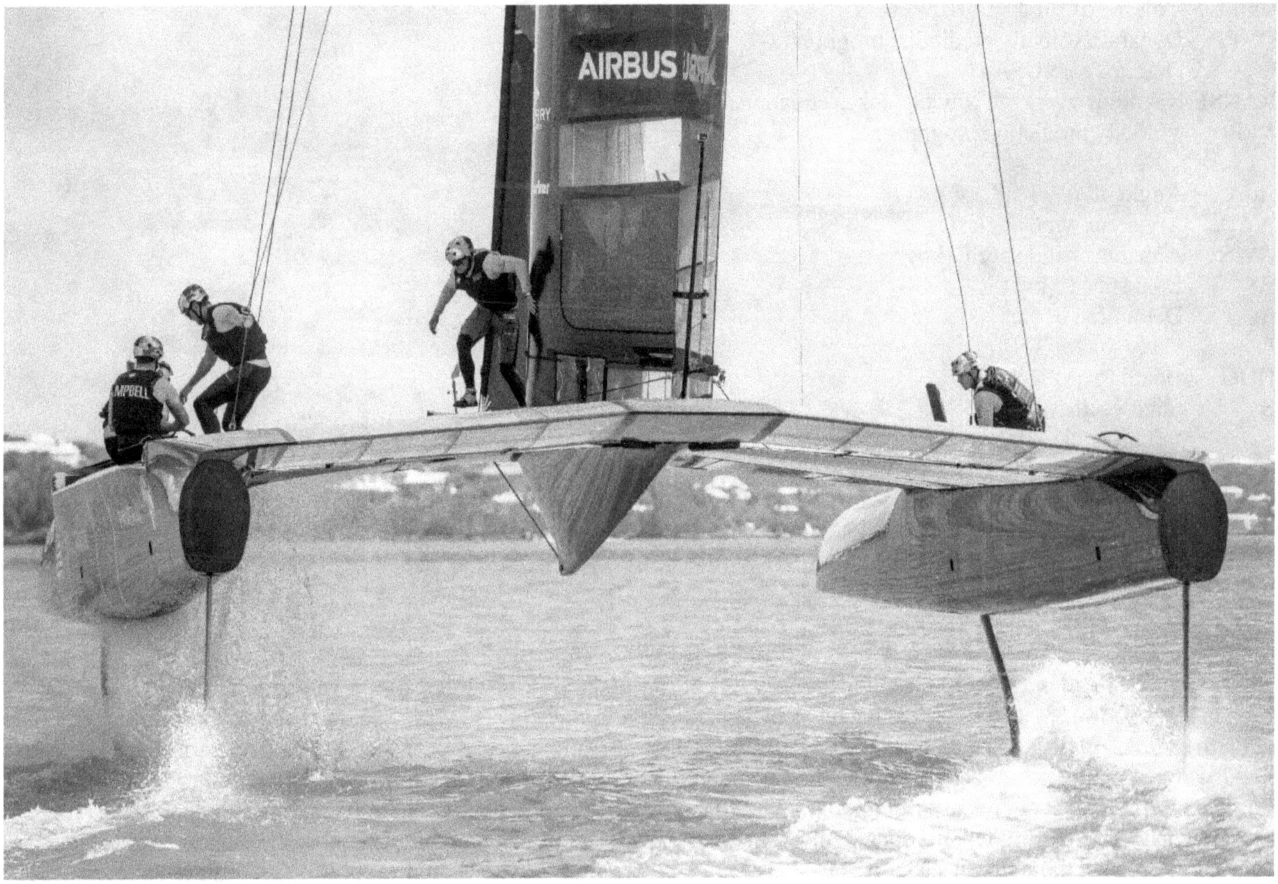

ABSTRACT

The stability and the dynamic behaviour is an integral part of designing hydrofoil supported sailing vessels, such as the America's Cup (AC) 50 class. The foil design and the control systems have an important influence on the performance and stability of the vessel. Both foil and control system design also drive the maneuverability of the vessel and determine maneuvering procedures. The AC50 class requirements lead to complex foil control systems and the maneuvering procedures become sophisticated and multifaceted.

Sailing and maintaining AC50 class yachts is a complex, expensive and time-consuming task. A dynamic velocity prediction program (DVPP) for the AC50 is therefore developed to assess the dynamic stability of different foil configurations and to simulate and optimize maneuvers. The goal is to evaluate certain design ideas and maneuvering procedures with this simulator so that sailing time on the water can be saved.

The paper describes the principal concepts of developing a AC50 model in the DVPP FS-Equilibrium. The force components acting on the yacht are defined based on physical principles, computational fluid dynamics (CFD) simulations and experimental investigations. The control systems for adjusting the aero- and hydrodynamic surfaces are modelled. Controllers are utilized to simulate the human behaviour of performing sailing tasks. Maneuvers are then defined as sequences of crew actions and crew behaviours.

In the paper examples of utilising the DVPP in preparation for the 35th America's Cup in Bermuda are described. The DVPP is for example used to investigate the effect of different boat set-ups on stability and handling during maneuvers. With the sailing team, maneuver procedures are developed and tested. Procedures such as dagger board and rudder elevator movement and crew position are investigated and evaluated to minimize the distance lost during tacking and gybing. The DVPP is also employed for trajectory optimization during maneuvers.

NOTATION

AC	America's Cup
AC50	America's Cup 50 class
CFD	Computational fluid dynamics
VOF	Volume of fluid
DVPP	Dynamic velocity prediction program
OTUSA	Oracle Team USA
RANSE	Reynolds averaged Navier-Stokes equations
VPP	Velocity prediction program

AoA	Angle of attack (°)
AWA	Apparent wind angle (°)
AWS	Apparent wind speed (kn)
BS	Boat speed (kn)
Δt_{step}	Delta time of step (s)
$\underline{F}$	Force vector (N)
HDG	Heading (°)
$\underline{S}$	State vector
$\underline{S_0}$	Initial values state vector
t	Time (s)
TWA	True wind angle (°)
TWS	Ture wind speed (°)
VMG	Velocity made good (kn)

INTRODUCTION

The America's Cup (AC) 50 class for the 35th AC in Bermuda in 2017 is the second generation of foiling catamarans used in the AC. In the campaigns leading up to the 34th AC, the foiling concept was developed and utilized for the first time with the AC72 class. When designing a hydrofoil supported sailing vessel, the stability and dynamic behaviour are important characteristics to consider. For being competitive in the 34th AC it was important to foil controllable and stable. For the 35th AC this is of course still important, but it must be done as efficiently as possible to be competitive.

Efficiency and stability tend to oppose each other for these types of sailing craft. The more efficient the foil design and configuration gets, the more unstable it gets as well. In addition to the foil design, the control systems are therefore a very important performance driver. A very responsive and accurate control system allows a less stable and more efficient foil design to be controllable. The AC50 class requirements drive the design to complex foil control systems.

Foil and control system design also impacts the maneuverability of the vessel and drives the maneuvering procedures. Depending on the control system design and set-up, maneuver procedures need to be developed and adjusted. For evaluating the overall performance of a configuration, maneuvering needs to be considered in addition to the straight-line performance.

Sailing and maintaining AC50 yachts is complex and expensive. Sailing time in preparation for an AC is very valuable, limited and needs to be prioritized. A dynamic velocity prediction program (DVPP) model is therefore set up to assist with assessing the dynamic stability of different foil configurations and to simulate and optimize maneuvers. The goal is to evaluate certain design ideas and maneuvering procedures with this simulator so that sailing time on the water can be saved.

The DVPP requirements for the yacht model and input data are quite different to a steady-state velocity prediction program (VPP). For dynamic simulations a wide range of sailing states needs to be covered by the model and input data. While for a steady-state VPP a much smaller range of sailings states is required, the model accuracy is paramount to enable the software to assist in making the correct design choices. High fidelity input data is important for a steady-state VPP, while a DVPP requires input data covering a wide range of sailing states.

The DVPP is set up as a second VPP in addition to the OTUSA VPP for steady-state velocity prediction. This enables cross-checking of steady-state results and allows both tools to be optimally developed and utilized for their specific purpose. The DVPP is interfaced with the OTUSA data acquisition, processing and analysis environment based on Bravo Systems[1] technologies. This enables efficient exchange, analysis and comparison of simulation results with measurements. DVPP simulations can be displayed in real-time when sailing and used as performance benchmarks for maneuvers.

DYNAMIC VELOCITY PREDICTION PROGRAM (DVPP)

FS-Equilibrium is an advanced workbench for the analysis of steady-states of equilibrium (VPP) and transient states of motion (DVPP). It is an open modular workbench for the analysis of floating rigid body equilibrium conditions and motions. Using a flexible architecture with so called force modules, specific set-ups can be built. The individual force components acting on the rigid body are modelled by force modules, which calculate the component forces and moments with theoretical, semi-empirical or numerical approaches or use data fed from external sources such as experiments or CFD.

The steady-state mode solves equilibrium conditions of external aero- and hydrodynamic forces and moments acting on yachts, ships and floating bodies for up to 6 degrees of freedom. The development started with the work of Hochkirch, 2000. Since then the software has continually been expanded and advanced.

Richardt et al., 2005, describes implementation of the transient mode to integrate the nonlinear differential equation of motion and compute the trajectory of a vessel so that maneuvers such as turns, stops, zig zag tests and real-time sailing scenarios can be simulated. Details of modelling unsteady motion and dynamics of flight are taken from Etkin, 2005. The DVPP is then used by BMW Oracle Racing in the 32nd AC campaign to build a sailing simulation for starting maneuver training as described by Binns et al., 2008.

Foiling sailing craft have since also been modelled with the workbench. Boegle et al., 2012, implement foil control systems of an International Moth and describe design considerations looking at the tradeoff between speed and stability. Paulin et al., 2015, discuss the performance assessment and optimization of a C-Class catamaran hydrofoil configuration.

DVPP YACHT MODEL

The core of the yacht model constitutes of the force modules set-up to model the forces acting on the yacht. Based on the state variables and the parameter settings, each force module returns the six force and moment components. In steady-state mode the forces/moments of all modules are added, and the residual force/moment of each component is minimized by adjusting the control variable associated with each degree of freedom. In addition, trim parameters can be defined to find the optimum with respect to an objective function, typically boat speed (BS) or velocity made good (VMG). A selection of balancing and optimization algorithms is available to achieve this.

In the DVPP mode the excess forces and moments are used to calculate the rigid body accelerations according to Newtons second law. The damping terms can be modelled by appropriate force modules and are then considered as part of the excitation force. The accelerations are fed into the numerical integration scheme to determine the values of the state variables for the next time step. A fourth order Runge-Kutta scheme, as well as a fifth order Runge-Kutta-Feldberg scheme, a variable time step scheme adapted from Cash-Carp and an explicit Euler scheme are available for time integration as described by DNV GL, 2018. The process flow of the time stepping procedure is schematically shown in Figure 1.

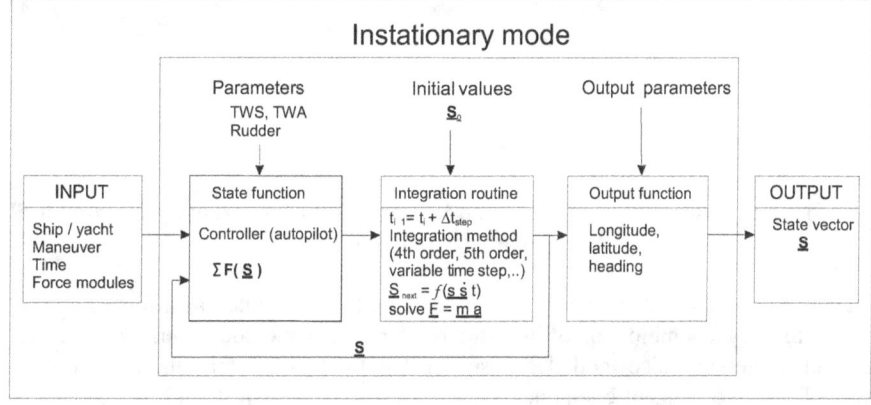

Figure 1 – Time stepping process flow

[1] Bravo Systems SL, www.bravosystems.es

Force Modules

The force components acting on the AC50 are defined by different types of force modules. Figure 2 shows an expletory list of force module types used to describe the AC50. Mass modules including inertia terms are specified for platform, wing and crew members. The mass of each crew member is defined in a separate force module so that individual crew movements to be modelled during maneuvers.

The wing and jib forces are modelled based on CFD data sets for upwind and downwind sail trims. The coefficients of the six force/moment components are modelled with an UniversalForce module as functions of the wing and jib trim parameters and the apparent wind angle (AWA), resulting in 8 independent variables. The UniversalForce module uses multidimensional regression models with freely definable independent variables to model an arbitrary force. The local velocity used to calculate the apparent wind speed (AWS) and AWA includes the dynamic components from the vessel's rotation.

During maneuvers such as tacks and gybes, the wing shape relative to the onset flow can be very different than in normal straight-line sailing conditions. A much larger range of operating conditions needs to be covered by the data set compared to the input data for a steady-state VPP. During a tack or gybe the AoA of the wing and the wing camber can become negative depending on the timing of the wing inversion and the wing trim. The wing inversion is crucial when simulating and investigating tacks and gybes, and the aerodynamic data must be included in the DVPP for these unusual operating conditions of the wing. To produce these extensive data sets, potential flow simulations of the wing and jib combination are conducted for a matrix of all relevant trim parameter variations and AWAs. Correction terms obtained from Reynold averaged Navier-Stokes equations (RANSE) volume of fluid (VOF) simulations are utilized to account for viscous effects.

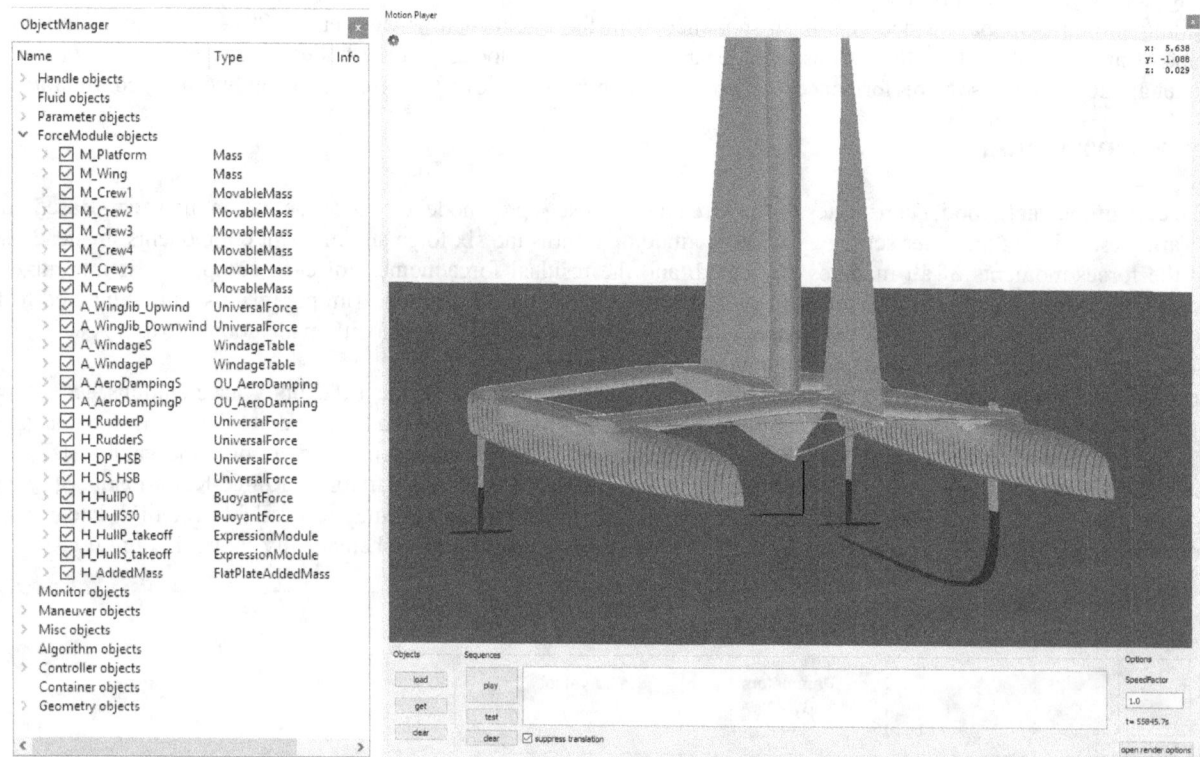

Figure 2 – List of force modules (left) and AC50 model in motion player (right) showing hull sections from BuoyantForce model in green

An aerodynamic damping model is developed from CFD data and implemented as a force module. The windage of the vessel's components is defined by a combination of high-fidelity CFD data and coefficient-based calculations.

The dagger boards and rudders are also modelled based on CFD data sets with UniversalForce modules. Each foil is modelled independently. The force/moment components are defined in a local foil coordinate system as functions of local flow and immersion parameters. While this approach disregards interaction between the foils, it allows to describe a wide range of operating conditions for each foil. This is required when simulating maneuvers in which each foil moves (changes position and/or orientation) and sees very different local flow and immersion depending on the maneuver characteristics.

The input data is predominantly developed from extended vortex lattice models with correctors from high-fidelity RANSE VOF simulations. The correctors account for viscous effects and apply penalties at the operational boundaries of the foils to emulate the effects of ventilation and cavitation. The BS during maneuvers is typically well below the maximum achievable BS so that significant cavitation or ventilation is not very likely.

The added mass of the principal hydrodynamic and aerodynamic components is calculated based on formulas for flat plates taken from Saunders, 1957.

While the primary focus is on modelling foiling states, a hydrodynamic model of the hulls is included to be able to simulate scenarios where the hulls briefly touch the water. The intention is not to model displacement sailing or take-off maneuvers. The BuoyantForce module generates and utilizes the hull sections shown in Figure 2. It calculates displacement, wetted surface area, frictional resistance and the resulting force/moment components at each time step. A viscous damping approximation is also computed. The added mass can be calculated from the Lewis transformation. An ExpressionForce module is created to account for additional forces and moments from hydrodynamic effects during takeoff. The mathematical expression is formulated based on CFD simulations.

Controllers

Controllers are defined in the DVPP to model the crew actions involved when sailing the yacht. With controllers, maneuvers can be repeated consistently and systematic studies can be carried out. The DVPP can also run in "simulator" mode where an operator controls the vessel through input commands.

Controllers are further utilized to model the control system of the boat. The control system design and implementation have great impact on the stability and controllability of the AC50 class yachts. The control system constitutes of hydraulic and mechanical actuators, and feedback control is limited by the class rules. The control system speed, responsiveness, accuracy and repeatability are all important parameters to consider and optimize. Key characteristics of the actuators installed on the boat are obtained from measurement data acquired when sailing. These characteristics are described by controllers in the DVPP to set up a simplified control system model. By changing the values of the controller parameters, modifications to the control system can be modelled and the effect on the controllability in straight-line sailing and in maneuvers can be investigated.

Besides typical linear time-invariant (LTI) relations, the controllers can use arbitrary mathematical expressions of time or state to define the rate of change and/or value of the respective actuator. Conventional proportional–integral–derivative (PID) dependencies as well as step changes and time delays can model the behaviour of the crew and the hydraulic and mechanical actuators of the control system.

Figure 3 shows, as an example, the ride height control implemented in the DVPP. One controller is defined for the human behaviour. The controller input is the ride height error. In an effort to minimize the error the controller determines the target rake. In this example the rake is adjusted by button pushes, which trigger step changes in the target rake. The target rake is the output of the human behaviour controller and becomes the input for the controller describing the rake actuator. This controller models the response time and speed of the rake actuator. The output of the rake actuator controller is the actual rake of the dagger board, which influences the boat response and affects the ride height.

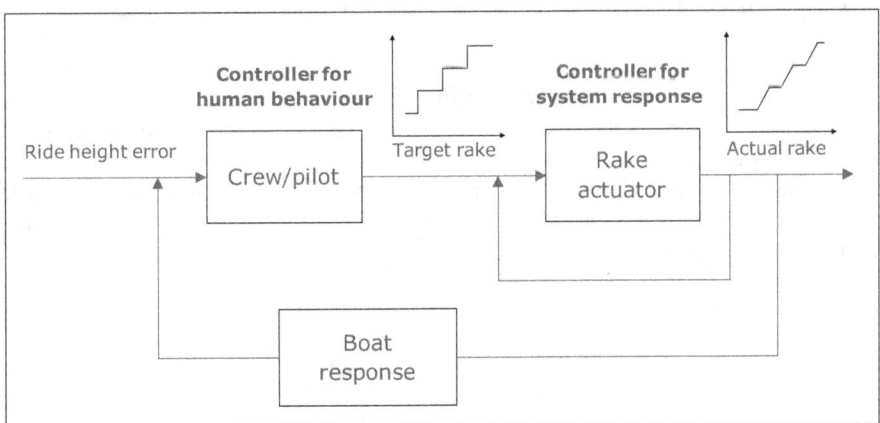

Figure 3 – Ride height controllers for human behaviour and system response

Crew

Controllers are implemented and employed to model the three main crew functions to control the vessel:

- Rudder angle: A controller simulates the helmsman changing the rudder angle to steer the vessel to a target TWA. The controller minimizes the error between target and actual TWA. For a maneuver the target TWA is prescribed by a sequence of target angles or a continuous function. The turning rate is utilized by the controller as the change rate of the error. The output of the controller is the rudder angle.

- Target wing rotation/twist: A controller simulates the wing trimmer changing the mast rotation or the wing twist to sail the vessel at a target heel angle. The controller minimizes the error between target and actual heel angle. The controller is disabled during certain parts of some maneuvers when controlling the heeling moment with the wing trim is not possible. This is for example the case in the middle of a tack or gybe while the wing is inverting. The heeling rate is used by the controller as the change rate of the error. The controller output is the target wing rotation/twist, which is the input for the wing rotation/twist actuator.

- Target dagger board rake: A controller simulates the pilot/flight controller changing the dagger board rake and thereby the AoA to sail at a target ride height. The controller minimizes the error between target and actual ride height. The vertical speed is utilized as the change rate of the error. The controller output is the target dagger board rake, which is the input for the rake actuator. During a maneuver when both dagger boards are in the water, the controller of both boards can be active at the same time or the active dagger board control is switched from one dagger board to the other at a certain point during the maneuver. This depends on the maneuver playbook. DVPP is for example also used to investigate approaches of controlling the dagger boards during maneuvers.

A reaction time of the crew and a sensitivity threshold to disturbances is not included because the sailors anticipate required actions and sense disturbances. The sailors comment is "if we only reacted once the boat starts heeling we would not be able to sail the boat. We anticipate the required actions based on the environment and the sailing condition." The parameter values are selected so that the controller response resembles the observed and measured crew response.

In addition to the three principle crew functions modelled by controllers, other crew functions are modelled without controllers (open-loop) including crew movement, dagger board dropping and raising, dagger board cant, dagger board rake drop pre-set, elevator rake and wing camber.

Actuators

Controllers are utilized to model the movement of actuators in the control system. The dagger board rake for example is modelled with a delay time for the hydraulic value to open and a rake speed.

The following actuator movements are modelled with controllers: dagger board rake, cant and extension, rudder elevator rake, wing rotation with different maximum speed for easing and bringing in, wing twist, wing camber and crew motion speed.

The rudder angle is modelled without an actuator since the input from the helmsman is directly transferred mechanically to the rudder blade. The actual rudder angle is therefore always equal to the target rudder angle.

Maneuvers

Parametric maneuvers with a series of actions are defined to model the maneuvers. The actions are triggered by conditions. A typical condition is a time delta to a reference time. Any other parameter than time can also be used to describe a condition, such as for example TWA, AWA and ride height.

Actions triggered by a condition include crew movement, dagger board dropping and raising, dagger board cant, dagger board rake drop pre-set, elevator rake, wing camber, changes in target TWA, target heel and target ride height, and activating and deactivating of controllers.

TACKING PROCEDURES

Together with the sailing team the DVPP is used to investigate tacking procedures to understand the influence of parameters on stability, controllability and performance. As a measure of performance, the distance lost during the tack is calculated. It is defined as the distance lost to windward relative to sailing continuously with the VMG at the beginning of the tack. The target TWA at the end of the maneuver is the inverse of the starting TWA. The boat set-up is symmetric so that the VMG will eventually be the same again after the tack. Provided the simulation time is sufficiently long, the distance lost is not affected by the time interval.

Assessing the distance lost in maneuvers from real-life measured data is often strongly dependent on the time interval due to fluctuations in TWS, TWA and BS. It is therefore very difficult to systematically evaluate different tacking procedures from measured data. Another reason is of course that it is impossible to achieve the required repeatability in the maneuver execution to allow a detailed assessment of variations in the maneuver procedures with respect to distance lost. The DVPP is therefore employed to investigate tacking procedures and suggest improvements, which can be tried on the water.

To investigate maneuver procedures, a baseline maneuver representing the current benchmark is typically set up in the DVPP. It is based on the analysis of measured data, videos, observations, the maneuver playbook and discussions with the sailors. The baseline maneuver is then used to systematically vary parameters to investigate their influence on the distance lost.

Crew Righting Moment

The effect of crew righting moment during a tack is one of the parameters investigated. Depending on the control system layout and the maneuver procedures, crew members are required to windward and leeward at different times during the maneuver. For making control system layout decisions and develop maneuver procedures it is therefore of interest to place a penalty on crew being to leeward during a maneuver.

Figure 4 shows the righting moment of the crew and BS against time for the baseline tack and for variations of crew movement during the maneuver. The baseline tack (black) is hidden first behind the green (circles) and then behind the blue (squares) crew righting moment curve. Each step in the righting moment indicates one or more crew members moving from one side to the other. It can be seen how sending crew to leeward earlier reduces the entry BS into the tack, which results in a reduced minimum (bottom) BS. The black vertical line indicates head to wind. Furthermore, it is evident that reduced righting moment from crew being to leeward when coming out of the tack slows down the speed build.

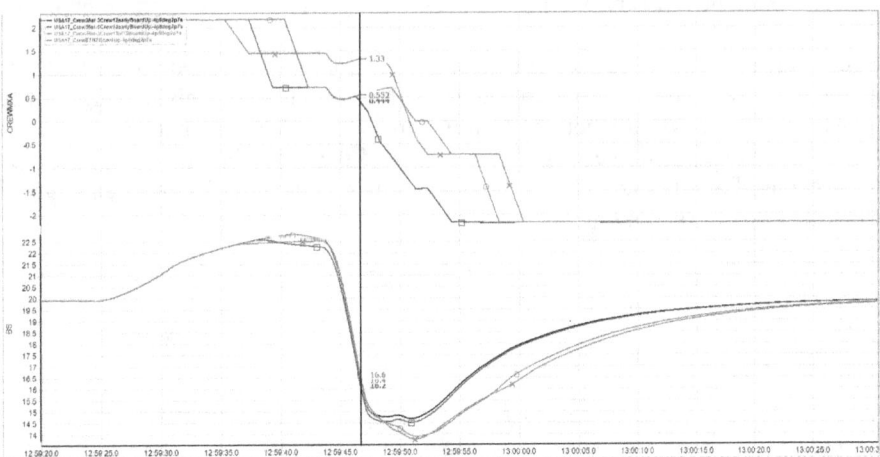

Figure 4 – Crew righting moment and BS against time during tack for different crew movement patterns

For each of the crew movement patterns the distance lost during the tack is computed and plotted in Figure 5 as the delta distance lost relative to the baseline tack. The x-axis shows the time that crew members are to leeward longer compared to the baseline tack. One scenario for example has the crew in positions 1 and 2 (sailors furthest forward) staying to leeward for 9.5s longer to raise the dagger board, which is a total of 18s crew is longer to leeward. Another scenario models the crew movement of a competitor, which is obtained from the analysis of video footage. From these and the other scenarios the relationship can be seen that one crew member being to leeward one second longer costs about 0.45m distance to windward. This number is used by the sailors to evaluate crew being the leeward against other parameters, such as having the windward dagger board in the water longer while the crew moves to the windward hull before raising the board.

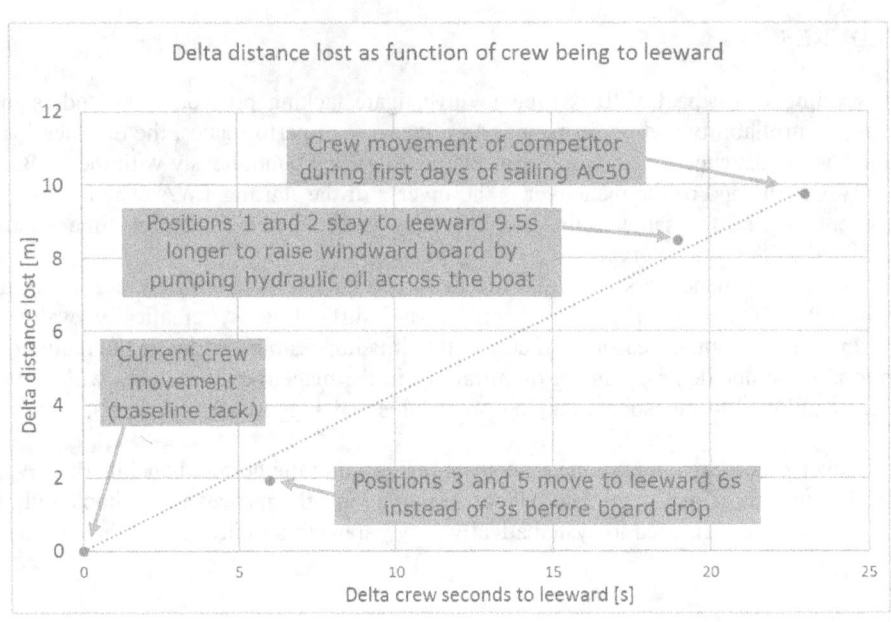

Figure 5 – Effect of crew position on distance lost in tack

Dagger Board Dropping and Raising

The effect of having two dagger boards in the water during a tack is also investigated. From a stability and controllability point of view, it makes sense to drop the windward board and let the boat settle for a moment before starting the turn. The dagger board behaviour during the immersion phase is very hard to predict and control consistently so that the board drop often unsettles the boat. On the other hand, the second board in the water adds additional drag. DVPP simulations are conducted with different delay times between board immersion and initiation of the turn into the tack. The turn is delayed by 1, 2, 3 and 4s compared to the baseline tack as can be seen by the heading (HDG) plot in Figure 6. The BS plot in Figure 6 shows that the tack entry BS reduces significantly the longer the turn is delayed. If the turn is delayed by 4s, the entry BS drops from around 23kn to 21kn. This results in a reduction of the minimum (bottom) BS by almost 1kn.

Figure 6 also includes the rake angle and the heave load of the dropped dagger board. The rake angle plot shows that the dagger board is stored at a rake of 2.87° and set to -4° before the drop. In this tacking procedure the ride height controller for the dagger board is activated when the turn is initiated. The controller increases the rake to the maximum angle. As the boat turns through the wind, the wing is inverted, and the boat weight is transferred to the dropped dagger board as shown in the heave load plot in Figure 6. As the boat reaches the bottom BS, the maximum rake angle is required to produce the required lift to stay on the foils.

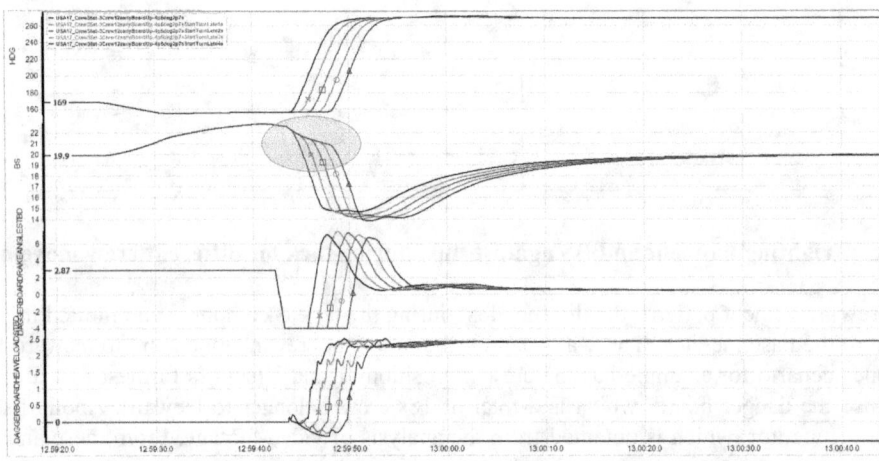

Figure 6 – Heading (HDG), BS, dagger board rake and heave force against time for different delay times before initiating turn into tack

In this tacking procedure the windward dagger board is set up to produce little heave force before and after the load transfer from and to the leeward dagger board. The heave load plot in Figure 6 shows that little force is produced by the windward dagger board after the drop until the rake angle increase.

The DVPP is also utilized to highlight that upwards heave force on the windward dagger board is unfavorable as it reduces the righting moment. Downward heave force increases the righting moment. DVPP simulations show that the drag of the dagger board outweighs the benefits of the additional righting moment. The distance lost increases. On the exit of the tack, the BS is at its minimum. Additional upwards force on the leeward dagger board to compensate downwards force on the windward board increases the required minimum BS to keep foiling. Consequently, the required tack entry BS is higher. A larger speed build before the tack is necessary, which further increases the distance lost.

The effect of delaying the turn by 1, 2, 3 and 4s shown in Figure 6 is expressed as the delta distance lost relative to the baseline tack in Figure 7. The relationship between delay and distance lost is close to linear. Delaying the turn after dagger board immersion increases the distance lost by around 3m per second delay.

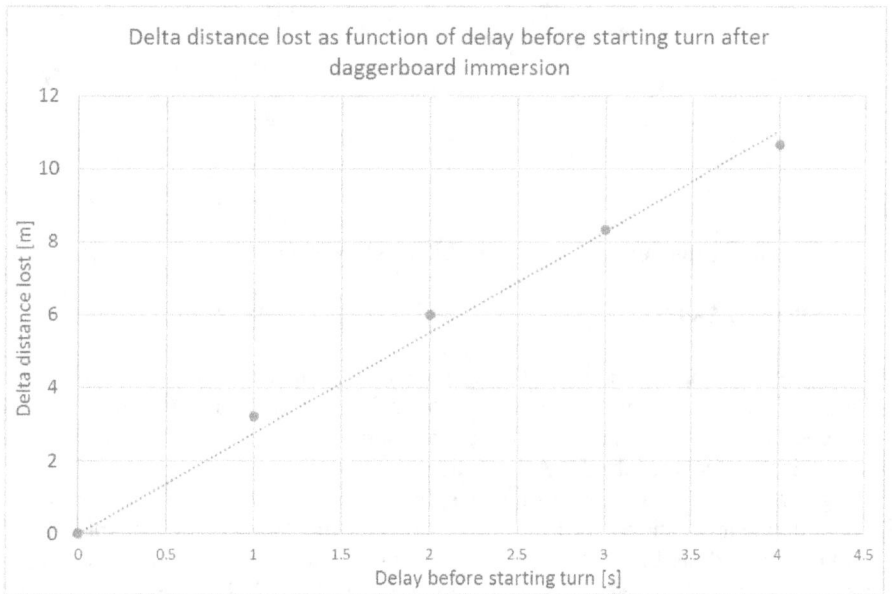

Figure 7 – Effect of delaying turn after dagger board drop on distance lost in tack

The effect of raising the dagger board slower is investigated analogously with the same baseline tack as reference. The time it takes to raise the dagger board is varied by adjusting the actuator speed with which the dagger board moves. The starting time of raising the dagger board is not changed. The motion starts as soon as the wing inversion is completed. At this point heeling moment to the new leeward side is generated and the heave load is transferred to the new leeward dagger board. The windward dagger board can be raised.

Figure 8 shows the effect of raising the dagger board slower. The distance lost is plotted against the delta time the dagger board is in the water relative to the baseline tack, which results from the slower dagger board retraction speed. The relationship is close to linear again. The distance lost increase by around 1.6m per second the dagger board is longer in the water.

These studies highlight that having the second dagger board longer in the water at the beginning of the tack (3m loss per second) has a larger negative effect than having the dagger board in the water longer when exiting the tack (1.6m loss per second). In the beginning of the tack the BS is much higher and the absolute reduction in BS and VMG is greater. The integral of the VMG change is the distance lost. In addition, a reduction in BS at the beginning of the tack is carried through the whole tack. Losses (or mistakes) in the beginning of a maneuver are therefore often costlier than mistakes later in the maneuver.

The influence of having the dagger board longer in the water when exiting the tack (1.6m loss per second) can also be evaluated against having crew to leeward longer (0.45m loss per second). If the crew crosses first to the new windward hull before raising the dagger board, it might be in the water 3s longer. This is approximately equivalent to having two crew members to leeward 5s longer to raise the board immediately.

The outcome of studies like these is used by the sailing team to analyse, evaluate and improve maneuvering procedures.

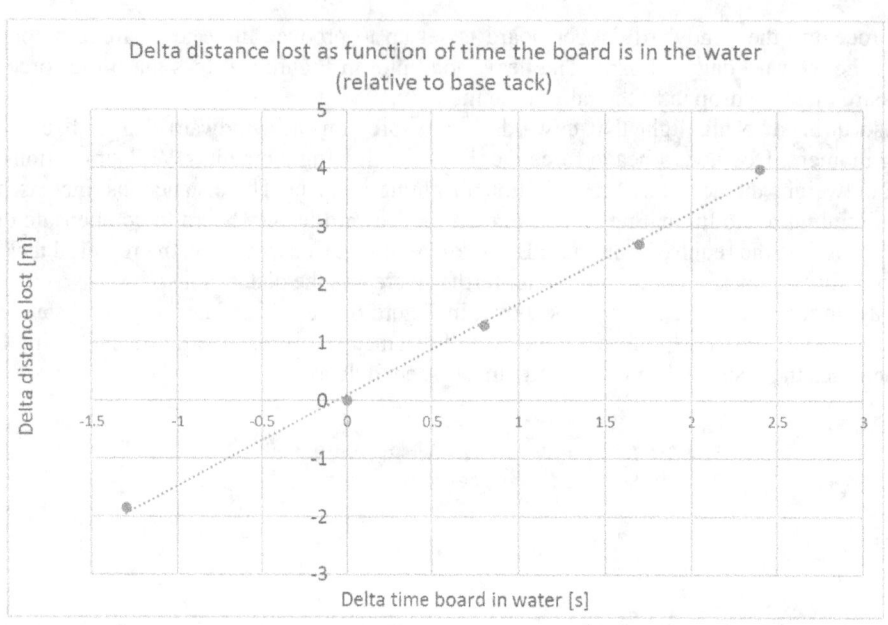

Figure 8 – Effect of raising dagger board slower on distance lost in tack

TACK TRAJECTORY OPTIMIZATION

Another aspect investigated to improve tacks is the trajectory. A formal optimization approach is utilized with the objective of minimizing the distance lost. The course through the tack is described by the change in heading as a function of time. From the change in heading the target TWA for the rudder controller simulating the helmsman is obtained. The target TWA at the end of the tack is the inverse of the TWA at the beginning of the tack.

The change in heading is defined as a function of time by spline elements with 7 free variables as shown in Figure 9. The build angle and build ratio define how quickly the turn is initiated. The turn rate specifies the maximum turn rate during the tack. The fade angle and fade ratio describe how abruptly the turn fades out. The overshoot angle defines how wide the boat comes out of the tack and the overshoot time expresses how long it takes to come back up to the target TWA. Tangentiality constraints are applied between the spline elements and the spline elements describing the overshoot shape are defined with fixed build and fade parameters.

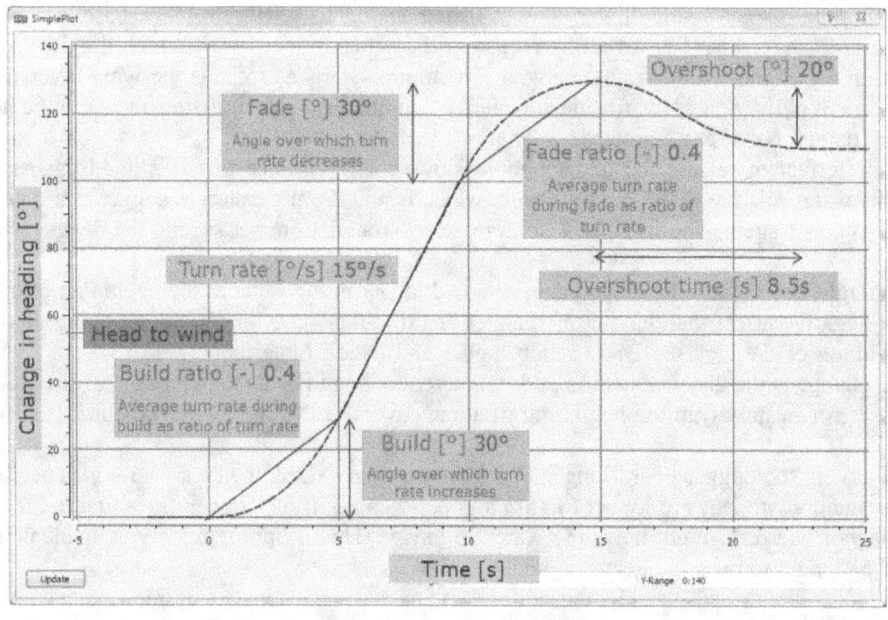

Figure 9 – Target heading definition as spline elements with 7 free variables

In the parametric maneuver, the dagger board drop is the main reference action. All other actions are defined relative to the board drop. The first crew members move to leeward before the board drop and the turn is initiated 1s after the board drop. Subsequent actions do not take place at fixed times but are triggered by conditions that account for different timing depending on the trajectory. The wing inversion is triggered by a threshold AWA, which is a function of turn rate. Further crew movement and dagger board raising are relative to completion of the wing inversion. The rudder elevator movement is controlled by rake changes on the dagger board.

For the baseline tack, the free variables of the target heading function are chosen to resemble a reference benchmark tack at the time. Figure 10 shows TWA, turn (yaw) rate and rudder angle against time of the measured reference benchmark tack (back dotted lines) and the baseline tack (red lines) from the DVPP. The baseline tack captures the characteristics of the reference tack well, including the overshoot angle. The baseline tack is defined to be symmetric, i.e. the TWA after the tack is the inverse of the TWA before the tack. The measured reference tack is not symmetric and therefore the TWAs do not match after exiting the tack.

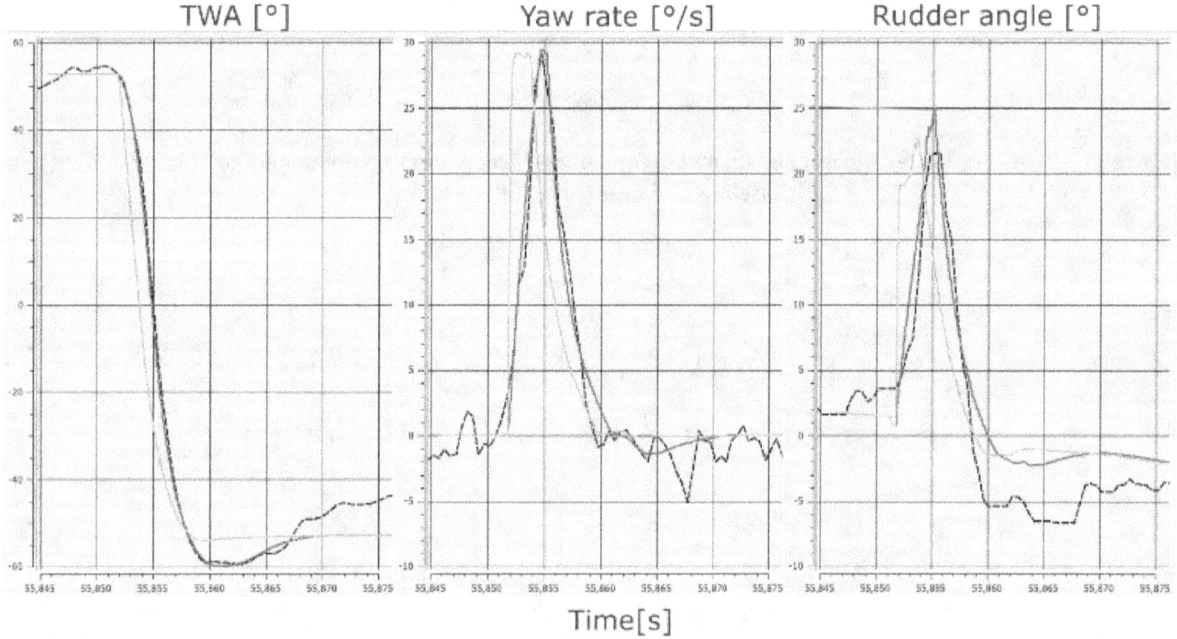

Figure 10 – TWA, yaw rate and rudder angle during reference (black, dotted), baseline (red, dark) and optimized (green, light) tack

For the trajectory optimization the DVPP is driven through the optimization tool FS-Optimizer. Sobol (quasi-random low-discrepancy sequence) and NSGA-II (Non-Dominated Sorting Genetic Algorithm) optimization algorithms with 7 free variables and the objective function to minimize the distance lost are employed. For each optimization at a TWS approximately 2000 trajectory variations are assessed.

Figure 11 shows optimization results for 14kn TWS. The distance lost objective is plotted against turn rate. The green circles denote fully foiling tacks while the orange circles are "touch and go" tacks where at least one hull touches the water. The baseline tack and the optimum tack with the minimal distance lost are highlighted with blue circles. The best tacks are all fully foiling tacks and the optimum turn rate is around 29°/s. The turn rate of the baseline tack and the optimum tack are similar.

It is conceivable that a higher turn rate is generally better since it reduces the time where wing and jib do not produce forward thrust. The distance lost increases as the turn rate decreases. With turn rates below 20°/s, fully foiling tacks are almost impossible. The distance lost also increases for turn rates above 30°/s. The time required to invert the wing and the dagger board rake actuator speed make tacks with higher turn rates difficult to control and less effective.

Figure 12 shows the same set of results for 14kn TWS. This time the distance lost objective is plotted against the overshoot angle. The best tacks have almost no overshoot angle (<1°). The distance lost increases with overshoot angle. Sailing at a wider TWA reduces the VMG out of the tack. Unlike heavy displacement sailing yachts, these lightweight and responsive catamarans have very little inertia so that bearing away to build speed out of a tack has no benefit.

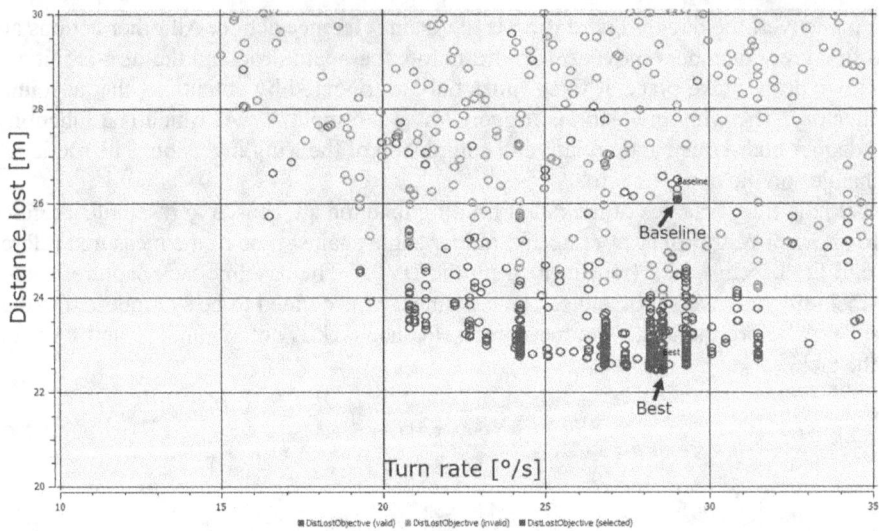

Figure 11 – Distance lost as function of turn rate for foiling (green, dark), "touch and go" (orange, light) and baseline and best (blue, filled) tacks

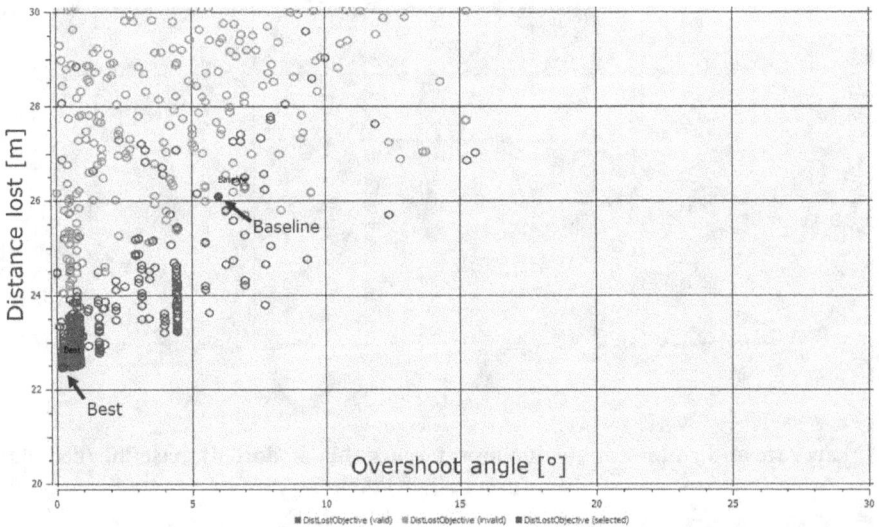

Figure 12 – Distance lost as function of overshoot angle for foiling (green, dark), "touch and go" (orange, light) and baseline and best (blue, filled) tacks

Figure 13 shows the distance lost objective plotted against the build angle when starting the turn. A small build angle means that the turn is initiated quicker, i.e. the turn rate increases more rapidly. Many of the best tacks have small build angles. Good tacks are also possible with lager build angles. But for these tacks the build ratio is high, which also results in a rapid increase in turn rate. These catamarans have very little inertia and lose momentum quickly as soon as wing and jib produce less forward thrust. It is therefore crucial to turn through the wind as quickly as possible. A rapid initiation of the turn reduces the turning time.

Figure 13 highlights that the optimum tack has a much smaller build angle than the baseline tack. The key differences between optimum and baseline tack are the more rapid initiation of the turn and the finer exit without overshoot. The optimum tack loses around 3.6m less than the baseline tack.

Figure 14 illustrates the trajectory of the optimum tack (green) compared to the baseline tack (red). The rapid initiation of the turn, the similar turn rate and the finer exit with virtually no overshoot can be seen. The catamaran symbols in Figure 14 mark the time when the boat in the baseline tack is head to wind. It is evident that the boat in the optimum tack has already gone through the wind at this time. The wind arrows in the upper left-hand corner of Figure 14 show that the boat on the optimized tack trajectory leads the boat on the baseline trajectory after exiting the tack.

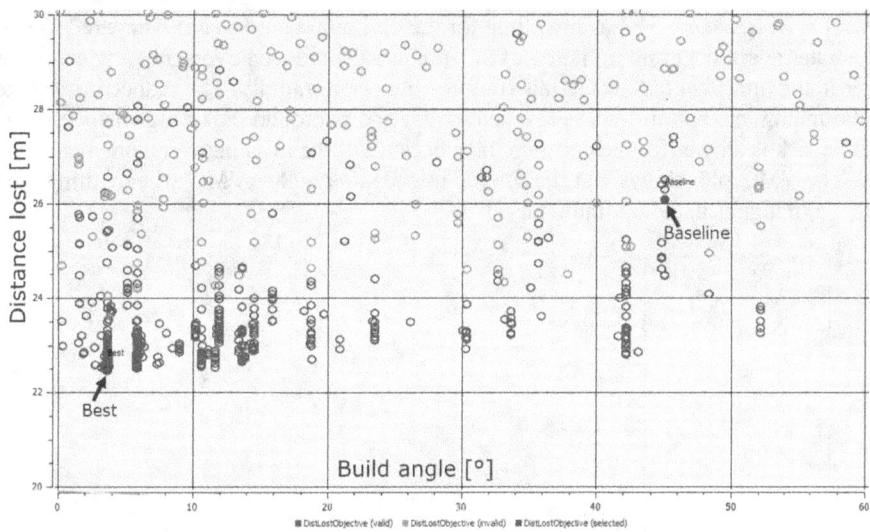

Figure 13 – Distance lost as function of build angle for foiling (green, dark), "touch and go" (orange, light) and baseline and best (blue, filled) tacks

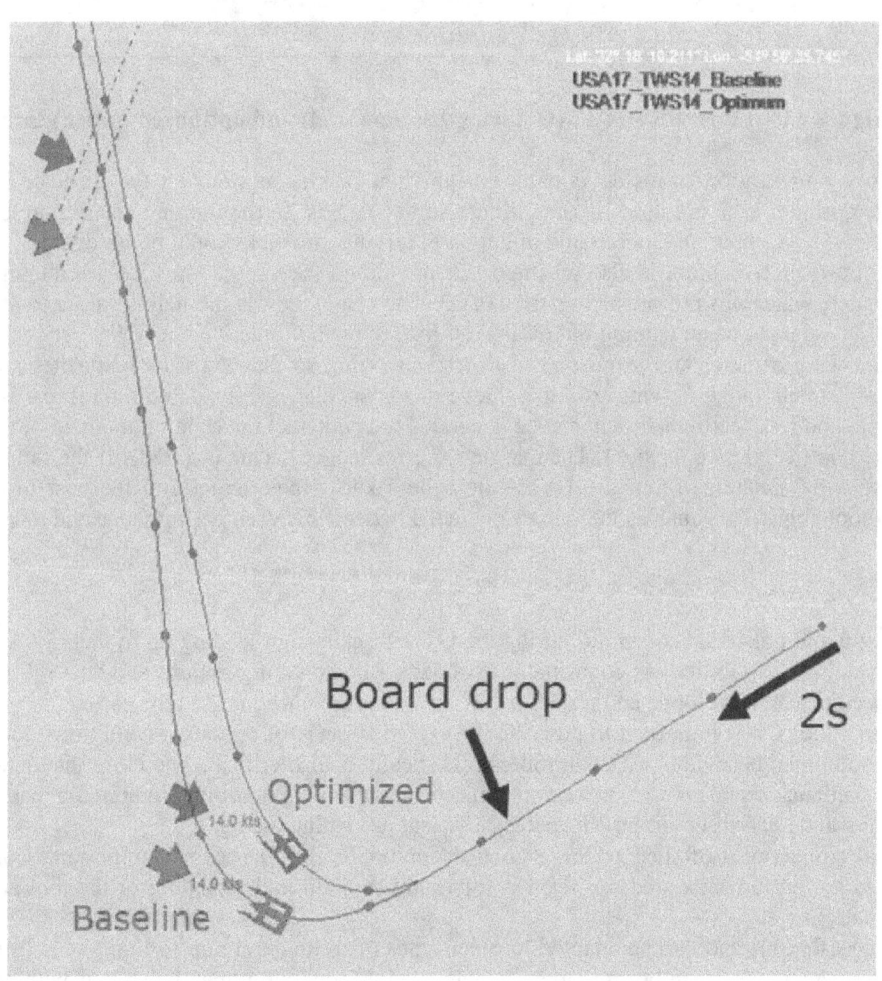

Figure 14 – Baseline (red) and optimized (green) tack trajectory

Figure 15 shows TWA, BS and VMG against time for the optimum tack (green) compared to the baseline tack (red). The TWA plot illustrates again the rapid initiation of the turn and the reduced overshoot angle when exiting the tack. The BS reduces quicker in the optimum tack when initiating the turn more rapidly. The reduced time taken to go through the wind means that the minimum (bottom) BS is reached earlier and is around 0.5kn higher than in the baseline tack. The speed build out of the tack is slower for the optimum tack because of the missing overshoot angle, resulting in a finer exit at a smaller TWA. The VMG plot shows that the finer exit leads to a higher VMG when exiting the tack. The minimum VMG is more than 1.5kn higher in the optimum tack.

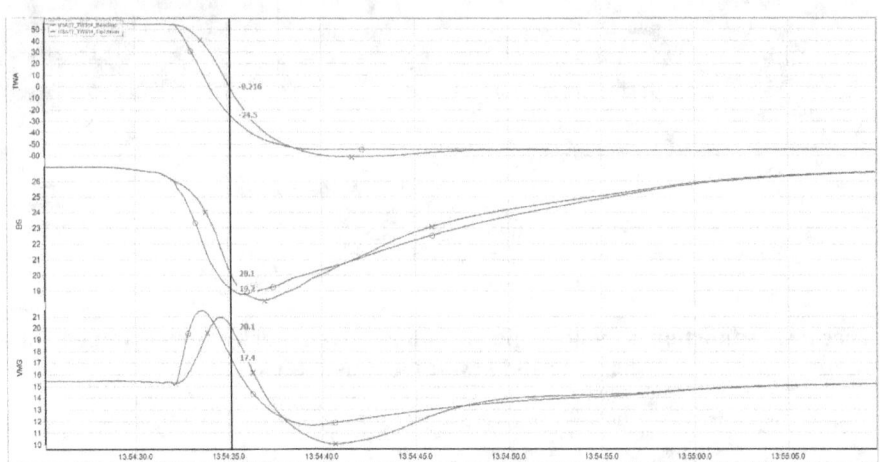

Figure 15 – TWA, BS and VMG during baseline (red) and optimized (green) tack

The tack trajectory optimization provides valuable insight into factors minimizing the distance lost in a tack. The characteristics of the optimum tack are used by the sailing team as targets. In the training leading up to the Louis Vuitton AC Qualifiers and the 35th AC more and more rapid initiation of turning into tacks could be observed.

A fine exit without much overshoot is also an important target and the sailors work on overcoming some additional challenges on the water, which are not present in the DVPP. The challenge for the helmsman is to steer the boat to the desired TWA without overshoot when coming out of a fast turn. True wind direction and TWS vary constantly, and waves disturb the boat. The wing trimmer and grinder need to trim the wing so that the flow reattaches after inversion. At a smaller TWA, the adjustment range in wing trim to achieve this is smaller, the available time frame is shorter due to the smaller rotation angle, and less heeling moment can be generated to counteract the righting moment of the foiling boat. It is harder to "catch" the boat with less overshoot. The margin for error reduces. This is a tradeoff the sailing team must make depending on the specific situation. From the DVPP the sailors know their target and the cost in distance lost when increasing the overshoot angle. This enables them to evaluate the tradeoff between performance and risk better.

CONCLUSIONS

The DVPP is set up and utilized successfully during the OTUSA campaign leading up to the 35th AC in Bermuda. It is employed for investigating the stability and controllability of dagger board configurations and boat set-ups. With the sailing team, maneuver procedures are developed and tested. A few examples are introduced in the paper.

For maneuver simulations, it is important to describe the extended range of operating conditions, such us inverted wing shapes. Modelling the human behaviour with controllers realistically is also critical. The close interaction with the sailing team allows direct feedback from on the water experiences into the DVPP set-up. Interfacing with the OTUSA data acquisition, processing and analysis environment enables efficient use within the team.

For the successful employment of such a tool, a good integration within the team is important. Constantly discussing results within the design, performance and sail team is important to build understanding of the tool and awareness of its capabilities and limitations.

The DVPP set-up is flexible and can be adapted to other types of foiling and non-foiling boats. In the future dynamic simulations including maneuvering can become more widely used. Modelling human behaviour with controllers produces repeatable results so that systematic parameter variations can be conducted. The DVPP model can also be utilized as the backbone of a simulator, where operator inputs drive the simulation. A simulator with a virtual reality environment for the sailors can then be used for training and development off the water.

ACKNOWLEDGEMENTS

The authors like to sincerely thank everyone in Oracle Team USA for supporting and contributing to this work. Many thanks also to Grant Simmer, Jimmy Spithill and Tom Slingsby for enabling and encouraging this topic.

REFERENCES

Binns, J.R., Hochkirch, K., De Bord, F. & Burns, I.A., "The Development and Use of Sailing Simulation for IACC Starting Manoeuvre Training", RINA 3rd High Performance Yacht Design Conference (HPYD), Auckland, New Zealand, 2008.

Boegle, C., Hansen, H. & Hochkirch, K., "Speed vs. Stability - Design considerations and velocity prediction of a hydro-foiled International Moth", RINA 4th High Performance Yacht Design Conference (HPYD), Auckland, New Zealand, 2012.

DNV GL, "FS-Equilibrium: User manual", Potsdam, Germany, 2018.

Etkin, B., "Dynamics of Atmospheric Flight", Dover Publications, ISBN 0-486-44522-4, Mineola, NY, USA, 2005.

Hochkirch, K., "Entwicklung einer Meßyacht zur Analyse der Segelleistung im Originalmaßstab (Design and construction of a full scale measurement system for the analysis of sailing performance)", Mensch & Buch Verlag, ISBN 3-89820-119-8, PhD-Thesis TU-Berlin, Berlin, Germany, 2000.

Paulin, A., Hansen, H., Hochkirch, K. & Fischer, M., "Performance Assessment and Optimization of a C-Class Catamaran Hydrofoil Configuration", RINA 5th High Performance Yacht Design Conference (HPYD), Auckland, New Zealand, 2015.

Richardt, T., Harries, S. & Hochkirch, K., "Maneuvering Simulations for Ships and Sailing Yachts using FRIENDSHIP-Equilibrium as an Open Modular Workbench", International Euro-Conference on Computer Applications and Information Technology in the Maritime Industries (COMPIT), Hamburg, Germany, 2005.

Saunders, H.E., "Hydrodynamics in Ship Design, Volume 1", The Society of Naval Architects and Marine Engineers (SNAME), New York, NY, USA, 1957.

6DOF behavior of an offshore racing trimaran in an unsteady environment

Paul Kerdraon[1], VPLP Design, Vannes, France, and Ecole Centrale Nantes, France

Boris Horel, Ecole Centrale Nantes, LHEEA res. dept. (ECN and CNRS), Nantes, France

Patrick Bot, Naval Academy Research Institute, Brest, France

Adrien Letourneur, VPLP Design, Vannes, France

David Le Touzé, Ecole Centrale Nantes, LHEEA res. dept. (ECN and CNRS), Nantes, France

ABSTRACT

While in recent years the use of hydrofoils has experienced a substantial growth, traditional design tools such as Velocity Prediction Programs (VPP) have proven inadequate to help architects and engineers with performance trade offs which now include specific stability issues related to these foils. The quest for performance also demands a better account of the unsteadiness of the environment in which the offshore yachts evolve.

Time-domain analysis and system-based modeling allow for an improved understanding of the controllability and dynamic stability of given geometries, enabling to adapt and refine the design. This paper presents such a dynamical unsteady model, based on the superposition of several loads components, computed from either numerical, empirical or analytical models. A test case and its results are presented to show the reliability and efficiency of the developed numerical tool, by comparing response amplitude operators of a reference hull form with experimental and numerical data.

Finally, the paper outlines two 6DOF dynamic simulations of an offshore trimaran. The first case shows a simple bearaway maneuver and compares two sail tuning strategies, while the second one presents the yacht evolution in unsteady wind demonstrating how in varying conditions the boat may reach attitudes that widely differ from the steady ones.

NOTATION

A	Wave amplitude	$\boldsymbol{\xi}$	Ship perturbation vector
$\mathbf{A}$	Added mass coefficients matrix	Φ_i	Potential of incoming waves
$\mathbf{B}$	Damping coefficients matrix	ρ	Water density
Fn	Froude number	ω	Frequency-domain variable
$\mathbf{F}_i$	Force component i		Wave circular frequency
g	Acceleration of gravity	$\boldsymbol{\omega}$	Yacht angular velocity vector
$\mathbf{I}$	Yacht inertia matrix		
$\mathbf{K}$	Retardation function matrix	CFD	Computational Fluid Dynamics
m	Yacht mass	DOF	Degree(s) of Freedom
$\mathbf{n}$	Outgoing body's normal vector	DVPP	Dynamic Velocity Prediction Program
p_i	Pressure of incoming waves	IMS	International Measurement System
S_w	Yacht wetted area	RAO	Response Amplitude Operator
$\bar{U}$	Yacht mean speed	TWA	True Wind Angle
$\mathbf{V}$	Yacht linear velocity vector	TWS	True Wind Speed
δ_R	Rudder angle	VPP	Velocity Prediction Program

---

[1]kerdraon@vannes.vplp.fr

INTRODUCTION

Velocity Prediction Programs are nowadays widely used in naval architecture offices to help architects and engineers in the design process of sailing yachts. Based on experimental, numerical or empirical data, they enable boats settings optimization to derive the reachable boat speeds in given steady conditions.
However, unlike inshore yachts such as the AC45 or the AC75, offshore vessels may encounter rough sea and wind conditions with short characteristic time of evolution. The study of the boat behavior in unsteady environment is therefore a key of the yacht real performance assessment and should be included in the design trade-offs.

In addition, recent years have seen a substantial growth of foiling technologies, leading to fully flying yachts and introducing specific stability issues with direct impact on speed. A better knowledge of the dynamic response of flying yachts is paramount for safe and sustained offshore flight. Non-linear couplings between the different degrees of freedom complicate further the study and traditional VPPs have proven inadequate to handle those matters with the accuracy required for high performance sailing.

In this context, numerical tools enabling time-domain analysis and including unsteady environment – often called Dynamic Velocity Prediction Programs (DVPPs) – have become a major research topic.

SAILING YACHT DYNAMIC SIMULATION

Ability to simulate maneuvers and especially tacking has long been the main subject of sailing dynamic studies (Keuning et al., 2005, Masuyama et al., 1995). In match racing, maneuvers are critical and simulations provide an efficient way for designers as well as crew to improve the on-water results (Binns et al., 2008). Most of those works are based on the usual maneuvering approach (Abkowitz, 1964) in which loads are described using hydrodynamic derivatives and enabling the modeling of the three or four degrees of freedom (DOFs) boat motion (surge, sway, yaw and sometimes roll). Nevertheless foiling greatly enhances the need to factor in the two other DOFs as heave (flight height) and trim (appendages angles of attack) are now at the core of boat stability (Heppel, 2015). Full 6 degrees of freedom modeling is therefore needed.

First works on adapting classical ship maneuvering theory to sailing yacht were undertaken in the late sixties, focusing mostly on steering and course stability. Introduction of time domain studies in the design of racing yachts occurred for the victorious 26th America's Cup challenger *Stars and Stripes* (see Oliver et al., 1987) using a quasi-steady approach. Based on the steady state models of the IMS VPP, Larsson (1990) and Ottosson et al. (2002) presented in the early nineties a four degrees of freedom simulator in which angular motions are accounted for by calculating induced velocities. Masuyama et al. (1993, 1995) developed a numerical tool based on hydrodynamic derivatives computed from tank tests and successfully compared their results to full scale measurements. A linear law was used to model the sail forces evolution at low apparent wind and simulate tacking maneuvers. Keuning et al. (2005) enlarged the possibilities and modularity of such tools by introducing the use of Delft Systematic Yacht Hull Series to compute the hydrodynamic coefficients. The same year, Richardt et al. (2005) proposed a full six degrees of freedom DVPP based on Masuyama's approach.

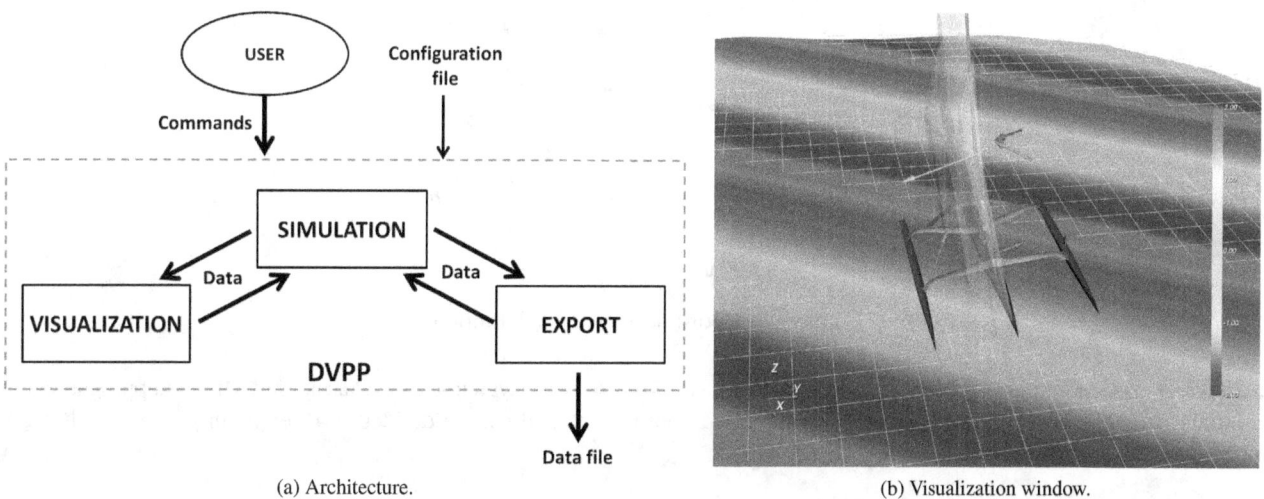

(a) Architecture.   (b) Visualization window.

Figure 1: Developed simulation tool.

Constant improvements of computational power have, on the other hand, opened the possibility to use CFD for time domain simulation by directly coupling the flow RANS solver with rigid body dynamic solvers (Jacquin et al., 2005, Lindstrand Levin and Larsson, 2017, Roux et al., 2008). Such approaches enable great accuracy while eliminating the need for empirical data or numerical pre-computations. The use of CFD as numerical VPP was achieved and work is now undertaken to add unsteady environment. Nevertheless, the computational time and costs of such techniques make them currently unavailable for naval architects when the comparison of several designs and configurations is needed.

System-based approaches, on the contrary, use empirical and theoretical models, experimental results or pre-computed numerical data to derive the hydrodynamic and aerodynamic loads and model the boat global behavior (Horel, 2016, 2017) with a computational efficiency that enables systematic studies, of appendages shapes and configurations for instance. Such a solution has therefore been selected and implemented (see Figure 1).

MATHEMATICAL MODELS OVERVIEW

Dynamics

The developed numerical tool is based on the resolution of the 6 degrees of freedom rigid body motion equations in the non inertial ship-fixed reference frame:

$$\begin{cases} m\left(\dfrac{d\mathbf{V}}{dt} + \boldsymbol{\omega} \wedge \mathbf{V}\right) = \mathbf{F} \\ \mathbf{I}\left(\dfrac{d\boldsymbol{\omega}}{dt} + \boldsymbol{\omega} \wedge \boldsymbol{\omega}\right) = \mathbf{M} \end{cases} \tag{1}$$

where $\mathbf{F}$, $\mathbf{M}$ are the external forces and moments and m, $\mathbf{I}$ the body mass and inertia.

The ship-fixed reference frame (X_b, Y_b, Z_b) is defined with X_b positive direction forwards, Y_b to port and Z_b upwards (see Figure 2). Its orientation with respect to the earth-fixed inertial reference frame (X_0, Y_0, Z_0) is expressed using the usual Cardan angle φ, θ, ψ (roll, pitch and yaw).

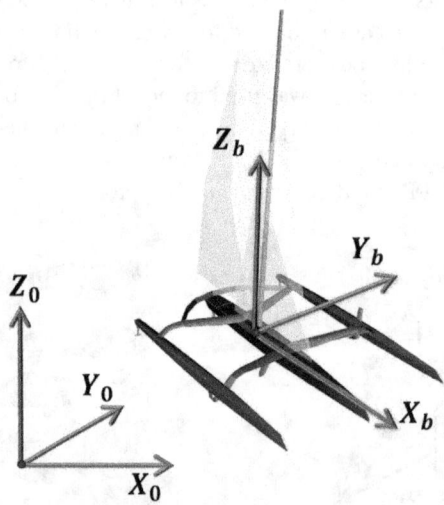

Figure 2: Coordinate systems definition.

Several explicit numerical integration schemes with various orders are available as well as adaptive time-stepping methods. External loads are given by the sum of all boat loaded components and can be divided in three main groups: hulls loads, appendages loads and aerodynamic loads.

Hulls loads

Hulls hydrodynamic loads are divided into low frequency (maneuvering) and high frequency (seakeeping: radiation and waves) loads, both are however expressed in the time domain.

• Maneuvering loads

Instead of the usual hydrodynamic derivatives approach (Abkowitz, 1964) for the maneuvering forces, the dynamic simulation tool uses polynomial response surfaces based on numerical viscous computations (RANS). The response surfaces are built on steady-state calculations for a wide range of hull attitudes, sinkages, leeway angles and speeds, which are then the input variables to the polynomial fit. It enables a full modeling of the six components of the hydrodynamic loads on each hull, including dependency to the boat possible changes of attitude and displacement due to the effect of appendages.

• Radiation force

The higher frequencies loads are based on the classical distinction between radiation, diffraction and Froude-Krylov forces. The former considers damping and added mass effects due to radiated waves generated by ship oscillations at the free surface. They are computed in the frequency domain using the Boundary Element Method (BEM) code Aquaplus (Delhommeau, 1993) developed at the Ecole Centrale de Nantes. Transformation of those frequency-domain coefficients to an impulse response function usable in the time-domain is carried out through Cummins equation (Cummins, 1962):

$$\mathbf{F}_{\text{Rad}} = -\mathbf{A}\left(\infty, \bar{U}\right) \ddot{\boldsymbol{\xi}} - \mathbf{B}\left(\infty, \bar{U}\right) \dot{\boldsymbol{\xi}} \\ - \int_0^t \mathbf{K}\left(t - \tau, \bar{U}\right) \dot{\boldsymbol{\xi}}\left(\tau\right) d\tau \tag{2}$$

where the retardation function $\mathbf{K}$ is given by the inverse Fourier transform of the frequency dependent damping component:

$$\mathbf{K}\left(t, \bar{U}\right) = \frac{2}{\pi} \int_0^\infty \left[\mathbf{B}\left(\omega, \bar{U}\right) - \mathbf{B}_\infty\left(\bar{U}\right)\right] \cos\left(\omega t\right) d\omega \tag{3}$$

• Wave loads

Diffraction loads are the first wave excitation force component. They originate in the reflection of the incident waves on the ship surface. Similarly, they are linearly modeled through the seakeeping code outputs, but directly using the frequency-domain expression :

$$\mathbf{F}_{\text{DF}} = A\left|\mathbf{F_d}\right|(\omega_e) \cos\left(kX - \omega t + \varphi_{\mathbf{d}}(\omega_e)\right) \tag{4}$$

where $\left|\mathbf{F_d}\right|$ and $\varphi_{\mathbf{d}}$ are the modulus and phase of the diffraction force, X the ship abscissa along the wave propagation axis and ω_e the frequency of encounter given by:

$$\omega_e = \omega - \frac{\omega^2}{g} U \cos\mu \tag{5}$$

with μ the angle between the ship track and the wave propagation.

Finally, Froude-Krylov force gathers the loads of the incident wave pressure field p_i on the instantaneous ship wetted surface:

$$\mathbf{F}_{\text{FK}} = -\iint_{S_w} p_i \, \mathbf{n} \, ds \tag{6}$$

where p_i is expressed through usual potential flow gravity waves theory:

$$p_i = -\rho \left[\frac{\partial \Phi_i}{\partial t} + \frac{1}{2}\left(\boldsymbol{\nabla} \Phi_i\right)^2\right] \tag{7}$$

While computing Froude-Krylov force, a correction of the hydrostatic loads to account for the deformation of the free surface is also performed.

Appendages loads

Appendages loads are in a first approach modeled using a Vortex Lattice Method with correction for viscous effects. Induced velocities, wave velocity field and appendages tuning angles are accounted for by computing effective angles between the appendage and the incoming flow following the Quasi-Steady Theory (QST) approach to compute the apparent flow velocity vector:

$$\mathbf{V}_{\mathbf{X}/\text{flow}} = \mathbf{V} + \boldsymbol{\omega} \wedge \mathbf{X} - \mathbf{V}^{\text{flow}} \tag{8}$$

where $\mathbf{X}$ is the coordinate vector of the considered location.

Hysteresis or fluid-structure interaction effects are neglected for the time being.

Aerodynamic loads

Similarly, the aerodynamic models are based on the usual Quasi-Steady Theory (QST) assumption (see Keuning et al., 2005, Richardt et al., 2005). Specifically, steady state sails polars are used while the apparent wind calculation accounts for the induced velocity due to the yacht angular motion. The IMS VPP (Claughton, 1999) approach that models sails de-powering through the *Flat* (sail lift reduction) and *Twist* (center of effort lowering) parameters is used.

A correction is added to model the aerodynamic inertia provided by the sails through the method described in Gerhardt et al. (2009). The sails added mass is approximated using a strip theory approach and integrating the potential flow expression of the added mass of an infinitely long, flat plate along the sail surface. The work of Tuckerman [1926] is used to derive a three-dimensional effect factor that proved rather consistent when compared to experimental measurements on a model sail by Gerhardt et al..

As was shown by Augier et al. (2013) and Fossati and Muggiasca (2011), such simple models do not reproduce fully the unsteady aerodynamic behavior of sails, and especially hysteresis phenomena. Research still needs to be carried out to integrate such aspects in DVPPs.

Control systems

Class rules on yacht control systems are a key issue in the future of high-performance sailing, especially offshore. Sportsmanship, human safety, energy consumption, financial costs are intimately linked to the decision to authorize them on board. For the time being, the Ultim Class 32/23 does not allow control system other than the helm autopilot. As this paper is concerned with the simulation of an offshore trimaran complying with those class rules, no control system of foils or centerboard is enabled and therefore no control of heel, pitch or ride-height. Unlike dinghies, Moths or America's Cup catamarans, Ultim trimarans take advantage of their substantial inertia which enable them to go through (limited) changes of the environmental conditions, wind gusts for instance, without immediate capsize.

The autopilot is an usual proportional-derivative controller:

$$\delta_R = K_p \left(\psi - \psi_T \right) + K_p T_D \dot{\psi} \tag{9}$$

where
    δ_R rudder angle,
    ψ_T targeted heading,
    $\psi, \dot{\psi}$ yacht heading and its first order derivative,
    K_p controller proportional coefficient,
    T_D controller derivative coefficient.

The coefficients used in this paper have been manually tuned on a one degree of freedom (yaw) test simulation. To this end, T_D is first set to zero and K_p adjusted by finding a compromise between the system's stability and settling time. Second, T_D is increased to reduce oscillations while avoiding excessive overshoots and instability.

VALIDATION: HEAVE AND PITCH OSCILLATIONS OF WIGLEY III IN WAVES

Case presentation

This test case is built on the study of four Wigleys in regular head waves by Journée (1992). Wigleys are fully parametric and mathematically defined hull forms. The presented work focuses on Wigley III. Its body plan and characteristics are respectively given in Figure 3 and Table 1.

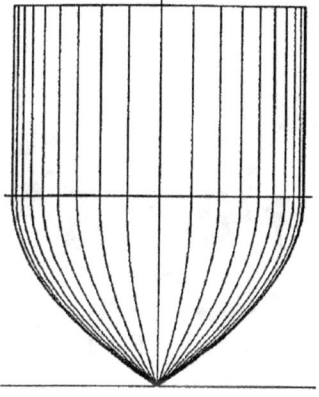

Length L	3.0 m
Breadth B	0.3 m
Amidship section coefficient C_m	0.667
Draft d	0.1875 m
Displacement ∇	0.078 m^3
Pitch radius of inertia k_{yy}	0.75 m
Center of rotation above base **KR**	0.1875 m
Center of gravity above base **KG**	0.1700 m

Figure 3: Wigley III body plan.   Table 1: Wigley III dimensions.

In his paper, Journée presents experimental results from tank testing at the Shiphydromechanics Laboratory of the Delft University of Technology as well as numerical ones computed through the 6 degrees of freedom program SEAWAY that implements both ordinary and modified strip theory methods. Hydrodynamic coefficients, wave loads, heave and pitch motions as well as added resistance results are presented.

Results

In the purpose of validation, results from the developed dynamic simulation tool are compared to the reference values provided by Journée. Simulations are run using the models previously presented. Infinite depth Airy waves are used, while one wave amplitude, $A = 0.02$ m, and four Froude numbers, 0, 0.2, 0.3 and 0.4, are considered.

The heave and pitch response amplitude operators are defined as:

$$\mathrm{RAO}_z\,(\omega) = \frac{z\,(\omega)}{A} \tag{10a}$$

$$\mathrm{RAO}_\theta\,(\omega) = \frac{\theta\,(\omega)\,L}{2\pi A} \tag{10b}$$

where z and θ are the heave and pitch response.

The moduli of the response amplitude operators for $Fn = 0$ are presented in Figure 4. The results correlate almost perfectly with both experimental and numerical results. The pitch motion peak is slightly lower than Journée's experimental data but is in excellent agreement with his own numerical results. Both low and high frequencies limits are consistent.

As the speed increases, a resonance peak of increasing intensity appears in both heave and pitch motions. The overall response and the peak position are rather well described, especially in pitch. The heave response shows however an underestimated resonance peak intensity. This discrepancy is due to the moderate quality of the speed dependency modeling of radiation and diffraction loads, which are here critical to properly track the resonance. In addition, the considered Froude numbers of the reference paper are relatively small compared to the usual values seen by high performance sailing yachts. For those reasons, tank tests have been planned to improve and validate our models in this regard.

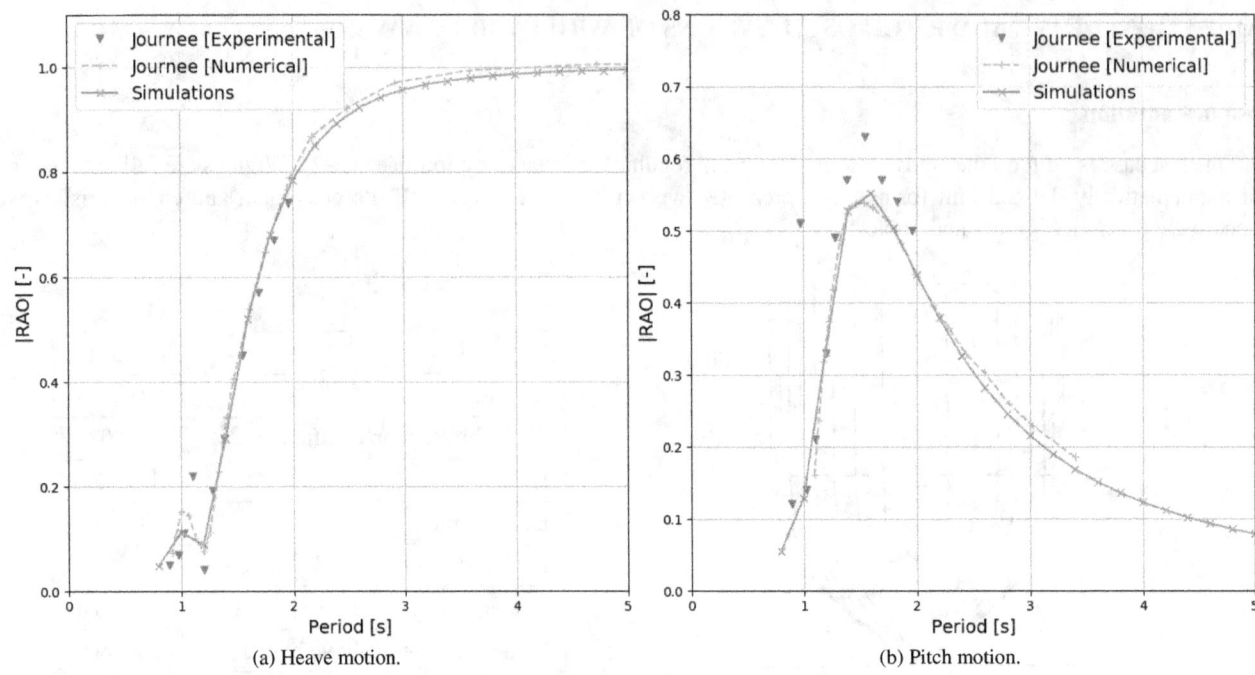

(a) Heave motion. (b) Pitch motion.

Figure 4: Wigley III response amplitude operators at $Fn = 0$.

OFFSHORE TRIMARAN SIMULATIONS

Considered yacht

This part presents two simulations examples which underline specificities and interests of dynamic studies compared to steady ones. The simulated yacht is *Macif 100*, an offshore trimaran of the Ultim class. Skippered by François Gabart, she has been the holder of the single handed sailing around-the-world record since 2017 (in 42 days 16 hours 40 minutes 35 seconds). Those simulations consider the appendages set that was used for the circumnavigation: two small L-foils, one centerboard and three T-rudders (one on each hull).

Length L	30.0 m
Breadth B	21.0 m
Max. draft D	4.5 m
Air draft H	35.0 m
Displacement m	14.5 t
Launched	2015
Architect	VPLP
Shipyard	CDK Technologies

From *www.macifcourseaularge.com/trimaran-macif/bateau/*.

Table 2: *Macif 100* characteristics.

Simple maneuver

As known by VPP engineers, a given configuration may lead to different equilibrium states (and thus different boat speeds), especially when the yacht has the ability to evolve in different modes (archimedean, fully flying, hybrid). The aim of this first simulation is to illustrate this specificity by comparing two sail trimming strategies and showing that even though the final configuration is the same in both cases, the final state largely differs, with a substantial speed delta.

Flat water conditions are considered and the yacht initially evolves upwind in 19 knots of wind at 50° True Wind Angle on port tack. The simulation is carried out in six degrees of freedom, with only an autopilot for the rudder angle. At $t = 50$ s, the target heading is increased by sixty degrees so that the yacht bears away (see Figure 5). No change of sail is allowed.

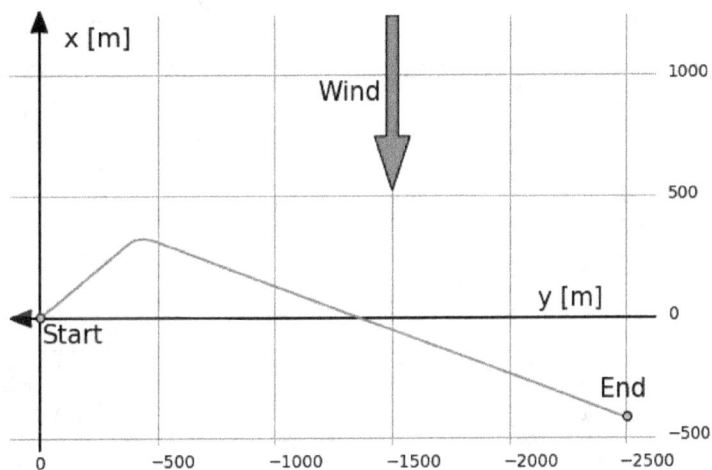

Figure 5: Yacht trajectory during the maneuver.

Initial and final states optimal configurations are known from steady VPP studies (board extension, rudder rake, *Flat* and *Twist*, etc.)., which show that in the final situation steady state optimal configuration sails must be depowered (twisted). The compared strategies focus on the timing of specific action: in the first one, this trimming is operated progressively, but directly after the rudder action, while in the second one, the sails are twisted only after the main hull has exited the water.

An important bias of the aerodynamic model must here be underlined. Steady state polars provide, for each wind angles, the lift and drag coefficients of the sails when optimally trimmed. Also, when used in dynamic situations, using such polars is equivalent to making the assumption that the sails are optimally trimmed at the same speed as the apparent wind angle increases. The strategies compared here only consider the sails twist, which enables the lowering of the center of effort at the cost of an increased drag coefficient. But in both cases, the aerodynamic model considers that the sheets are eased as the yacht bears away.

With the second strategy the yacht maintains a strong heeling moment which makes her heel and decreases the dynamic displacement of the main hull. The yacht can then accelerate, increasing thus the lift force of the foil in a virtuous circle that sees the hulls dynamic buoyancy and drag replaced by the foil action. Finally, the sails need to be twisted to enable the heeling moment to be balanced (as the boat accelerates, the aerodynamic heeling moment keeps on increasing otherwise). Acting too soon, as in the first strategy, prevents the yacht, under-powered, from lifting the main hull (Figure 6c), with an implicated lack of speed of more than 4.5 knots (Figure 6a).

Pushed by the inertial forces after the start of the turn, the yacht keeps briefly a non negligible speed component in the direction of her initial motion, that is to windward. This explains the negative leeway peak shown in Figure 6b.
The ratio of the main hull wetted surface area to its nominal value is shown in Figure 6f. Its evolution is close to the heel angle behavior and, while it is blocked at about 25% when using the first strategy, it indeed tends to zero in the second case. The pitch angle evolution is visible in Figure 6g. Its evolution is driven by two main phenomena. First, after the maneuver, the speeds are globally higher in both strategies. As the speed increases, the aerodynamic pitching moment increases, leading to a greater pitch angle (bow down). However, another aspect is at stake in the second strategy: the main hull leaves the water and its pitching moment component cancels out. That is why the second boat shows a lower pitch angle than the first one.

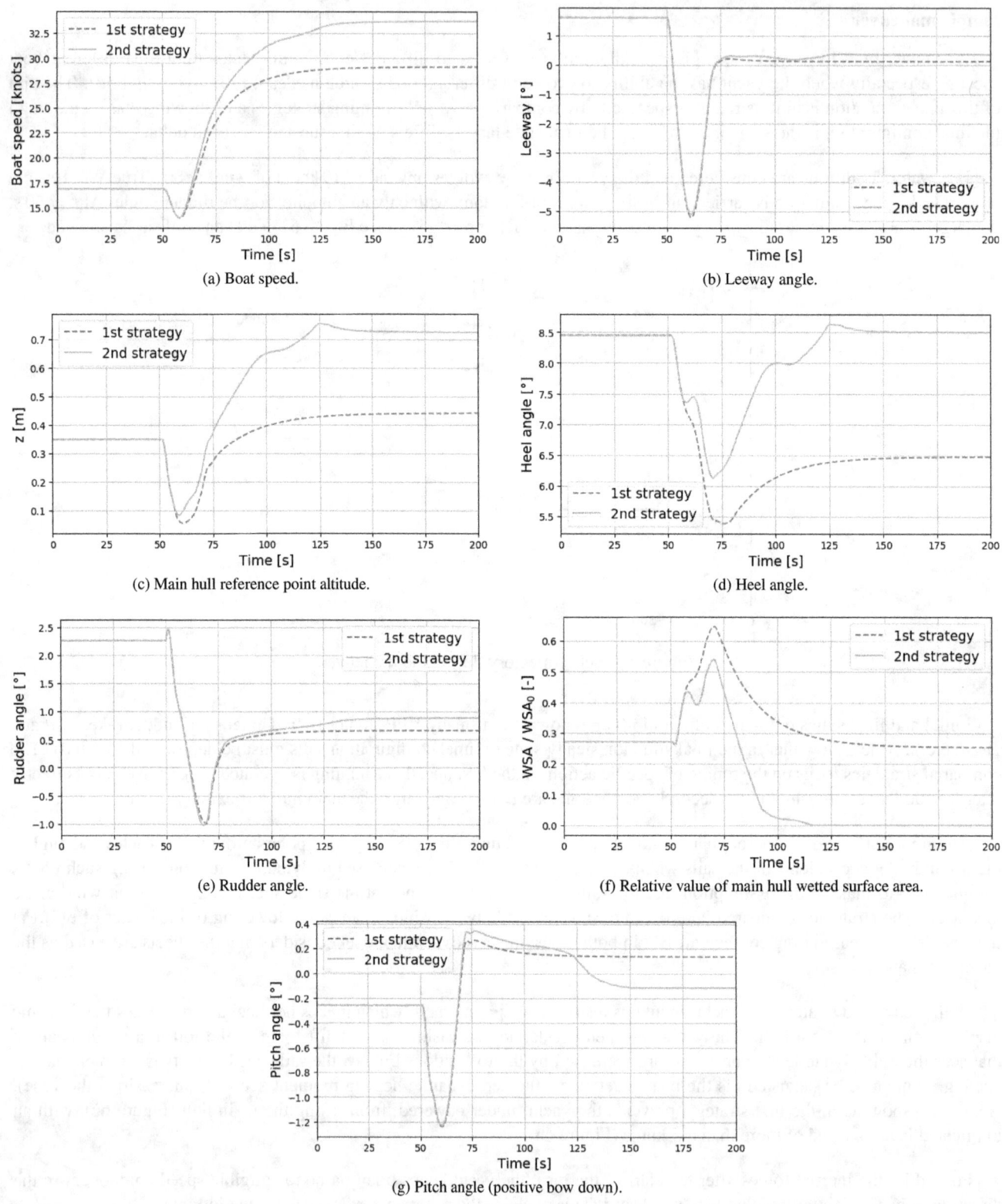

(a) Boat speed.

(b) Leeway angle.

(c) Main hull reference point altitude.

(d) Heel angle.

(e) Rudder angle.

(f) Relative value of main hull wetted surface area.

(g) Pitch angle (positive bow down).

Figure 6: Comparison of the two trimming strategies.

Behavior in unsteady wind conditions

This second simulation case aims at showing the interest of dynamic studies to predict potentially critical situations when evolving in unsteady conditions. The consequences on the yacht behavior of an irregular wind are studied. Unsteadiness is modeled by adding harmonic components to the mean True Wind Direction TWD_0 while the True Wind Speed is kept constant at 18 knots:

$$TWD(t) = TWD_0 \left[1 + \sum_i k_i \sin\left(\frac{2\pi}{T_i}t\right) \right] \qquad (11)$$

The periods T_i and the intensity factors k_i of the chosen harmonic components are given in Table 3. Periods are chosen so that they are not multiples of each other, in order to increase the time necessary to observe a periodic behavior. Time evolution of the True Wind Direction is visible in Figure 7.

i	Period [s]	Frequency [Hz]	Intensity factor
1	41	0.0244	0.04
2	17	0.0588	0.03
3	7	0.143	0.02
4	6.5	0.154	0.005
5	5	0.200	0.01

Table 3: Wind harmonic components.

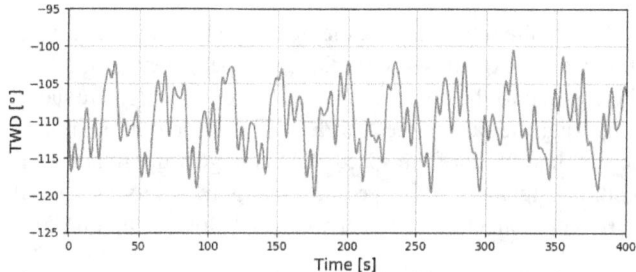

Figure 7: True Wind Direction evolution.

The boat is released from a VPP optimized equilibrium corresponding to TWA 110° / TWS 18 kn at $t = 0$. It is interesting to notice that, in such a configuration the boat speed is about 30 knots, and therefore a major component of the apparent wind. Fluctuations are thus much smaller in apparent wind than in true wind. During the 400 s simulation the standard deviation of the True Wind Direction is 4.3° and the amplitude between its extrema is 19.5°, the corresponding measures for the Apparent Wind Angle give respectively 1.4° and 6.1°. Some of the simulation outputs are shown in Figure 8. The yacht is free to evolve in 6 degrees of freedom, while an autopilot with a constant heading target ($\psi_T = 0°$) controls the rudders.

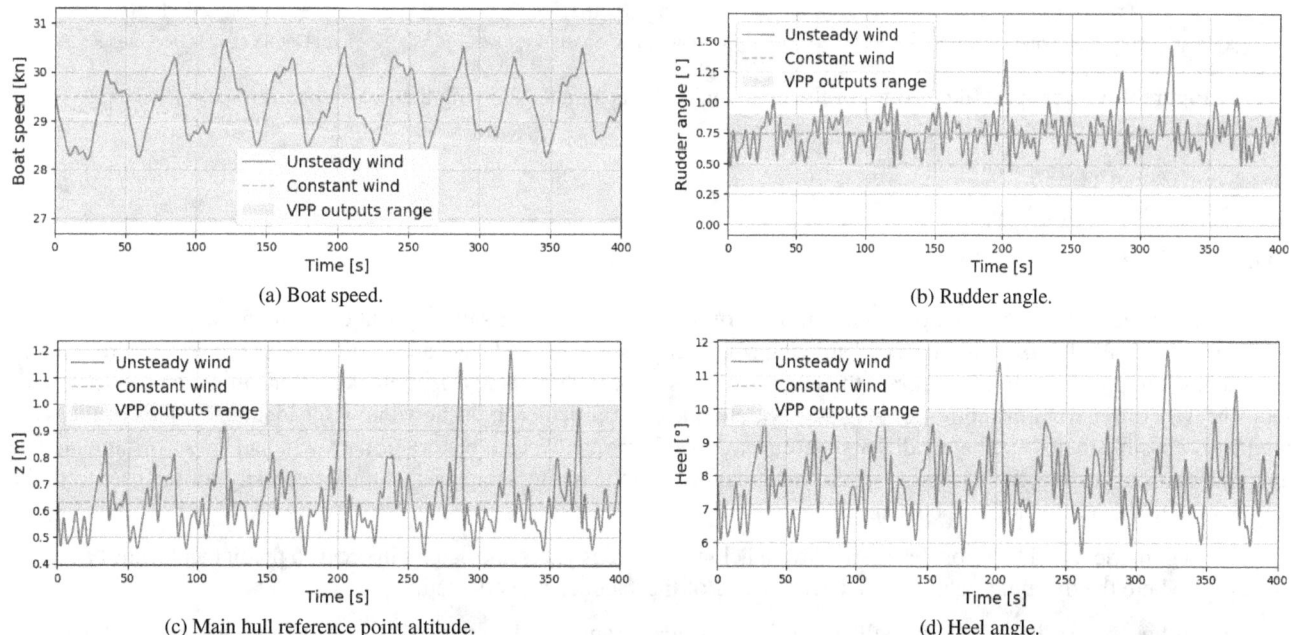

(a) Boat speed.

(b) Rudder angle.

(c) Main hull reference point altitude.

(d) Heel angle.

Figure 8: Dynamic simulation results compared to steady VPP optimization on the same True Wind Angle range.

One can notice three peaks where the heel angle almost reaches 12°. Similar peaks are also visible on the altitude response. Consistently, Figure 7 shows that they correspond to situations where the wind heads (TWD maxima as the yacht heads North on port tack). However, such values of the True Wind Angle are reached several times without resulting in such situations. This demonstrates how the wind sequence has a strong impact on the yacht instantaneous behavior and proves how necessary time-domain simulations are for the complete understanding of the yacht behavior.

Similarly, Figure 8 shows in green the limits reached in a steady VPP when the yacht configuration is kept constant and the True Wind Angle ranges from 100° to 120°, the minimal and maximal values of Figure 7. From time to time those values are largely exceeded. This shows, as could be expected, that steady state analysis can hide critical situations. On the contrary, the simulated boat speed is well contained within the VPP limits, as the wind oscillates too quickly for the speed to settle to the steady equilibrium values seen in VPP. The average speed during the simulated sequence is 29.3 knots, which is slightly below the speed reached in steady wind (29.5 knots).

Unlike the other outputs, the boat speed presents a low frequency harmonic strongly dominating the high frequency ones. This can be explained by the loads' dependency to the boat speed that tends to damp the response and by the low-pass filtering due to the yacht inertia, that filters the high frequency components of the excitation signal. This can be verified by comparing the power spectral densities of the output signals with the input one (Figure 9), which consistently shows that the first harmonic of the speed PSD is largely dominant over the other components. On the contrary, the linear and angular positions signals show a spectrum that is relatively close to the input one, with some additional very small harmonics. The fourth component being rather weak and close to the third one, it is hardly distinguishable in the shown spectra.

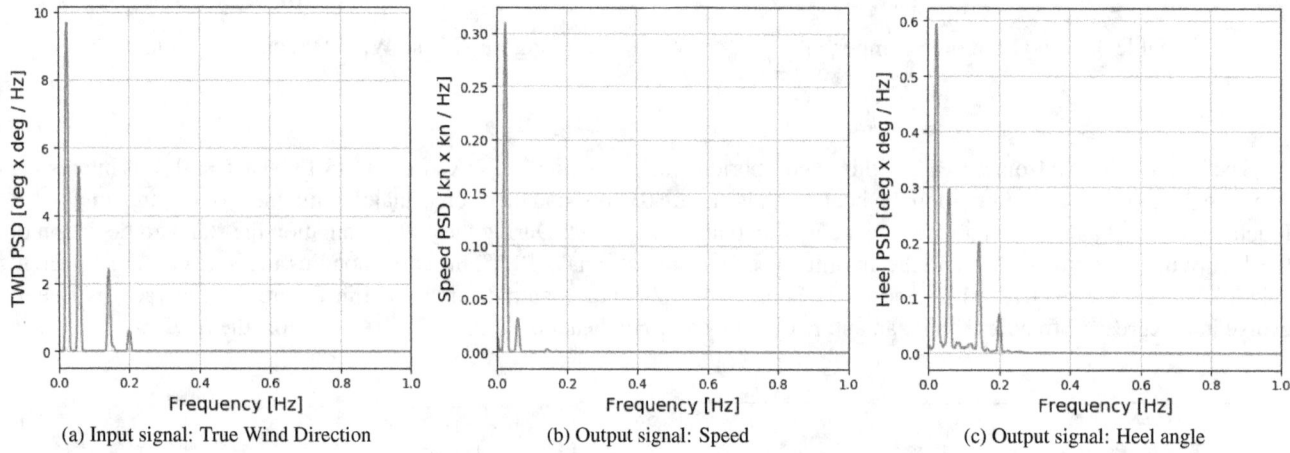

(a) Input signal: True Wind Direction   (b) Output signal: Speed   (c) Output signal: Heel angle

Figure 9: Power spectral densities of the input and outputs signals showing the boat speed low-pass filtering.

CONCLUSION

This paper describes the development and models of a numerical tool for sailing yacht dynamic behavior analysis. Validation case on a Wigley III hull shows encouraging consistent results for the heave and pitch response in waves. Preliminary simulations are presented to demonstrate the DVPP abilities, especially to study dynamic situations in which the yacht attitudes heavily differ from the corresponding steady state ones. In particular, it is shown how for identical values of the wind angle it is possible to observe highly different attitudes depending on the recent history of the yacht and her environment. The example cases also show the sensibility of the yacht behavior to the tuning strategies, and the necessity for the user to find a tuning path to the VPP optimized boat speed.

Such a numerical tool therefore brings a relevant help to yacht designers and sailors in order to predict and identify critical situations where the dynamic stability and performances of the yacht can be affected.

Work is however still needed, especially regarding appendages and aerodynamic loads, to handle a wider range of dynamic situations. Tuning strategies and appendages control are also major issues to be able to pilot the boat optimally.

PERSPECTIVES

Future work will aim at improving the previously mentioned points. Model tests in the towing tank facility of the LHEEA laboratory are also planned to improve the transient loads modeling of hulls and appendages. Such experiments will also enable to validate additional aspects of the implemented dynamic models.

ACKNOWLEDGMENTS

This work is supported and funded by VPLP Design and the ANRT (National Association for Research and Technology).

REFERENCES

M. A. Abkowitz. Lectures on ship hydrodynamics - Steering and manoeuvrability. Report no. hy-5, Hydrodynamics Department, Hydro- and Aerodynamics Laboratory, Lyngby, Denmark, May 1964.

B. Augier, P. Bot, F. Hauville, and M. Durand. Dynamic behaviour of a flexible yacht sail plan. *Ocean Engineering*, 66: 32–43, 2013.

J. R. Binns, K. Hochkirch, F. De Bord, and I. A. Burns. The development and use of sailing simulation for IACC starting manoeuvre training. In *3rd High Performance Yacht Design Conference*, pages 158–167, Auckland, New Zealand, 2008.

A. Claughton. Developments in the IMS VPP formulation. In *The 14th Chesapeake Sailing Yacht Symposium*, Annapolis, MD, USA, 1999.

W. E. Cummins. The impulse response function and ship motion. Report 1661, Navy Department, David Taylor Model Basin – Hydromechanics Laboratory, MD, USA, 1962.

G. Delhommeau. Seakeeping codes Aquadyn and Aquaplus. In *19th WEGEMT School on Numerical Simulation of Hydrodynamics: Ships and Offshore Structures*, Nantes, France, Sept. 1993.

F. Fossati and S. Muggiasca. Experimental investigation of sail aerodynamic behavior in dynamic conditions. *Journal of Sailboat Technology*, 2011.

F. C. Gerhardt, D. Le Pelley, R. Flay, and P. Richards. Tacking in the wind tunnel. In *The 19th Chesapeake Sailing Yacht Symposium*, pages 161–175, Annapolis, MD, USA, Mar. 2009.

P. Heppel. Flight dynamics of sailing foilers. In *5th High Performance Yacht Design Conference*, pages 180–189, Auckland, New Zealand, 2015.

B. Horel. *Modélisation physique du comportement du navire par mer de l'arrière*. PhD thesis, Ecole centrale de Nantes, France, 2016.

B. Horel. System-based modeling of a foiling catamaran. In *The 4th International Conference on Innovation in High Performance Sailing Yachts*, Lorient, France, 2017.

E. Jacquin, P.-E. Guillerm, Q. Derbanne, L. Boudet, and B. Alessandrini. Simulation d'essais d'extinction et de roulis forcé à l'aide d'un code de calcul Navier-Stokes à surface libre instationnaire. In *10èmes Journées de l'Hydrodynamique*, Nantes, France, Mar. 2005.

J. M. J. Journée. Experiments and calculations on 4 Wigley hull forms in head waves. Report 0909, Ship Hydromechanics Laboratory, Delft University of Technology, Delft, The Netherlands, May 1992.

J. A. Keuning, K. J. Vermeulen, and E. J. De Ridder. A generic mathematical model for the maneuvering and tacking of a sailing yacht. In *The 17th Chesapeake Sailing Yacht Symposium*, pages 143–163, Annapolis, MD, USA, Mar. 2005.

L. Larsson. Scientific methods in yacht design. *Annual Review of Fluid Mechanics*, 22(1):349–385, 1990.

R. Lindstrand Levin and L. Larsson. Sailing yacht performance prediction based on coupled CFD and rigid body dynamics in 6 degrees of freedom. *Ocean Engineering*, 144:362–373, 2017.

Y. Masuyama, I. Nakamura, H. Tatano, and K. Takagi. Dynamic performance of sailing cruiser by full-scale sea tests. In *The 11th Chesapeake Sailing Yacht Symposium*, pages 161–179, Annapolis, MD, USA, Jan. 1993.

Y. Masuyama, T. Fukasawa, and H. Sasagawa. Tacking simulation of sailing yachts–numerical integration of equations of motion and application of neural network technique. In *The 12th Chesapeake Sailing Yacht Symposium*, Annapolis, USA, 1995.

J. C. Oliver, J. S. Letcher, Jr., and N. Salvesen. Performance predictions for Stars & Stripes. In *Transactions Society of Naval Architects and Marine Engineers*, volume 95, pages 239–261, New York, NY, USA, Nov. 1987.

P. Ottosson, M. Brown, and L. Larsson. The effect of pitch radius of gyration on sailing yacht performance. In *High Performance Yacht Design Conference*, Auckland, New-Zealand, Dec. 2002.

T. Richardt, S. Harries, and K. Hochkirch. Maneuvering simulations for ships and sailing yachts using Friendship-Equilibrium as an open modular workbench. In *International EuroConference on Computer Applications and Information Technology in the Maritime Industries, COMPIT*, pages 101–115, Hamburg, Germany, May 2005.

Y. Roux, M. Durand, A. Leroyer, P. Queutey, M. Visonneau, J. Raymond, J.-M. Finot, F. Hauville, and A. Purwanto. Strongly coupled VPP and CFD RANSE code for sailing yacht performance prediction. In *3rd High Performance Yacht Design Conference*, pages 215–226, Auckland, New Zealand, 2008.

Recent Advances in Downwind Sail Aerodynamics

Jean-Baptiste R. G. Souppez, Warsash School of Maritime Science and Engineering, Solent University, Southampton, UK.

Abel Arredondo-Galeana, School of Engineering, Institute for Energy Systems, University of Edinburgh, Edinburgh, UK.

Ignazio Maria Viola, School of Engineering, Institute for Energy Systems, University of Edinburgh, Edinburgh, UK (corresponding author, i.m.viola@ed.ac.uk).

ABSTRACT

Over the past two decades, the numerical and experimental progresses made in the field of downwind sail aerodynamics have contributed to a new understanding of their behaviour and improved designs. Contemporary advances include the numerical and experimental evidence of the leading-edge vortex, as well as greater correlation between model and full-scale testing. Nevertheless, much remains to be understood on the aerodynamics of downwind sails and their flow structures. In this paper, a detailed review of the different flow features of downwind sails, including the effect of separation bubbles and leading-edge vortices will be discussed. New experimental measurements of the flow field around a highly cambered thin circular arc geometry, representative of a bi-dimensional section of a spinnaker, will also be presented here for the first time. These results allow to interpret some inconsistent data from past experiments and simulations, and to provide guidance for future model testing and sail design.

NOTATION

AoA	Angle of Attack
CFD	Computational Fluid Dynamics
Cp	Pressure Coefficient
DES	Detached Eddy Simulation
FEA	Finite Element Analysis
FSI	Fluid-Structure Interaction
LE	Leading Edge
LES	Large Eddy Simulation
LEV	Leading-Edge Vortex
LSB	Laminar-Separation Bubble
NACA	National Advisory Committee for Aeronautics
PIV	Particle Image Velocimetry
RANS	Reynolds-Averaged Navier Stokes
Re	Reynolds Number
TKE	Turbulent Kinetic Energy
VLM	Vortex Lattice Method

INTRODUCTION

Sailing has been a central part of History, and has heavily influenced the development of humanity, with evidence of sailing vessels as early as the 6th millennium BC (Carter, 2006). While sailing downwind has benefited from millennia of evolution, the very first instance of a highly cambered and dedicated downwind sail, termed spinnaker, did not occur until 1865, as reported by King (1981), and was not popularized until the 1970s and 1980s; primarily thanks to the development of symmetric spinnakers for the America's Cup.

Asymmetric spinnakers were then introduced in the 1980s in the 18ft fleet in Sydney, before being popularised on offshore racing yachts in the 1990s. These new sails were promptly adopted in many significant sailing events; firstly in offshore races such as the Vendée Globe and the Whitbread 60, and later in the America's Cup (Fallow, 1996; Richards *et al*, 2001; Viola & Flay, 2009). The significant advances made in terms of spinnaker design and analysis during this particular decade can be related to the greater part that downwind legs took in the 1995 America's Cup (Fallow, 1996), thus motivating further research and development.

The 1990s also coincide with a fast increase in accessible computational power, allowing advanced numerical methods to be used in sail design (Hedges, 1993; Hedges *et al*, 1996), particularly for downwind sails. Upwind sails, where the flow remains largely attached, have been successfully analysed using inviscid codes since the 1960s, with the pioneering work of Milgram (1968) on Vortex Lattice Method (VLM), and later Gentry (1971), to eventually be extensively utilized in America's Cup sails development (Gentry, 1988). Conversely, for downwind sails, where the flow is largely separated, the use of Reynolds-Averaged Navier-Stokes (RANS) simulations is necessary (Lasher *et al*, 2005). The first instances of RANS occurred in 1996 for downwind sails (Hedges *et al*, 1996) and 1999 for upwind sails (Miyata & Lee, 1999). The complexity of downwind sail flow also prompted the development of dedicated experimental facilities, namely twisted flow wind tunnels (Flay & Vuletich, 1995), the need for which was highlighted a few years before by Flay & Jackson (1992).

One of the benefits of experimental testing is the ease to achieve the effective flying shape of the sail from the moulded one. In order to achieve the flying shape from numerical simulations, Computational Fluid Dynamics (CFD) is coupled with Finite Element Analysis (FEA). This approach has led to major advances in the field of Fluid-Structure Interaction (FSI) of downwind sails (Richter *et al*, 2003; Renzsch *et al*, 2008; Durand, *et al*, 2014; Sacher *et al*, 2015).

With the continuous growth of computational power (Viola, 2009), leading in 2011 to over a billion cells being used for yacht sail simulations for the first time (Viola & Ponzini, 2011), and with the wide adoption of asymmetric spinnakers, there is more than ever a strong incentive to employ numerical methods to further the understanding of downwind sails design. This has recently enabled the discovery of the Leading-Edge Vortex (LEV). Indeed, the first evidence of the presence of a stable LEV on a downwind yacht sail was provided numerically in 2014 (Viola *et al*, 2014), before being confirmed experimentally three years later (Viola & Arredondo-Galeana, 2017; Arredondo-Galeana & Viola, 2018), prompting new interpretations of full-scale pressure measurements on downwind sails (Richards & Viola, 2015).

This paper first introduces the background to the LEV, and the numerical and experimental work demonstrating its presence and impact on sailing performance. Then, the correlation between full-scale experiments, numerical simulations and model-scale testing are presented, focusing on recent findings and discrepancies. Successively, novel results are introduced to understand anomalies observed in pressure distributions between various experiments at model-scale. The findings of this experiment will be discussed, eventually concluding on the recent advances in downwind sail aerodynamics, and suggesting refined wind tunnel testing practice for future experimental work.

LEADING-EDGE FLOW

The LEV identified on yacht sails has significant similarities with that on delta wings. Following the work undertaken by the National Advisory Committee for Aeronautics (NACA) during World War II, and driven by the will to achieve supersonic planes, a vast amount of research was undertaken on delta wings. The leading-edge separation highlighted in the 1950s (Marsden *et al*, 1958; Harvey, 1959) then resulted in the leading-edge vortex theory developed by Hall (1961) and applied by Earnshaw (1962) in the early 1960s. The LEV on delta wings (pictured in Figure 1a) provides most of the lift at high angles of attack (AoA). Anecdotally, Bethwaite (1993) anticipated that highly swept back asymmetric spinnakers could function in a similar fashion as delta wings on high performance dinghies. It was subsequently hypothesised by Viola & Flay (2012) that an LEV, analogous to that of delta wings, was present at the head of spinnakers on both the leading and trailing edges, provided that sailing occurs at sufficiently high apparent wind angles.

Vortex lift, such as the lift due to the LEV, was modelled successfully by Polhamus (1966) for slender delta wings through the 'suction force analogy'. Despite the LEV being inherently tri-dimensional, the contribution of the vortex to the sectional lift can be modelled as a 2D effect using this analogy, which is based on the leading-edge suction associated with potential-flow leading-edge singularity. Saffman & Sheffield (1977) further explored theoretically the effect of a 'trapped', two-dimensional vortex near the leading edge of a flat plate through inviscid potential flow. They found that at suitable locations the vortex might remain stationary relative to the flat plate and that the vortex provides a significant lift contribution. At the leading-edge, however, the vortex was sensitive to flow disturbances and no stable positions were found. Huang & Chow (1982) expanded the study to circular arcs and Joukowski airfoils and found consistent results.

More recently, a tri-dimensional LEV was found to be the reason for the Hawkmoth (*manduca sexta*), and more generally insects, being able to fly thanks to the lift contribution of the LEV. This pioneering work and flow visualization realized in 1996 (Ellington *et al*, 1996) was validated using Particle Image Velocimetry (PIV) in 2005 (Bomphrey *et al*, 2005). Around the same time, in 2004, evidence of LEV on bird wings was also provided (Figure 1b, Videler *et al*, 2004) suggesting lift enhancement. This seminal work on insect and bird flight led to significant studies of the LEV on oscillating and revolving wings (Figure 1c, Taira & Colonius, 2009) and has been of paramount importance to understand the different stabilization mechanisms that allow a vortex to remain stably ('trapped') near a wing (Eldredge & Jones, 2019). In the case of spinnakers

(Figure 1d), the LEV has similarities with that of bird wings (Figure 1b) and translating wings (Figure 1c) due to the comparable sweepback angle and strong interaction with the tip vortex.

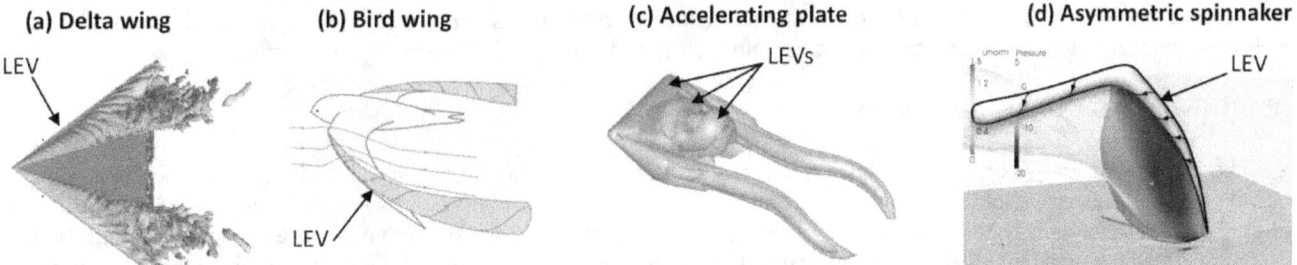

Figure 1 - The LEV on different lifting surfaces. (a) The steady LEV of a delta wing (Mitchell *et al*, 2006); (b) the steady LEV on a gliding bird's wing (similar to the periodic LEV on flapping wings) (Videler *et al*, 2004); (c) the unsteady LEV of an accelerating plate after having convected 5 chord lengths (Taira & Colonius, 2009); and (d) the intermittently steady LEV on an asymmetric spinnaker.

Remarkably, the LEV has been identified across a wide range of Reynolds numbers (Re). In laminar flow conditions, it has been found on auto-rotating seeds (Lentink *et al*, 2009) and on the wings of insects (Muijres *et al*, 2008) and small birds (Lentink *et al*, 2007). In transitional and turbulent flow conditions, it has been detected on larger bird wings (Hubel & Tropea, 2010), fish fins (Borazjani & Daghooghi, 2013) and delta wings (Gursul *et al*, 2005, Gursul *et al*, 2007).

In the examples above, the LEV provides an essential source of lift augmentation. However, the LEV is not always desirable. In helicopter rotors (Corke *et al*, 2015) and wind turbines (Larsen *et al*, 2007) the LEV is a powerful but dangerous flow feature, since it generates large load oscillations. When the LEV is shed downstream, it leads to a lift overshoot above the quasi-static maximum lift, as well as an abrupt and dangerous change in the pitching moment.

A key characteristic of the LEV is that it is a feature of the instantaneous flow field, and not of the time-averaged one. This is relevant to yacht sails, where a distinction can be made between the recirculating flow at the leading edge of upwind sails, such as jibs and genoas, and that of downwind sails.

Vortices that exists only in the time averaged sense include those on upwind sails, where the leading-edge bubble is similar to those of flat plates at incidence characterized by Newman & Tse (1992), and those of plates with a blunt leading edge (Figure 2a and 2b). Another time-averaged vortex is that occurring as a result of a detached boundary layer, giving a region of recirculating reverse flow (O'Meara & Muller, 1987). This type of bubbles typically occurs on thin foils and reattachment is due to the laminar-to-turbulent transition of the separated shear layer (Crabtree, 1957); in this case they are called Laminar-Separation Bubbles (LSB) (Figure 2c). Bubbles and their impact on the flow field and resulting pressure distribution was extensively discussed by Ward (1963), with the characteristic plateau in the pressure coefficient (*Cp*) indicating the presence of a bubble. This will be emphasized in the following sections, when discussing the pressure distribution over asymmetric spinnakers.

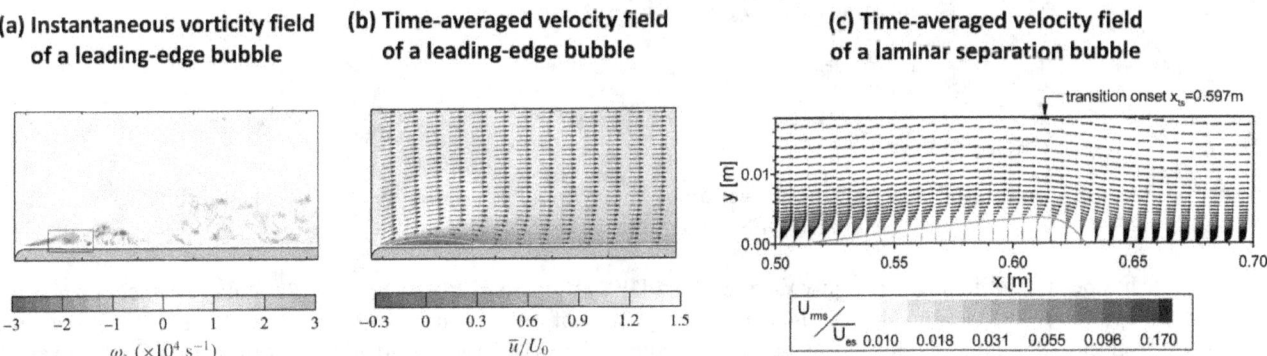

Figure 2 – Leading-edge bubble and laminar-separation bubble. The vorticity (a) and velocity field (b) of a leading-edge bubble measured by Stevenson *et al.* (2016) and the velocity field of a laminar-separation bubble (c) measured by McAuliffe & Yaras (2010).

On the other hand, spinnakers generate a much more coherent vortex structure that can be identified in the instantaneous flow field. This vortex is formed by the roll of vorticity at the leading edge, as identified on, and exploited by both biological flyers and delta wings, with a significant increase in performance. The underlying stabilization mechanisms that allows the

LEV to be more coherent on spinnakers than on upwind sails are still to be fully understood. Arredondo-Galeana (2019) suggested that the sweep back of the leading edge, and the strong tip vortex, are the main elements that stabilize the LEV on downwind sails. Finally, it is interesting to note that the interference of the mast on the mainsail results in two counter-rotating vortices on the windward and leeward side of the mast (Fossati, 2009; Larsson *et al*, 2013). The vortex lift due to these two coherent vortices is unlikely to cancel each other, and their net contribution has never been estimated.

THEORETICAL CONSIDERATIONS

The mechanism by which the LEV increases lift is as follow. Let assume that the sail is trimmed at the ideal angle of attack, *i.e.* that the flow velocity is tangent to the sail at the leading edge and no LEV occurs. The integral of the vorticity in the sail boundary layer is equal to the 'bound' circulation, which represents the strength of a vortex inside of the sail ('bound to the sail'). The Kutta-Joukowski's theorem states that the lift is proportional to the bound circulation. If the sail is trimmed at higher angles of attack, the sharp leading edge leads to flow separation. Without the LEV, the sail would stall and both circulation and lift would decrease. With the LEV, the loss in bound circulation is accounted for, at least in part, by unbound circulation which is contained within the LEV. Its circulation contributes to the lift of the sail to a different extent depending on its position and velocity (Li & Wu, 2018). Hence, the LEV allows to retain some of the circulation that would otherwise be lost. In some cases, such as in Figure 3, it also enables flow reattachment. In these cases, the LEV is said to be 'trapped' by the streamlines that keep it attached to the sail. The sum of the bound circulation and the circulation in the trapped vortex is equal to the circulation that the sail would have if the boundary layer was attached without the presence of the vortex (DeVoria & Mohseni, 2017).

LEADING-EDGE VORTEX OF DOWNWIND SAILS: NUMERICAL SIMULATIONS

The use of Detached Eddy Simulation (DES) revealed the presence of an LEV on a model-scale sail. The vortex size was shown to increase spanwise towards the head of the sail. The visual representation of the numerical evidence of the LEV, related to the work of Viola *et al* (2014), is depicted in Figure 3. The LEV can be identified as the region of high vorticity, whose contour is shows by the isoline of vorticity. Remarkably, the isoline of axial velocity reveals that, inside the core of the LEV, the flow velocity u_a along the axis of the vortex (out of the plane of the figure) is even higher than the free stream velocity U. This region of very high velocity is associated with a low pressure, with a pressure coefficient Cp ranging from -2 to -4. The time-averaged streamline shows the reattachment downstream of the LEV.

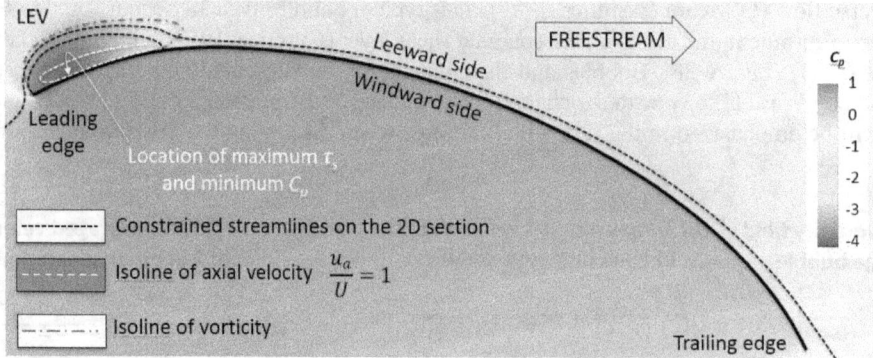

Figure 3 - The LEV computed on a horizontal section of an asymmetric spinnaker. Unpublished results from the time-averaged DES simulations of Viola *et al.* (2014).

The LEV has been identified with a highly time resolved DES (Viola *et al*, 2014) at lower Reynolds numbers than at full scale. In the opinion of the authors, despite the continuous growth of the computational resources, it is still not possible to use more advanced turbulence models such as Large Eddy Simulations (LES). The grid and time resolution required to apply an LES model is significantly higher to that of DES, and is unachievable even by very large supercomputers. However, some under-resolved simulations have been performed with some degree of success. LES have been applied to upwind sails and flat plates (Nava *et al*, 2018), the latter being based on the low-Reynolds number experimental work of Crompton & Barrett (2000) that has often been used as a benchmark for low-camber sails (Collie, 2006; Collie *et al.*, 2008). A key advantage of LES over RANS was originally shown by Sampaio *et al* (2014) and confirmed by Nava *et al* (2017) on upwind sails: the ability to predict the second recirculation bubble typical of long leading-edge bubbles. Modelling correctly this secondary bubble, which sits between the leading edge and the main bubble core, is critical to accurately predict the direction of the separated shear layer and thus the reattachment point of the bubble. Therefore, while the use of LES for yacht sails is expected to significantly improve the accuracy of the solution, the required computational power is still unaffordable.

LEADING-EDGE VORTEX OF DOWNWIND SAILS: EXPERIMENTAL MEASUREMENTS

Model-scale testing of a solid asymmetric spinnaker, identical to that of the previous numerical work (Viola *et al*, 2014), was conducted in the current flume at the University of Edinburgh, utilizing PIV to provide flow visualization. This experiment confirmed the existence of the LEV on the upper half of an asymmetric spinnaker. Vorticity formed at the leading edge rolls up and it is extracted by axial flow at the top of the sail, providing at least 25% of the total sectional lift. It is suggested that the overall effect on the whole sail could be significantly more than 10% (Arredondo-Galeana & Viola, 2018). Due to the forward position of the LEV, an even higher impact on drive force can be speculated.

The LEV was observed to be stable only intermittently (Figure 4a and 4b). When unstable, it was continuously formed and, once it reached a critical strength, convected downstream (Figure 4c). The aerodynamic forces depend on the position and velocity of the LEV with respect to the sail. Hence the convection of the LEV could lead to mild load fluctuations. However, the distance between two convecting LEVs is a fraction of the sail chord, and hence several LEVs are simultaneously present on the leeward side of the sails (Figure 4c). This mitigates the load fluctuations associated with the convection of each LEV. Moreover, the mean aerodynamic forces were similar for the stable and unstable mode of the LEV.

Figure 4a shows the experimental time averaged-streamlines of a section of the model spinnaker (Arredondo-Galeana & Viola, 2018) at 75% of the span from the foot. Figures 4b and 4c show a stable and a shedding LEV respectively, modelled through a potential flow approach, where the circulation and position of each vortex was informed by the PIV measurements. This modelling approach informed by the experiments allows the quantification of the contribution of the LEV to the sectional lift of the spinnaker.

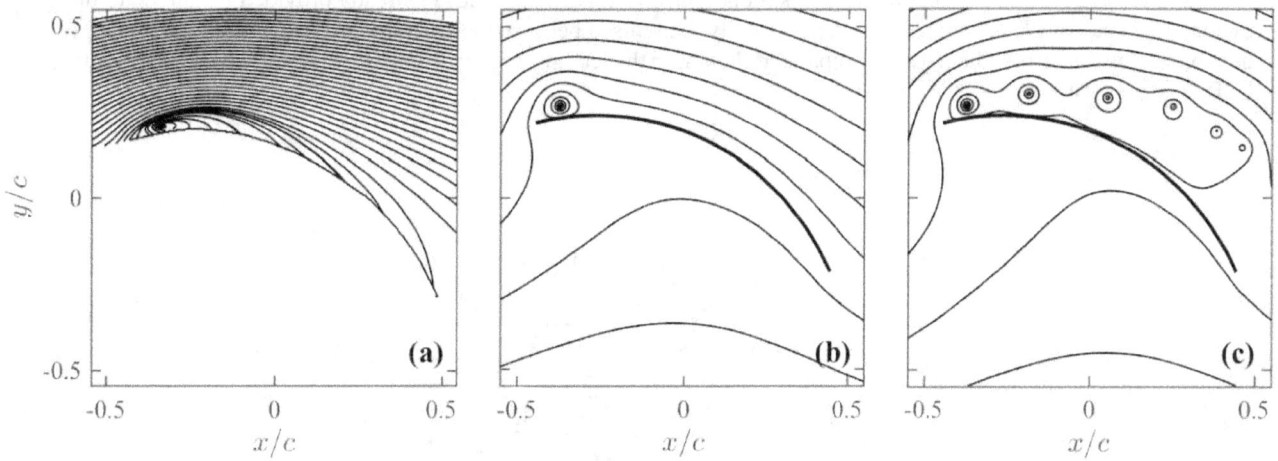

Figure 4 – The LEV computed and modelled on a horizontal section of an asymmetric spinnaker. The experimental time-averaged streamlines (a) measured by Arredondo-Galeana & Viola (2018); the complex potential model of the LEV when steady (b) and when unsteady (c).

MODEL-SCALE / FULL-SCALE CORRELATION

Background on Full-Scale Testing

Despite the earliest report of full-scale testing on yacht sails dating back to 1923 (Marchaj, 1979), the absence of correlation between the measurements and sail shape has been a major limit in the study of sail aerodynamics. With the growing interest for numerical modelling, advances in sensors, full-scale testing and the validation of numerical simulations and model-scale tests, there is a strong demand for full-size benchmark cases.

With the development of dedicated wind tunnels in the 90s, the demand for validation data and benchmarks led to full-scale measurements to be performed on a 35-footer by Milgram *et al* (1993). Subsequently, similar experiments were performed on 33-footers by Masuyama & Fukasawa (1997) and Hochkirch & Brandt (1999) respectively. The former primarily focused on sail forces for the purpose of velocity prediction, and achieved a good agreement between the experimental data gathered (on both tacks) and numerical methods such as vortex lattice and RANS CFD, based on the sail shapes recorded by on-board cameras. This experiment primarily tackled upwind sails and the numerical validation with full-scale data.

Hansen *et al* (2003) targeted downwind sails, with a validation focused on comparison between full-size and wind tunnel data. More recently, load and position sensors were fitted to the spinnaker of a 26-footer (J-80) by Augier *et al* (2012), this time with a stronger emphasis on the more realistic unsteady fluid-structure interaction. Indeed, the greater availability of computer power now allows to run more cost-effective virtual wind tunnel tests with FSI, modelling the changes between the moulded and flying shape depending on the point of sail, wind speed and trim. An area where more research is certainly needed is, for example, luff flapping (Viola & Flay, 2009; Deparday, 2016; Aubin *et al*, 2018).

Full-scale pressure measurements have, in some instances, provided evidence of the presence of the LEV. Viola & Flay (2011) identified a suction peak at the trailing edge of full-scale downwind sails at high apparent wind angles. It is argued that this is evidence of delta wing-like vortex formation on the top section of the spinnaker.

Motta *et al* (2015) also performed full-scale pressure measurement, detecting low pressure peaks convecting chordwise; a phenomenon assimilated to the shedding of an LEV. This is also the argument brought forward by Richards & Viola (2015): the inability to sustain an LEV leads to its shedding in the upper sections of asymmetric spinnakers.

On Water, Wind Tunnel and Computational Measurements

Viola & Flay (2011) compared the forces and pressures measured at full-scale on water, at model-scale in the wind tunnel, and numerically using RANS. This was performed on both upwind and downwind sails, the latter being of primary interest in this instance. In this study, the drive and side force coefficients were shown to be within 0.5% between the wind tunnel and numerical models. However, significant differences were observed on the pressure distributions. Due to their free leading edge, spinnakers tend to be trimmed tighter in full-scale sailing conditions to prevent the luff from collapsing. Conversely, in the steadier conditions of the wind tunnel, the spinnaker can be eased closer to the flapping point. This is revealed in Figure 5, where the single suction peak for the full-scale spinnaker suggests that it is trimmed too tightly. Conversely, the wind tunnel and the numerical simulation feature two suction peaks, suggesting a lower angle of attack. Here the sails were trimmed for the maximum drive force.

It is interesting to observe that the RANS simulations of Viola & Flay (2011) already provided insights into the presence of a tri-dimensional LEV, but these could not be fully recognised because the simulations were not time-resolved. Hence, the helicoidal flow pattern in the region of separated flow near the leading edge was interpreted as a feature of the time-averaged LSB.

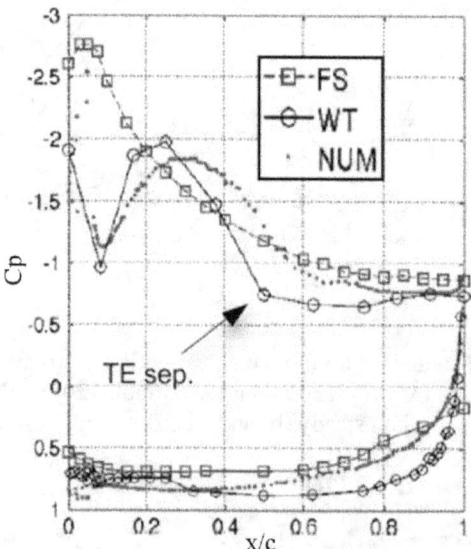

Figure 5 – Pressure distributions on the mid-span section of an asymmetric spinnaker, measured at full-scale (FS), in a wind tunnel (WT) and computed numerically (NUM) by Viola & Flay (2011).

LEADING-EDGE-SEPARATION BUBBLE AND LAMINAR-SEPARATION BUBBLE

As previously highlighted, when the LEV of downwind sails is unstable, it is shed downstream and convected along the surface of the sail (Richards & Viola, 2015), ultimately resulting in a similar time-averaged flow field to that of leading-edge bubbles. A similar phenomenon occurs on upwind fore sails, such as jibs and genoas.

Ota *et al* (1981) showed that the flow regime of leading-edge separation bubbles can be equally laminar, transitional (*i.e.* laminar-to-turbulent transition occurs on the separated shear layer), or turbulent (with either turbulent separation, or transition immediately downstream of the separation point). The instantaneous vorticity field and the time-averaged velocity field of leading-edge bubbles can be found in Figures 2a and 2b respectively. The time averaged and instantaneous flow fields of an LSB, shown in Figure 2c, are similar to those of a leading-edge bubble. Following the separation of the boundary layer due to an adverse pressure gradient, a laminar to turbulent transition occurs, with a reattachment downstream, thus forming a laminar-separation bubble. The LSB can be identified thanks to the presence of a plateau in the pressure distribution (Ward, 1963). Increasing the Reynolds number and the background turbulence, the transition and reattachment points move upstream, hence the length of the bubble decreases (O'Meara & Mueller, 1987). Conversely, increasing the AoA, the separation point moves downstream and the length of the bubble increases.

In this paper, the term LSB is used differently from several other papers, including those of the very same authors. Here the LSB is used to identify a separation bubble where the key mechanisms of reattachment are due to the laminar to turbulent transition. In fact, one of the aims of this paper is to emphasise that the leading-edge bubble of model-scale sails may remain laminar. Consequently, the reattached boundary layer is laminar. In this case, the sail curvature may lead to separation of the laminar boundary layer, resulting in two possible outcomes: either an LSB is formed and hence the flow reattaches forming a turbulent boundary layer; or reattachment does not occur. It will be shown in the following section how these two outcomes leads to very different pressure distributions and lift forces.

The presence of the leading-edge bubble on yacht sails have been shown on model-scale downwind sails in wind tunnels (Viola & Flay, 2009), on circular arcs in CFD (Brault, 2013), and on circular arcs in wind tunnel (Flay, et al, 2017). These results are sometimes contradictory because of misinterpretations of the role of the vortex in generating lift, and in the assumption that it would have always triggered laminar-to-turbulent transition.

In some of these past experiments, the leading-edge bubble was laminar and the boundary layer downstream of the reattachment was also laminar. As noted by Flay et al (2017), this observation is supported by evidence of a LSB farther downstream along the chord. For example, Martin (2015), Flay et al (2017) and Nava et al (2017) have shown a clear LSB towards the rear of a circular arc, especially for a low AoA (lesser than the ideal one), and low Reynolds number. The sudden change occurring below and above a specific critical Reynolds number was discussed by Flay et al (2017).

For greater Reynolds numbers, the flow behaviour is closer to that of a streamlined profile. In these conditions, transitions occurs upstream of trailing edge separation, either within the leading-edge bubble or between the reattachment point and the LSB. Consequently, the boundary layer remains attached longer, decreasing both the wake and the drag, while the improved suction region results in higher lift, as is the case, for instance, of Nava et al. (2016). The LSB is also observed at angles of attack below the ideal one, where there is no leading-edge bubble.

These discrepancies motivated further work to be undertaken on a circular arc, which has the same camber of a typical spinnaker. It will be shown in the next section that a laminar leading-edge separation bubble may indeed occur on model scale sails, resulting in an unrealistic laminar trailing-edge separation. Importantly, the next section will describe the conditions at which model-scale sails must be tested to prevent this effect.

HIGHLY-CAMBERED, THIN, CIRCULAR ARC

Background

The use of a highly cambered (over 20% camber) thin circular arc with a sharp leading edge has been used extensively in recent years to further the understanding of the flow field past downwind sails. Indeed, the circular arc represents a typical cross section through a modern asymmetric spinnaker. The geometry was originally employed by Velychko (2014) in wind tunnel tests, followed by numerical work (Brault, 2013) and by water tunnel experiments (Lebret, 2013; Lombardi, 2014; Martin, 2015; Thomas, 2015; Couvrant, 2015). These research works, together with other wind tunnel tests on the same geometry, are reported by Flay et al (2016).

From these results, it was unclear which was the minimum AoA at which the leading-edge bubble occurs (*i.e.* the ideal angle of attack), and if the appearance of the bubble resulted in a lift increase or decrease. It was also unclear if, and where, laminar-to-turbulent transition occurred and its effect on the lift.

Method

To investigate, a carbon fibre circular arc was built, with a chord of 200 mm and a camber of 22.32% (as per the literature). Pre-preg was employed to achieve the thinnest possible geometry, more representative of the thin membrane than spinnakers are. The final thickness of the tested arc was 1.8 mm.

The force measurements were recorded at 1000 Hz for 6 seconds using potentiometers, at speeds equivalent to Re found in the literature, namely 53k, 68k, 150k and 220k. This allowed to validate the accuracy of the forces measured against Velychko (2014) and demonstrated an abrupt increase in lift and decrease in drag at a critical AoA for a given Re. Additional work on flow visualization was then carried out using PIV on the same geometry, in order to provide a physical explanation to the abrupt change in lift coefficient. Transition was detected by quantifying the Turbulent Kinetic Energy (TKE) from one hundred pairs of PIV flow fields.

Results

The experiments reveal that, at low Re, the leading-edge bubble is laminar! This demonstrates that at model scale there might be a non-realistic laminar boundary layer, which is more prone to separation than the turbulent boundary layer at full scale. These experiments showed there is an AoA at which the reattached, laminar, boundary layer turns into turbulent before trailing edge separation occurs. The authors called this angle the 'critical AoA'. In order to correctly scale the point where trailing-edge separation occurs on a scaled model, tests must be performed at an AoA higher than this angle. This angle decreases with the Re.

Figure 6a shows a sub critical AoA, where trailing-edge separation is laminar and transition occurs in the separated shear layer. Conversely, Figure 6b shows a super-critical AoA, where transition occurs in the boundary layer and trailing-edge separation is turbulent. The turbulent boundary layer is more resilient to separation and, hence, trailing edge separation occurs further downstream.

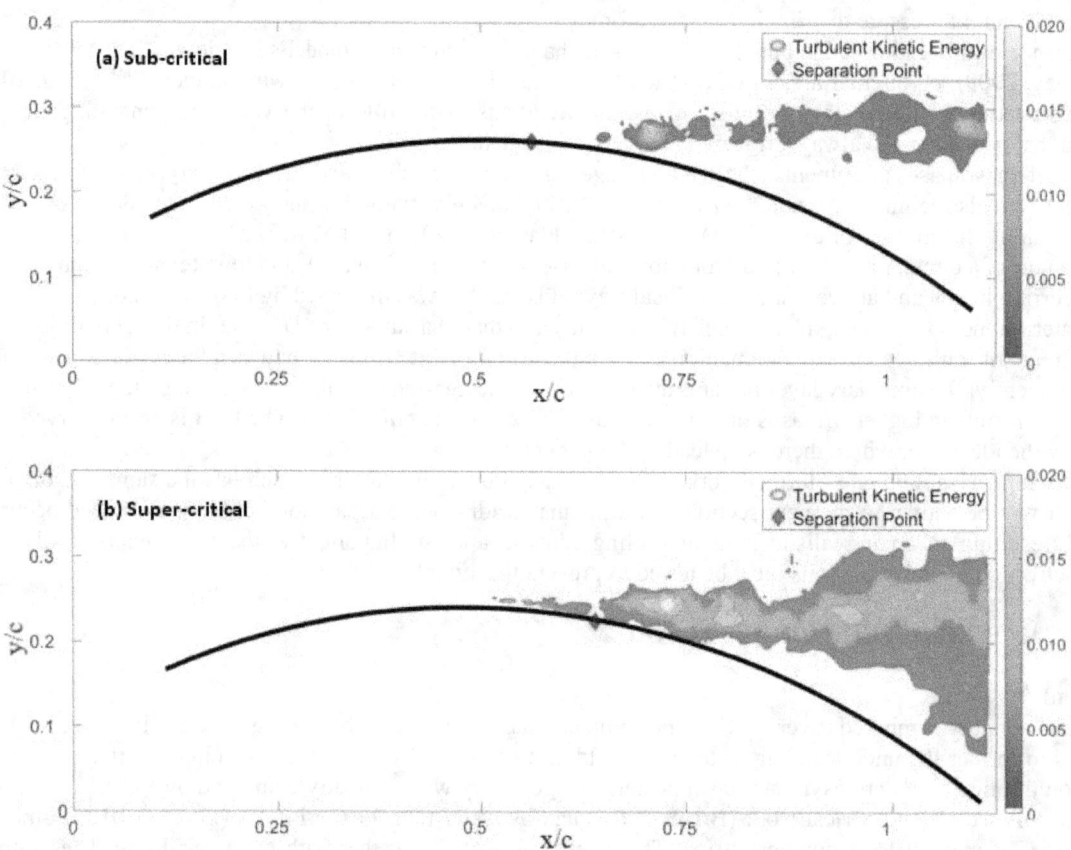

Figure 6 – Flow measurements over a circular arc. Contours of turbulent kinetic energy and position of the separation point over a circular arc at Re = 68k and at: a) sub-critical AoA and b) super-critical AoA.

In order to investigate the effect of the leading-edge bubble on the lift force, the minimum AoA at which the bubble occurs is investigated. Figure 7 highlights the position of the stagnation point near the leading edge (LE) of the arc for different AoA. This figure clearly reveals that 11 degrees (deg) is the ideal AoA for the tested circular arc: the leading-edge bubble must occur at 11 deg and cannot occur at 10 deg.

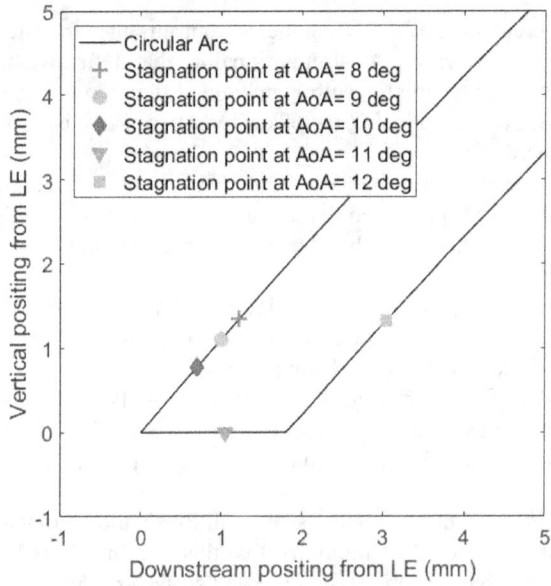

Figure 7 - Location of the stagnation point at the leading edge of a circular arc for different AoAs.

Wind tunnel tests of flexible sails are performed only at AoA higher than the ideal AoA. Hence, the Re at which the critical AoA is 11 deg was searched for. The authors called this Reynolds number the 'critical Re'. If a circular arc is tested at the critical Re or higher, there is no risk of laminar trailing-edge separation. It was ascertained that, for the tested model, the critical Re is 144k (±2k). It must be remarked that the critical Re and the critical AoA depend on the specific geometry.

Finally, by combining the existing literature data and devised experiments, a schematic diagram of how the lift coefficient varies with the AoA was produced, as depicted in Figure 8. A low lift coefficient curve corresponds to subcritical AoA and Re. Past a Re of 218k, the flow will be turbulent for any AoA. However, for each lower Re, there is a critical AoA, where the lift abruptly increases due to the transition in the boundary layer.

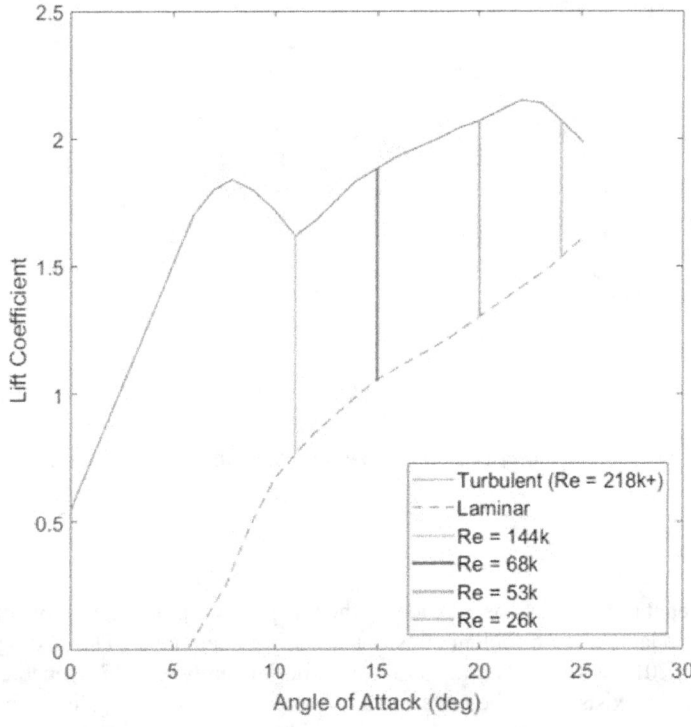

Figure 8 – Schematic diagram of the lift curve and critical AoAs for different Re.

Discussion

The recent research findings inherent to the highly cambered thin circular arc with sharp leading edge yielded four significant results. Firstly, there are pairs of critical AoA and critical Re that define when the laminar to turbulent transition occurs upstream of the trailing-edge separation point, resulting in higher lift and lower drag. On the other hand, for the critical Re ≈ 220k, this occurs for any AoA. At the critical Re = 68k, the critical AoA is between 14 deg and 15 deg, which is consistent with previous work (Lombardi, 2014).

The leading-edge bubble is not correlated with the discontinuous lift jump, which instead is due to the transition in the boundary layer. For example, Martin (2015) speculated that the growth of the leading-edge bubble was causing the jump in lift. However, the present work shows that this is not the case, and the previous conclusion drawn were a coincidence of the tested Re.

Thirdly, the ideal AoA for the circular arc, previously defined as 8 deg by Martin (2015) based on the peak in lift coefficient has been shown to be erroneous. Indeed, PIV revealed that the stagnation point was located on the leading edge at an AoA of 11 deg This is consistent with the PIV measurements of Thomas (2015).

Lastly, the present results show that, for this model, the critical Re = 144k (±2k) is associated with a critical AoA that coincides with the ideal AoA, 11 deg. Let assume, in first approximation, that this is the critical Re of the ideal AoA also for a model-scale sails. For Re higher than 144k (±2k), transition would occur upstream of trailing-edge separation, as on a full-scale sail.

These results suggest that, for a highly cambered model-scale sail, Re should be much higher than 144k (±2k) so that the model flow field is as per the full-size. It should be reminded that this limiting Re is likely to strongly depend on the sail curvature, and probably on the twist, the background turbulence and surface roughness. For example, the tests by Bot *et al* (2014) on a rigid model-scale sail suggest that Re = 230k was insufficient for their model. In fact, reviewing their sail pressure distributions in light of these findings, the presence of the LSB can clearly be identified (Figure 9). Bot *et al* (2014) tested at an average chord-based Re of 230k. This represents an example of a test where too low a Re was employed, with the critical AoA occurring after the ideal AoA.

Similarly, circular arcs at low Re have showcased an LSB, the evidence of which was provided either numerically with the average velocity field (Brault, 2013), or experimentally with pressure taps (Flay *et al*, 2017). The latter also reported both an AoA and Re dependency. Conversely, no evidence of an LSB can be found in the literature for full size spinnakers, including the experiments reported by Viola and Flay (2011), Motta *et al* (2014) and Deparday *et al* (2017).

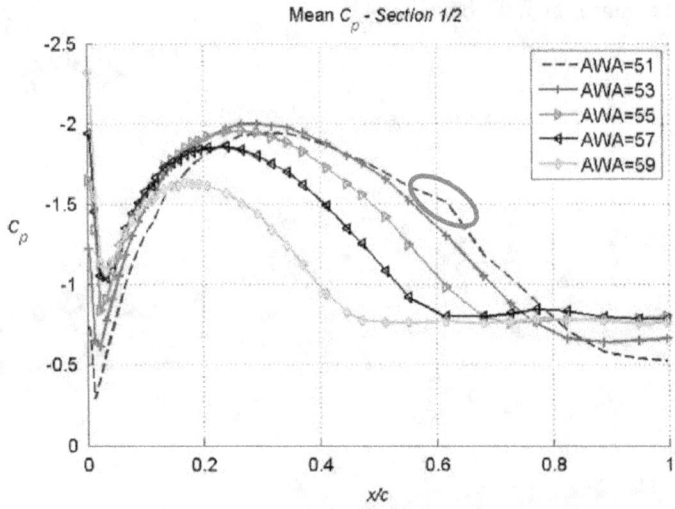

Figure 9 - Pressure coefficient along the mid-section of a spinnaker at different AoA (Bot *et al*, 2014).

CONCLUSIONS

In the last decade, significant progress has been made on the understanding of the aerodynamics of sails, and particularly on flow at the leading edge and laminar-to-turbulent transition in the boundary layer. The leading-edge vortex was first shown to be present on spinnakers in 2014 numerically and confirmed experimentally in 2017. Nonetheless, more research is needed to confirm that the LEV actually exists in a coherent structure at full-scale and in a realistic unsteady flow condition, which are the sail types and apparent wind angles at which it occurs, and how it can be exploited by design.

The last decade also saw significant advances in the correlation between the surface pressure distributions measured in wind tunnel, computed numerically or measured on full-scale sails. This now leads to a better understanding of the flow features affecting the design and testing of asymmetric spinnakers. Anomalies in some pressure distributions measured at model scale prompted further work to be conducted on highly cambered thin circular arcs resembling a 2D section of an asymmetric spinnaker. In this paper, some new key results of this model are presented. These experiments demonstrated that several wind tunnel tests were performed at too low Reynolds numbers. At the low Reynolds numbers, the leading-edge bubble is laminar, and hence the reattached boundary layer is laminar, and trailing edge separation might occur upstream of where a turbulent boundary layer would have separated. Transition was shown to be governed by the combination of a critical Reynolds number and a critical AoA. The analysis of previous work in light of these new results suggests that model-scale tests should conservatively be tested at Re > 230k.

These results further supported recent findings on the importance of the leading-edge vortex in lift generation. Indeed, noting the recent experimental evidence suggesting that 25% of the sectional lift can be provided by the LEV, it is envisaged that future work will see a stronger emphasis on how to promote a sustainable leading-edge vortex by design. Moreover, the leading-edge vortex being located at the luff of the sail, the relative increase in driving force will be substantially higher than the increase in lift. Hence, downwind sail design, and more precisely modern asymmetric spinnakers, can be tremendously refined by fully exploiting the effect of the LEV as part of the sail design process.

REFERENCES

Arredondo-Galeana, A., *A study of the vortex flows of downwind sails*, PhD Thesis, University of Edinburgh, UK, 2019.

Arredondo-Galeana, A. & Viola, I.M., *The leading-edge vortex of yacht sails*, Ocean Engineering, vol. 159, pp. 552-562, 2018.

Aubin, N., Augier, B., Deparday, J., Sacher, M., & Bot, P., Performance enhancement of downwind sails due to leading edge flapping: A wind tunnel investigation. *Ocean Engineering*, 169, 370-378. 2018

Augier, B., Bot, P., Hauville, F., & Durand, M., *Experimental validation of unsteady models for fluid structure interaction: application to yacht sails and rigs*, Journal of Wind Engineering and Industrial Aerodynamics, vol. 101, pp. 53-66, 2012.

Bethwaite, F., *High performance sailing*, International Marine Publication Company, 1993.

Bomphrey, R.J., Lawson, N.J., Harding, N.J., Taylor, G.K. & Thomas, A.L.R., *The aerodynamics of Manduca sexta: digital particle image velocimetry analysis of the leading-edge vortex*, The Journal of Experimental Biology, volume. 208, pp. 1097-1094, 2005.

Borazjani, I. & Daghooghi, M., *The fish tail motion forms an attached leading edge vortex*, Biological Sciences, The Royal Society, vol. 280, pp. 2012-2071. 2013.

Bot, P., Viola, I.M. & Flay, R.G.J., & Brett J.S., *Wind-tunnel pressure measurements on model-scale rigid downwind sails*, Ocean Engineering, vol. 90, pp. 84-92, 2014.

Brault, E., *Aerodynamics on Very Thin Airfoil*. Newcastle: MSc Thesis, Newcastle University, 2013.

Carter, R., *Boat remains and maritime trade in the Persian Gulf during the sixth and fifth millennia BC*, Antiquity, vol. 80(307), pp. 52-63, 2006.

Collie, S., *Application of computational fluid dynamics to two-dimensional downwind sail flows*, Auckland: PhD thesis, The University of Auckland, 2006.

Collie, S., Gerritsen, M., Jackson, P., *Performance of Two-Equation Turbulence Models for Flat Plate Flows with Leading Edge Bubbles*, Journal of Fluid Engineering – Transactions of ASME, vol. 130, issue 2, pp. 0212011-02120111, 2008.

Corke, T.C. & Thomas, F.O., *Dynamic stall in pitching airfoils: aerodynamic damping and compressibility effects*, Annual Reviews in Fluid Mechanics, vol. 47, pp. 479–505, 2015.

Couvrant, H., *Reynolds number and angle of attack effects on a leading edge vortex (LEV) for a highly cambered thin profile*, Brest: IERNAV, 2015.

Crabtree, L. F., *The formation of regions of separated flow on wing surfaces*, Aeronautical research council reports and memoranda, 1957.

Crompton, M. & Barrett, R., *Investigation of the separation bubble formed behind the sharp leading edge of a flat plate at incidence*, Proceedings of the Institution of Mechanical Engineers, Part G, vol. 214, pp. 157-176, 2000.

Deparday, J., *Experimental studies of fluid-structure interaction on downwind sails*, Brest: PhD thesis, Universite de Bretagne Occidental, 2016.

Deparday, J., Bot, P., Hauville, F., Augier, B. & Rabaud, M., *Modal Analysis of Pressures on a Full-Scale Spinnaker*, Journal of Sailing Technology, 2017.

DeVoria, A.C., & Mohseni, K., On the mechanism of high-incidence lift generation for steadily translating low-aspect-ratio wings, Journal of Fluid Mechanics, vol. 813, pp. 110-126, 2017.

Durand, M., Leroyer, A., Lothodé, C., Hauville F., Visonneau, M., Floch, R. & Guillaume, L., *FSI investigation on stability of downwind sails with an automatic dynamic trimming*, Ocean Engineering, vol. 90, pp. 129-139, 2014.

Earnshaw, P.B., *An experimental investigation of the structure of a leading-edge vortex*, Aeronautical Research Council Reports and Memoranda, 1962.

Eldredge, J.D. & Jones, A.R., *Leading-Edge Vortices: Mechanics and Modelling*, Annual Reviews of Fluid Mechanics, vol. 51, pp. 75–104, 2019.

Ellington, C., van den Berg, C., Willmott, A. & Thomas, A., *Leading-edge vortices in insect flight*, Nature, vol. 384, pp. 626-630, 1996.

Fallow, B.J., *America's Cup sail design*, Journal of Wind Engineering and Industrial Aerodynamics, vol. 63, pp. 183-192, 1996.

Flay, R.G.J. & Jackson, P. S., *Flow simulation for wind tunnel Studies of sail aerodynamics*, Journal of Wind Engineering and Industrial Aerodynamics, vol. 41-44, pp. 2703-2714, 1992.

Flay, R.G.J., Piard, A. & Bot, P., *Aerodynamics of a highly cambered circular arc aerofoil: experimental investigations*, International Conference on Innovation in High Performance Sailing Yachts, Lorient, France, pp.150-162, 2017.

Flay, R.G.J. & Vuletich, I.J., *Development of a wind tunnel test facility for yacht aerodynamic studies*, Journal of Wind Engineering and Industrial Aerodynamics, vol. 58, pp. 231-258, 1995.

Fossati, F., *Aero-Hydrodynamics and the Performance of Sailing Yachts*, London: Adlard Coles Nautical, 2009.

Gentry, A., *The aerodynamics of sail interaction*, The Ancient Interface III, 3rd AIAA Symposium on Sailing, Redondo Beach, California, US, 1971.

Gentry, A., *The application of computational fluid dynamics to sails*, Newport, Rhode Island, Proceedings of the Symposium of Hydrodynamic Performance Enhancement for Marine Applications, 1988.

Gursul, I., Gordnier, R., Visbal, M., *Unsteady aerodynamics of nonslender delta wings*, Progress in Aerospace Sciences, vol. 41, pp. 515–557, 2005.

Gursul, I.,Wang, Z. & Vardaki, E., *Review of flow control mechanisms of leading-edge vortices*, Progress in Aerospace Sciences, vol. 43, pp. 246–270, 2007.

Hall, M., *A theory for the core of a leading-edge vortex*, Journal of Fluid Mechanics, vol. 209, pp. 209-228, 1961.

Hansen, H., Jackson, P.S. & Hochkirch, K., *Comparison of wind tunnel and full-scale aerodynamic sail force measurements*, International Journal of Small Craft Technology, RINA Transactions, vol. 145(B1), pp. 23-31, 2003.

Harvey, J., *Some measurements on a yawed slender delta wing with leading edge separation*, A.R.C. R.&M. 3160, 1959.

Hedges, K.L., *Computer modelling of downwind sails*, ME Thesis, University of Auckland, New Zealand, 1993.

Hedges, K., Richards, P. & Mallison, G., *Computer modelling of downwind sails*, Journal of Wind Engineering and Industrial Aerodynamics, vol. 63, pp. 95-110, 1996.

Hochkirch, K. & Brandt, H., *Full-scale hydrodynamic force measurement on the berlin sailing dynamometer*, Proceedings of the 14th Chesapeake Sailing Yacht Symposium, January 30th, Annapolis, MD, pp. 33-44, 1999.

Huang, M.K. & Chow, C.Y., *Trapping of a free vortex by Joukowski airfoils*, AIAA Journal, vol. 20, pp. 292–298, 1982.

Hubel, T.Y. & Tropea, C., *The importance of leading edge vortices under simplified flapping flight conditions at the size scale of birds*, Journal of Experimental Biology, vol. 213, pp. 1930–1939, 2010.

King, R., *Spinnaker*, London: Granada Publishing, 1981.

Larsen, J.W., Nielsen, S.R.K. & Krenk, S., *Dynamic stall model for wind turbine airfoils*, Journal of Fluids and Structures, vol. 23, pp. 959–982, 2007.

Larsson, L., Eliasson, R.E. & Orych, M., *Principles o Yacht Design*, International Marine Publishing Co, 4th edition, 2013.

Lasher, W.C., Sonnenmeier, J.R., Forsman, D.R. & Tomcho, J., *The aerodynamics of symmetric spinnakers*, Journal of Wind Engineering and Industrial Aerodynamics, vol. 93, pp. 311-337, 2005.

Lebret, C., *High Cambered thin profile study*, Auckland: The University of Auckland, 2013.

Lentink, D., Dickson, W.B., van Leeuwen, J.L., Dickinson, M.H., *Leading-edge vortices elevate lift of autorotating plant seeds*, Science, vol. 324, pp. 1438–1440, 2009.

Lentink, D., Müller, U.K., Stamhuis, E.J., de Kat, R., van Gestel, W., Veldhuis, L.L.M., Henningsson, P., Hedenström, A., Videler, J.J. & van Leeuwen, J.L., *How swifts control their glide performance with morphing wings*, Nature, vol. 446, pp. 1082–1085, 2007.

Li, J. & Wu, Z.N., *Vortex force map method for viscous flows of general airfoils*, Journal of Fluid Mechanics, vol. 836, pp. 145-166, 2018.

Lombardi, A., *Experimental analysis of a highly-cambered thin profile*, Strathclyde: MSc Thesis, University of Strathclyde, 2014.

Marchaj, C., *Aero-hydrodynamics of sailing*, Dodd Mead and Company, New York, 1979.

Marsden, D., Simpson, R. & Rainbird, W., *An investigation into the flow over delta wings at low speeds with leading edge separation*, Cranfield: College of Aeronautics, Cranfield Report No. 114. ARC 20,409, 1958.

Martin, V., *Reynolds number and angle of attack effects on a highly cambered thin profile flow topology*, Brest: IRENAV Laboratory, 2015.

Masuyama, Y. & Fukasawa, T., *Full-Scale Measurements of Sail force and Validation of Numerical Calculation Method*, Proceedings of the 13th Chesapeake Sailing Yacht Symposium, January 25th, Annapolis, MD, pp. 23-36, 1997.

Mcauliffe, B.R. & Yaras, M., *Transition mechanisms in separation bubbles under low and elevated-freestream turbulence*, Journal of Turbomachinery, vol. 132 (1), 2010.

Milgram, J.H., *The Aerodynamic of sails*, Proceedings of the 7th Symposium on Naval Hydrodynamic, pp. 1397-1434, 1968.

Milgram, J.H., Peters, D.B. & Eckhouse, D.N., *Modelling IACC sail forces by combining measurements with CFD*, Proceedings of the 11th Chesapeake Sailing Yacht Symposium, January 29th-30th, Annapolis, MD, pp. 65-73, 1993.

Mitchel, A.M., Morton, S. A., Forsythe, J.R. & Cummings, R.M., *Analysis of delta-wind vortical substructures using detached-eddy simulation*, AIAA Journal, vol. 44 (5), pp. 964-972, 2006

Miyata, H. & Lee, Y., *Application of CFD simulation to the design of sails*, Journal of Marine Science and Technology, vol. 4, pp. 163-172, 1999.

Motta, D., Flay, R.G.J., Richards, P.J., Le Pelley, D.J. & Deparday, J., *Experimental Investigation of Asymmetric Spinnaker Aerodynamics Using Pressure and Sail Shape Measurements*, Ocean Engineering, vol. 90, pp. 104-118, 2014.

Motta, D., Flay, R.G.J., Richards, P.J., Le Pelley, D.J., Bot, P. & Deparday, J., *An investigation of the dynamic behaviour of asymmetric spinnakers at full-scale*, 5th High Performance Yacht Design Conference, Auckland, New Zealand, 2015.

Muijres, F.T., Johansson, L.C., Barfield, R., Wolf, M., Spedding, G.R. & Hedenström, A., *Leading-edge vortex improves lift in slow-flying bats*, Science, vol. 319, pp. 1250–1253, 2008.

Nava, S., Bot, P. & Carter, J., *Modelling the lift crisis of a cambered plate a 0 degrees angle of attack*, 20th Australasian Fluid Mechanics Conference, Perth, Australia, 2016.

Nava, S., Cater, J. & Norris, S., *Modelling leading edge separation on a flat plate and yacht sails using LES*, International Journal of Heat and Fluid Flow, vol. 65, pp. 299-308, 2017.

Nava, S., Cater, J. & Norris, S., *Large Eddy Simulation of an asymmetric spinnaker*, Ocean Engineering, vol. 169, pp. 99-109, 2018.

Newman, B.G. & Tse, M.C., *Incompressible flow past a flat plate aerofoil with leading edge separation bubble*, Aeronautical Journal, vol. 96, pp. 57-64, 1992.

O'Meara, M.M. & Mueller, J.T., *Laminar separation bubble characteristics on an airfoil at low Reynolds numbers*, AIAA Journal, vol. 25(8), pp. 1033-1041, 1987.

Ota, T., Asano, Y. & Okawa, J.I., *Reattachment length and transition of the separated flow over blunt flat plates*, Bulletin of the JSME, vol. 24(192), pp. 941-947, 1981.

Polhamus, E.C., *A concept of the vortex lift of sharp-edge delta wings based on a leading-edge suction analogy*, NASA Technical Report TND-3767, 1966.

Renzsch, H., Muller, O. & Graf, K., *FLEXSAIL – a fluid structure interaction program for the investigation of spinnakers*, Proceedings of the International Conference on Innovations in High Performance Sailing Yachts, Lorient, France, 2008.

Richards, P.J., Johnson, A. & Stanton A., *America's Cup sails – vertical wind of horizontal parachutes?* Journal of Wind Engineering and Industrial Aerodynamics, vol 89, pp 1565-1577, 2001.

Richards, P.J. & Viola, I.M., *Leading Edge Vortex Dynamics*, 17th Australian Wind Engineering Society Workshop, Wellington, New Zealand, 2015.

Richter, H.J., Horrigan, K.C. & Braun, J.B., *Computational fluid dynamics for downwind sails*, Proceedings of the 16th Chesapeake Sailing Yacht Symposium, Annapolis, Maryland, USA, 2003.

Sacher, M., Hauville, F., Bot, P. & Durand, M., *Sail trimming FSI simulation - comparison of viscous and inviscid flow models to optimise upwind sails trim*, 5th High Performance Yacht Design Conference, HPYD 2015, pp. 217-228, Auckland, New Zealand, 2015.

Saffman, P.G. & Sheffield, J.S., *Flow over a Wing with an Attached Free Vortex*, Studies in Applied Mathematics, vol. 57, pp. 107–117, 1977.

Sampaio, L.E.B., Rezende, A.L.T., & Nieckele, A.O., *The challenging case of the turbulent flow around a thin plate wind deflector, and its numerical prediction by LES and RANSE models*, Journal of Wind Engineering and Industrial Aerodynamics, vol. 133, 52-64, 2014.

Stevenson, J.P.J., Nolan, K.P., & Walsh E.J., *Particle image velocimetry measurements of induced separation at the leading edge of a plate*, Journal of Fluid Mechanics, vol. 184, pp. 278-297, 2016.

Taira, K. & Colonius, T., *Three-dimensional flows around low-aspect-ratio flat-plate wings at low Reynolds numbers*, Journal of Fluid Mechanics, vol. 623, pp. 187-207, 2009.

Thomas, G., *Flow phenomenology around a highly cambered thin profile*, Brest: IRENAV, 2015.

Velychko, N., *Study of highly cambered aero foil using JR3 sensor*, Auckland: The University of Auckland, 2014.

Videler, J., Stamhuis, E. & Gde, P., *Leading-edge vortex lift swifts*, Science, vol. 306, pp. 1960-1962, 2004.

Viola, I.M., *Downwind sail aerodynamics: a CFD investigation with high grid resolution*, Ocean Engineering, vol. 36, pp. 974-987, 2009.

Viola, I.M. & Arredondo-Galeana, A., *The leading-edge vortex of yacht sails*, Innovsail International Conference on Innovation in High Performance Sailing Yachts, Lorient, France, pp. 115-126, 2017.

Viola, I.M., Bertesaghi, S., Van-Renterghem, T. & Ponzini, R., *Detached eddy simulation of a sailing yacht*, Ocean Engineering, vol. 90, pp. 93-103, 2014.

Viola, I.M. & Flay, R.G.J., *Force and pressure investigation of modern asymmetric spinnakers*, International Journal of Small Craft Technology, RINA Transactions, vol. 151 (B2), 2009.

Viola, I.M. & Flay, R.G.J., *Sail pressures from full-scale, wind tunnel and numerical investigations,* Ocean Engineering, vol. 38(16), pp. 1733-1743, 2011.

Viola, I.M. & Flay, R.G.J., *Sail aerodynamics: on-water pressure measurements on a downwind sail*, Journal of Ship Research, vol. 56 (4), pp. 197-206, 2012.

Viola, I.M. & Ponzini, R., *A CFD investigation with high-resolution grids of downwind sail aerodynamics*, International Conference on Developments in Marine CFD, London, UK, 2011.

Ward, J.W., *The behaviour and effects of laminar separation bubbles on aerofoils in incompressible flow*, Journal of the Royal Aeronautical Society, vol. 67, pp. 783-790, 1963.

An Energy Aware Autopilot for Sailboats

Mathilde Tréhin, Université de Bretagne Sud/Madintec, Lorient, France
Johann Laurent, Université de Bretagne Sud, Lab-STICC UMR6285, Lorient, France
Hugo Kerhascoët, Madintec, Lorient, France
Jean-Philippe Diguet, CNRS, Lab-STICC UMR6285, Lorient, France

ABSTRACT

In this paper, we propose a new control method for the next generation of autopilots. These new systems will need to manage more actuators to control the hydrofoils, which is going to significantly increase the energy requirements. So, this method is aware of the autopilot power consumption. It uses a model predictive controller to manage the actuators (position control appendage angle control). This controller uses a dynamic model of the actuator, running in real time, to anticipate the future behavior of the system. Once the predictions are made, it determines the future control sequence to apply in order to follow the reference trajectory. To do so, it minimizes a cost function which includes the quadratic error according to the behavior prediction and the associated energy consumption. So, it takes into account two criteria: the precision/rapidity of the system and the energy. With the proposed control method, skippers can weight each criterion in order to focus on one or the other depending on their goals and the boat's energy balance. We apply this method to one of the autopilot's subsystems, namely the rudder control. The electric actuator intervening in this control loop and the load representing the force opposed to its motion are modelled to design the control law. The first results of that method are compared with a standard autopilot. We increase by 40% the precision level and we are able to reduce the consumption by at least 20%. This work provides the first necessary components of a future autopilot that will control the whole appendages to a three dimensional piloting. Moreover, this type of management is a first step towards possible fossil fuel free sailboats.

INTRODUCTION

Over the years, the sailboats designed for ocean races have evolved a great deal in order to gain in performance. The hulls changed to get more powerful profiles and are now composed of the latest generation composite materials. Foils equip them and thanks to these improvements, the sailboats, initially in an Archimedean mode, became "flying machines". Embedded systems are part of these most recent technological developments and evolve into key components to optimize the boat settings "Douguet (2013)". They already use and analyze more than one hundred data sensors and are becoming increasingly complex under great performance and safety pressure.

For example, the autopilot, which is basically a control process running on a microcontroller, assists the skipper to steer the sailboat more than 95% of a solo race. Thanks to the autopilot system, the skipper can focus all his attention on sail trim and free up more time to refine tactical elements or for media communications. So, it has become an indispensable element of a solo race. An autopilot is composed of two control loops, the heading/attitude control and the rudder angle control (Figure 1).

To enable a quasi permanent use of the autopilot, the system needs a reliable energy supply around the clock. However, energy resources are limited on board. They have a few renewable energy producers (solar panel, wind turbine, etc.), but they can meet the boat needs only under specific weather conditions. They have a limited quantity of fuel which can recharge a battery set but they try to gain in performance by minimizing the weight which means reducing the fuel volume.

With the arrival of the new hydrofoil sailboats, the next generation autopilot will need to manage more actuators to control all these appendages. It is going to increase the computing and energy requirements. So, taking into account the energy criterion in the control laws and optimize its use is becoming mandatory to increase the number of managed actuators, to move towards a sailboat without fossil fuel and to guarantee skipper safety.

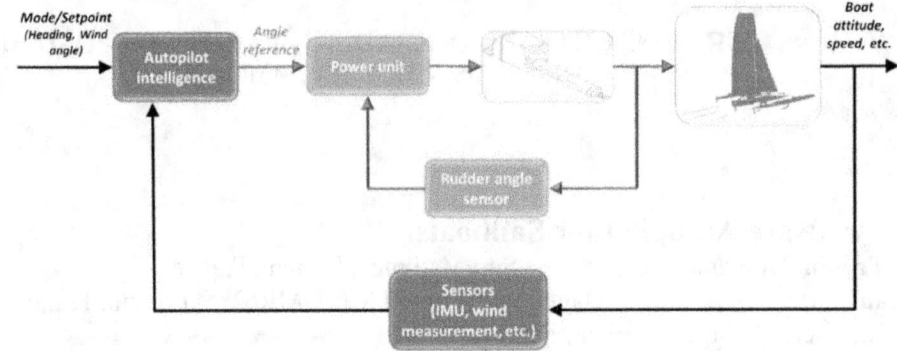

Figure 1: Autopilot control loops

Currently, the methods available in the scientific literature don't pay attention to the autopilot power consumption "Shi, et al., (2017)" "Tomera (2010)". They focus only on the boat performance and safety by improving the reactivity, the robustness or the adaptability of the system in a complex environment. As seen previously, to be able to manage the new hydrofoil sailboats, what is really needed is an autopilot which can adapt its control law to the energy constraints linked to the offshore sailing race environment. So, in this paper, we propose a new control method aware of the autopilot power consumption.

The first section of this paper includes a set of today's autopilot control laws. The second section explains our control method which is aware of the power consumption, its principle and setting up. In the last section, the rudder control case study is entirely detailed (models, control tuning and results). Finally, a summary of the main concepts of the paper and a concise presentation of the future prospects end this document.

AUTOPILOT CONTROL LAWS

Many laws exist in the field of drive control and a lot of them have been tested for autopilot systems. Some of them are presented in this section.

In the sailboat industry

According to some skippers' feedback about the autopilots currently on the market, it is regularly necessary to adjust the gains/settings in function of the sailing conditions. In fact, the implemented control laws used in the sailing domain remain quite simple and only one tuning can't adapt to every sailing condition. These are usually PID controllers. As a reminder, a PID controller (Proportional Integral Derivative) acts in function of the error value as the difference between a set point and a measured signal. It applies a correction based on Proportional, Integral and Derivative terms (Figure 2). So, this method is based only on measuring and calculating the error and not on a knowledge of the system behavior. It can perform well within some industrial environments and appears to be conceptually intuitive but its tuning can be difficult if opposing objectives such reactivity and high stability are to be achieved "Ang, et al., (2005)". Some PID configuration methods exist (Ziegler Nichols, Nyquist, etc.) but they can be difficult to apply to a sailboat because they need to have an observation and a characterization of the system answer to an excitement. Moreover, the boat operates in a complex and uncertain environment, the skipper often needs to modify the controller gains in navigation to match the sailing conditions.

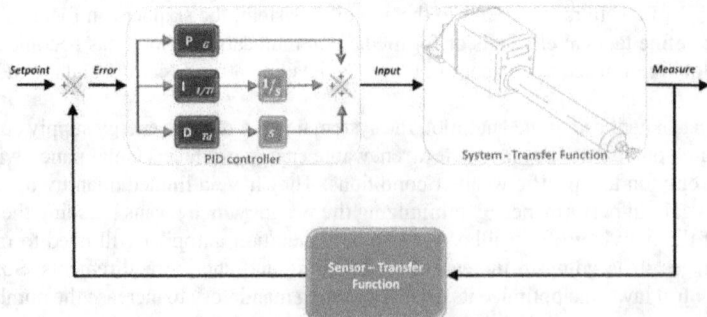

Figure 2: PID controller

The autopilot control is a complex problem and various scientific studies looked into the problem to improve it. The next section details some of them.

In the areas of research

Most of the scientific studies focus on ships or sailing robots which don't have the same steering responses under stress. Only a few are interested in the sailboat, a complex system severely disrupted by the two uncertain environments where it operates, water and air environments.

A significant part of these studies tries to reproduce the human behavior, its sensations at the helm that enable to adapt to any situation by using artificial intelligence methods. For instance, "Tiano, et al., (2001)" draws on neural networks to design a course keeping autopilot for a sailboat. Neural networks are based on the concept of black boxes. They need a large amount of data to derive a representative set of the sailboat behavior. However, it can be difficult to obtain these needed data in our case. It is expensive and time consuming to sail a hydrofoil sailboat, so to log data 24 hours a day, 7 days a week. Moreover, they often face the same sailing conditions, the data may not be really diversified. This option, at this stage, is not conceivable because it is not supervised. An unknown situation, which has not been noticed until now by the neural network, can lead to unsafe behavior. For instance, if the wind increases sharply, the boat can be over powered, and the skipper must release the sheets to lighten the boat. If he doesn't have the time because he is sleeping or engaged in some other task and if the autopilot behaves in the same way as usual, the boat can lay on its side, it can even turn over if it is a multi hull.

Other studies focus on fuzzy and adaptive fuzzy control "Velagic, et al., (2003)", a method widely used in the field of control systems and beyond (robotics, air control, meteorology, etc.). This control method is based on fuzzy logic which enables to approach to the flexibility of human reasoning. It allows the events to be partially true or false, it is not binary. Alternative methods, neural networks as noted above or genetic algorithms, can adopt the same behavior as fuzzy logic in many cases. But its key advantage is that its expression is easily understandable by a human, so he can put into the control his experience. The difficulty of this approach lies in the definition of the membership functions and the defuzzification process (data fusion and parameters transformation into digital data). "Zimmermann (1996)" explains that in his book.

Sliding mode controllers are also frequently used in the state of the art because they are known to provide a robust control. They are based on a variable structure control method and can switch from one continuous law to another in order to adapt to the dynamic behavior of nonlinear systems. However, the robustness, inherent in this method in order to face model uncertainties and external perturbations, is heavily dependent on the parameter tuning which must adapt to any sailing condition. It is tedious and time consuming and there is no guarantee that the result is an optimal solution. To address this issue, "McGookin, et al., (2000)" proposes using genetic algorithms to optimize the tuning of a sliding mode controller of a ship autopilot.

The H infinity control uses optimization techniques to get a robust control. This method minimizes the effect of the system's input/output. It minimizes the maximum possible loss in the frequency domain, or in other words, its control input enables the system to amplify the energy of the input signal as much as possible to achieve the desired state. This control method is used for rudder roll stabilization "Robert (2008)" and track keeping design for ships "Alfi, et al., (2015)".

Model Predictive Controller (MPC) also uses optimization techniques. The main advantage of MPC control is its capacity to make a prediction on the future trajectory in order to optimize the current control input. "McGookin, et al., (2008)" compares the sliding mode control with a MPC controller in another study and demonstrates that MPC is more efficient and reliable. By minimizing the cost function which represents the heading error, it performs superiorly than the sliding mode control. Moreover, this control method enables to take into account actuators and trajectory constraints to address the physical limits of the system (e.g. rudder's mechanical stop). The study's team notices a safer steering behavior and an increase in actuator life.

All the reviewed studies focus on the improvement of tracking and control algorithms. The impact of these control laws on the energy consumption is rarely addressed. The power allocated in control is sometimes even oversized to assure track keeping. Only a few studies about sailing robots take an interest about energy resource "Briere (2011)" but the scale and the constraints are far from comparable to real sailboats. The control method explained in this paper proposes a first approach to answer this lack.

PROPOSED CONTROL LAW

The proposed control law is based on MPC. This control method is selected because:
the main idea is intuitive and easy to understand;
it expresses optimality concerns;
it enables to handle constraints on the state and on the control input;
it prevents any excessive fluctuations on the manipulated variables, the control output is smoother. It enables a better use of the actuators (rams, valves, motors), so their life time is increased;
in case of measurable disturbances, the system can automatically adapt.

MPC control can be used to control complex systems with multiple inputs/outputs. This method is particularly interesting for systems with significant delays, inverted responses or with a large number of external disturbances "Lewis, et al., (2012)". This corresponds to the sailboat case which has delays due to the boat inertia and a lot of external disturbances as wind and sea state. The fundamental idea of this method is that it is built on a mathematical model of the process to control. It enables to predict the behavior of the system. It is a feedback implementation of optimal control using finite prediction horizon and on line computation. At each sampling period (Figure 3), the controller:

computes the predictions of the controlled variables for a given time horizon thanks to the internal model, it is based on measurements obtained at time t0;

develops a reference trajectory to follow;

following this trajectory, determines the next control sequence to apply to the system to reach the set point. In order to ensure this, the controller seeks to minimize a cost function which differs depending on the application. Usually, this function includes quadratic errors between the reference trajectory and the predictions for the control horizon as well as the control input variations to reduce the risk of instability;

applies the beginning of the control strategy until the next decision instant.

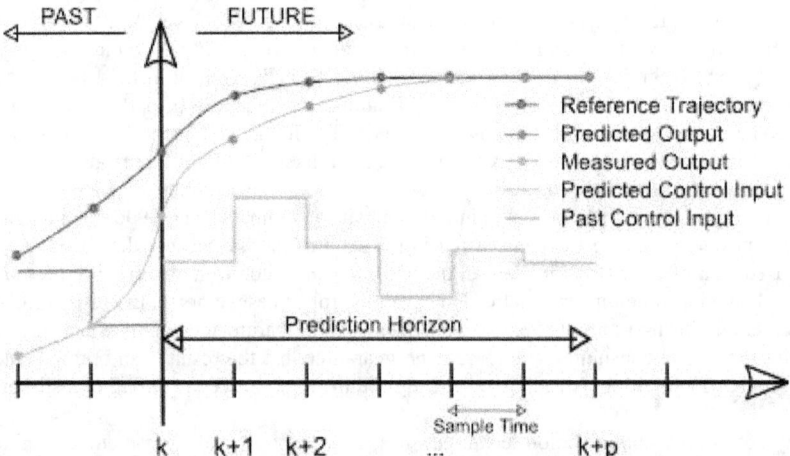

Figure 3: Principle of Model Predictive Control (By Martin Behrendt via Wikimedia Commons)

The main users of the predictive control are oil refineries, agri food industries, aerospace and car industries, etc. MPC also shows a good level of performance in ship's applications "Li, et al., (2012)". It enables a greater freedom to define control objectives and constraints. The hard task in the case of sailing is to adapt this control method to a much more complex behavior and to define a cost function to minimize which achieves a good level of precision and performance while taking into account the power consumption.

The cost function enables to have a scalar representation of an event. Thus, it transcribes again one or more costs associated with this event that the controller is going to minimize through this representative function.

The optimization problem to solve is defined as:

$$\forall k \in \mathbb{N}, \quad x(k) \in \mathbb{R}^3 \qquad P\big(x(k)\big) : \min_{u}\big\{ V(x(k), u) \mid u \in U\big(x(k)\big) \big\}$$

$V(x(k), u)$: a cost function to minimize
u : the decision variable
U : a set of admissible values

So, the function V is minimized by modifying the variable u under the constraint $u \in U$. This control shape could suggest a classic LQR (Linear Quadratic Regulator) problem except that it takes into account inequalities constraints and a finite horizon. So, it is more complex.

The discretization of the system enables to define a given prediction horizon N_p. The following cost functions take into account two separate criteria, the precision and the energy used by the system:

Criteria 1   Precision: square error minimizing

$$V(x(k),p) = \sum_{n=1}^{N_p} ||y(k+n,p) - y_r(k+n)||^2_{Q_1}$$

Criteria 2   Energy: power consumption minimizing / power total over a time horizon

$$V(x(k),p)\quad \sum_{n=1}^{N_p} ||i(k+n,p).u(k+n\quad 1,p)||^2_{Q_2}$$

y : system output $x(k)$: system state
y_r : system set point p : system input
i : electric current (A) Q_1, Q_2 : load factor
u : voltage (V)

The variables y, i and u are, in practice, estimated using the mathematical model of the process. In order to assess and validate this approach, a case study of a controller which minimizes this shape of cost function is covered in the following section.

CASE STUDY: RUDDER CONTROL

With a bottom up approach, the first step to obtain an energy aware control is to optimize the rudder motions (Figure 1: Autopilot control loops). In fact, the performance of a sailboat is highly dependent on how the rudder is controlled in order to move with minimum disturbances (wind shift, unnecessary rudder movements, jolts, etc.). So, before going more into the details of the heading control, it appears important to optimize the rudder movements in order to go into the lowest possible power mode, while also fulfilling precision, robustness and reactivity criterions. Let's introduce a case study of a model predictive energy aware control, the rudder control.

System model

As described in the previous section, the controller is based on a mathematical description of the process. The system model consists of two subsets (Figure 4): an electric actuator model (the electric ram) and a load model (the opposing force to the ram motion). It should be noted that the controller can be designed without this second model, considering the load as a system disturbance.

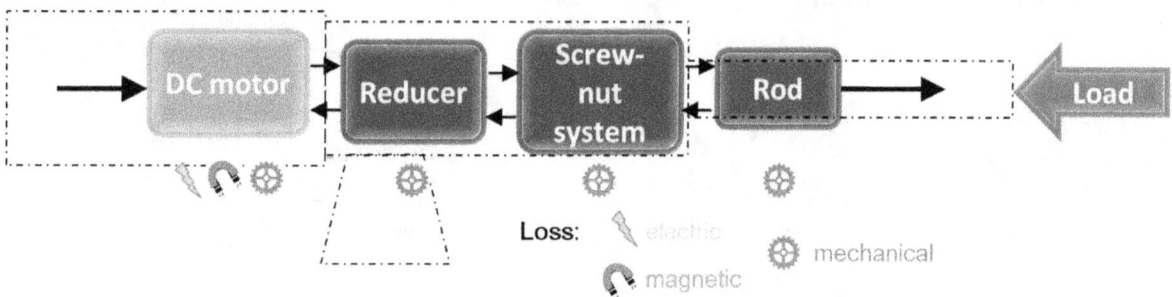

Figure 4 : System model

Electric actuator model

The electric actuator is composed of four subsets (Figure 4):

a DC motor (Direct Current motor) which is characterized by its inertia J_m, torque constant K_c, EMF (Electromagnetic Field) constant K_e, inductance L, resistance R, friction coefficient b "Aung (2007)"

$\Rightarrow$ $\begin{cases} J_m\ddot{\theta}_m = -b\dot{\theta}_m + K_c i - C_{res} \\ \quad Li + Ri = -K_e\dot{\theta}_m + u \end{cases}$ $\dot{\theta}_m$ rotation speed, i electric current, C_{res} resistive torque, u voltage

a reducer modelled by its reduction ratio r, efficiency N_r

$$\Rightarrow \quad \begin{cases} \dot{\theta}_r = r\dot{\theta}_m \\ C_{res\_out} = \dfrac{r}{N_r} C_{res\_in} \end{cases} \qquad \dot{\theta}_r \text{ rotation speed, } C_{res} \text{ resistive torque}$$

a screw nut system which is represented by the screw inertia J_v, screw thread p and efficiency N_v

$$\Rightarrow \quad \begin{cases} \vartheta = \dfrac{p}{2\pi}\dot{\theta}_r \\ C_{res} = J_v\ddot{\theta}_r + \dfrac{p}{2\pi N_v} F_{res} \end{cases} \qquad \vartheta \text{ linear speed, } F_{res} \text{ resistive force}$$

a rod characterized by its mass M_c

$$\Rightarrow \quad F_{res} = M_c\gamma + F_{ext} \qquad \gamma \text{ linear acceleration, } F_{ext} \text{ resistive force}$$

The accumulation of mechanical slacks in the linkage are represented by a slack on the lifting motion of the rod. It is modelled by a hysteresis which is only applied for a certain range of force and motion conditions. These conditions can boost or reduce even erase the mechanical slacks.

Moreover, the system is irreversible. That means only the motor can act on the rod of the actuator, a load force on the rod alone cannot cause a motion. This irreversible nature is represented thanks to torques' adjustments.

Load model

The load model enables to estimate the mechanical power that the actuator will need to generate to move the rudder instead of experiencing this load as a disturbance. This mechanical power is directly related to the electric power consumed by the motor. By predicting the needed effort, the controller can anticipate the motion, it enables to improve the energy performance of the control significantly. This load model is composed of three sub functions:

a function of heel correction

When the boat is heeling, as shown in the left part of Figure 5, the "neutral" angle which enables the boat to go straight ahead varies. In Figure 5 you can see that the boat has a constant heading with a rudder angle which oscillates around 6 degrees. It strikes a balance with a rudder angle different from 0 degree. This variation greatly affects the load calculation. So, this function enables the determination of the influence of the heel on the load calculation and to apply a correction offset.

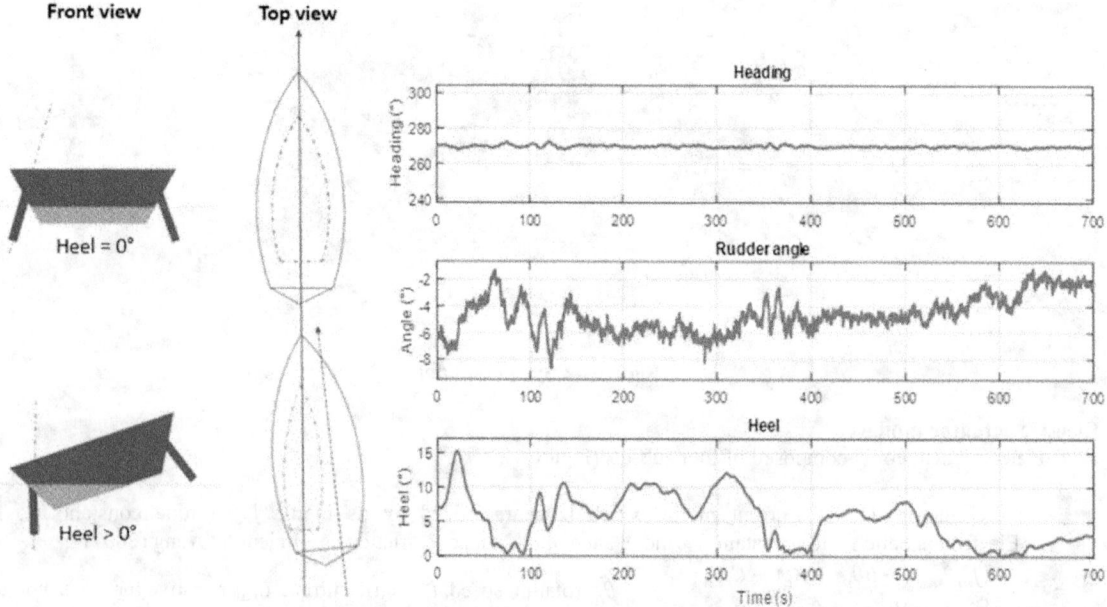

Figure 5: "Neutral" angle illustration

a function of load calculation

The function of load calculation is mainly a computation of hydrodynamic forces. The load related to the added mass is considered to be negligible. The thin foil theory is behind this computation.

$$\vec{F_{hydr}} = \vec{F_p} + \vec{F_t} \quad \text{with} \quad \begin{cases} F_p = \frac{1}{2}\rho V^2 S C_p \\ F_t = \frac{1}{2}\rho V^2 S C_t \end{cases}$$

The lift C_p and drag C_t coefficients are defined as follows:

$$\begin{cases} C_p = k_\lambda \alpha \\ C_t = \frac{C_p^2}{\pi \lambda e} \end{cases} \quad \text{with} \quad k_\lambda = \frac{2\pi}{1+\frac{2}{\lambda}} \quad \text{and} \quad \lambda = \frac{E^2}{S}$$

F_{hydro}: hydrodynamic force (N) λ: aspect ratio
F_p: lift force (N) k_λ: proportional coefficient
F_t: drag force (N) α: angle of attack (rad)
ρ: water density (kg/m³) V: speed of the incident stream (m/s)
S: Surface (m²)   e: planform efficiency factor (Oswald coefficient)
E: span (m)

In this model, the angle of attack α and the speed V of the incident stream depend on the boat speed, the rudder angle and the rotation speed of the rudder.

a function of force transmission

This function enables to calculate the force that acts on the actuator rod from the moment applied on the rudder's arm and the kinematic chain between these two points. It is based on solid mechanics. Figure 6 shows the rudder mechanism. The load is applied at points O'' or O''' of the rudder and transmitted to the electric actuator via the connecting bars. The vector "Force" represents the image of this load applied at point A (the point where the actuator is fixed).

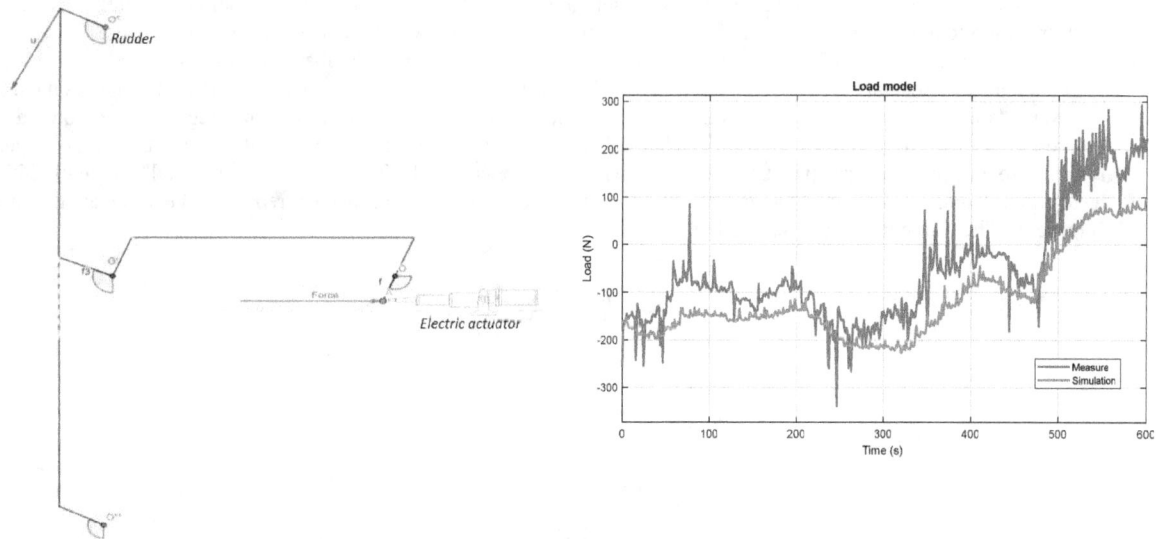

Figure 6: Rudder mechanism   **Figure 7: Results of the load model**

This load model is validated thanks to the instrumentation of a sailboat. As shown in the graph above (Figure 7), the load simulation from the boat's log enables a precision of +/ 20%. The actual load corresponds to a load cell measurement installed on the boat. This precision is sufficient because the load dynamic of our model corresponds to the reality. The margin of error is treated as a disturbance. This model could be improved by adding a sea state representation.

Movements representation

Moreover, all angular movements of the rudder are converted into linear movements of the rod (Figure 8) using the following function:

$$d = \sqrt{l^2 + L^2 - 2lL\cos(\theta + \gamma_0)}$$

$$\text{with} \qquad \gamma_0 = \cos^{-1}\left(\frac{l^2 + L^2 - d_0^2}{2l}\right)$$

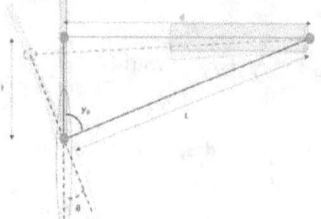

Figure 8: Angular/linear conversion

In fact, the control is carried out on the linear position of the rod and not on the angular position of the rudder to evade the non linearities related to the boat installation.

So, the mathematical model for the controller design is stated. This controller is implemented into a digital system; therefore, it is necessary to discretize the model to proceed to the controller design. The selection of the discretization step must take into account the dynamic of the system to control as the selection of the controller settings.

Settings

Two tuning descriptions are detailed in this section: the prediction horizon and the control frequency. They enable the controller to properly understand the system behavior and to not miss any important event. The mis set of these parameters can change the dynamics of control completely and cause system instabilities.

The tuning of the prediction horizon

A too short horizon can lead to a system instability. To ensure stability, there is a need to respect a minimum horizon or to reduce the state vector. However, the latter solution means a suboptimal control to initial requirements. On the contrary, a too long horizon can lead to an overload of the device which won't met the real time conditions. To offset this problem, reduced dimensional parametrization methods can conciliate a low number of degrees of freedom and a long prediction horizon that may be necessary for stability "Lewis, et al., (2012)". In order to not oversize the prediction horizon to ensure stability and to best adapt to every boat, frequency studies made for each sailboat category enable to guide the horizon selection. These studies show a correlation between the appropriate horizon length and the boat's inertia ie. its response to an event. For instance, an IMOCA (60ft) needs a longer prediction horizon than a "small" trimaran (24ft) to have an optimal control. In fact, the graph shows that the IMOCA has lower range of frequency due to its greater inertia (Figure 9). The trimaran is more responsive.

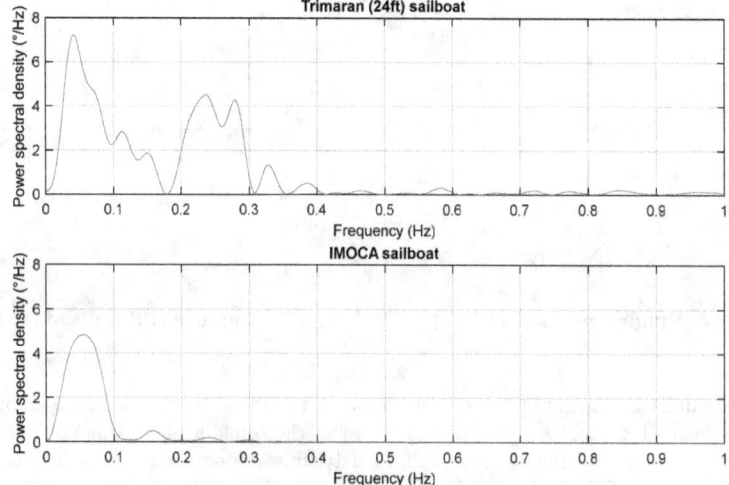

Figure 9: Power spectral density

The tuning of the prediction/control frequency

The right choice of prediction/control frequency enables not to miss any system event and to respond accordingly. In this case study, the events relating to the electric behavior are the most critical because they have the highest dynamics. So, this frequency must be adjusted according to the electric time constant of the motor in order to be able to track the current peaks. This time constant is defined as:

$$\tau_{elec} = \frac{L}{R} \qquad L \text{ motor inductance, } R \text{ motor resistance}$$

So, the prediction/control frequency must be of the same order of magnitude as this electric time constant depending on the motor characteristics.

Control results

In order to test the control method, the performance of various controller tunings is compared in a realistic simulation environment (Figure 10). As a reminder, the objective is to design and test a rudder control. The simulation environment is based on the models which are described in the first part of this case study.

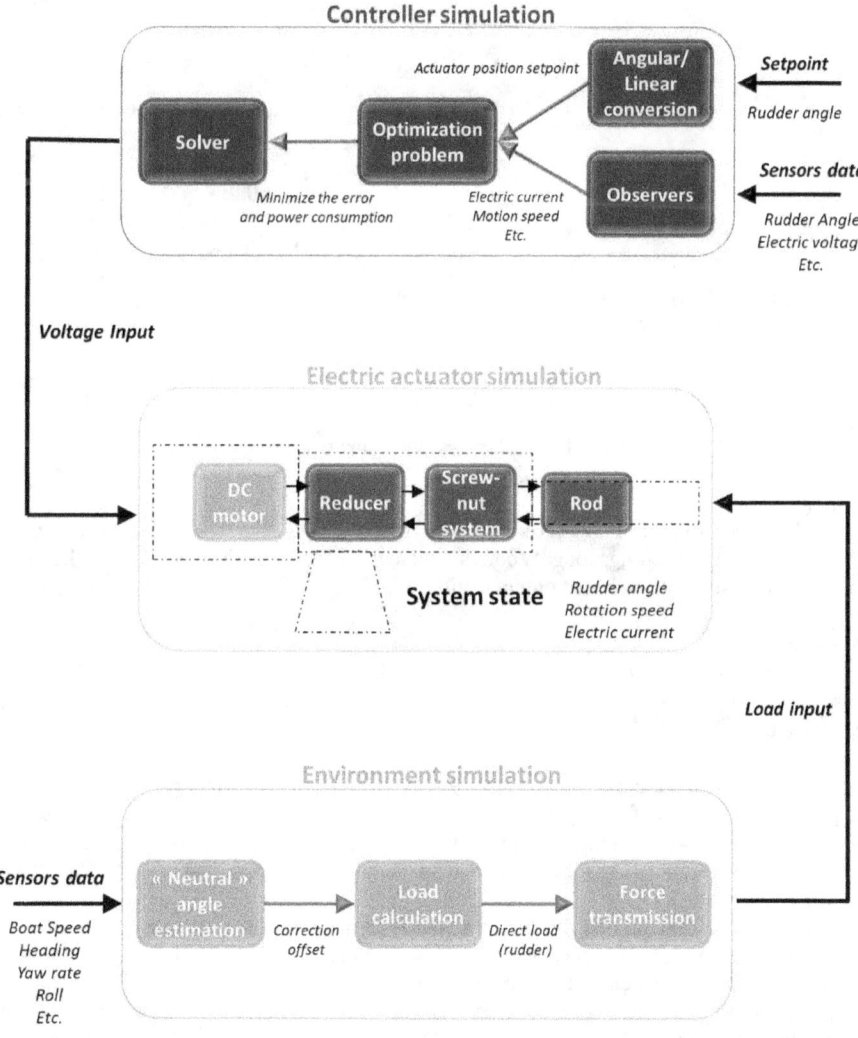

Figure 10: Simulation environment

The simulation inputs are derived from log coming from real navigations. Series of simulation of navigation (5min) are run with various load factor values Q_1, Q_2. The Q_1 load factor corresponds to the importance which is given to the precision and the Q_2 load factor to the power consumption. As a reminder, the cost function to minimize is defined as:

$$V(x(k),p) = \sum_{n=1}^{N_p} ||y(k+n,p) - y_r(k+n)||^2_{Q_1} + \sum_{n=1}^{N_p} ||i(k+n,p).u(k+n \quad 1,p)||^2_{Q_2}$$

y : system output
y_r : system set point
i : electric current (A)
u : voltage (V)

$x(k)$: system state
p : system input
Q_1, Q_2 : load factor

Tuning results

The Figure 11 shows the results for various prediction horizons and frequencies. Changing the load factor enables to draw a Pareto Front for each tuning. A prediction horizon sets to 0.05s and a frequency to 400Hz seems to give the best performance results for this sailboat. As underlined in the table, this tuning gives the best precision for about the same energy consumption.

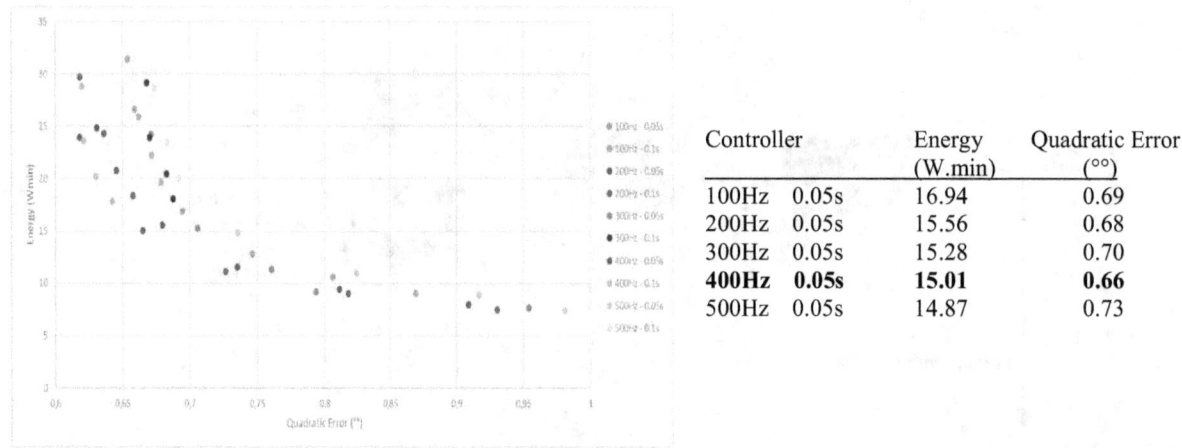

Controller		Energy (W.min)	Quadratic Error (°°)
100Hz	0.05s	16.94	0.69
200Hz	0.05s	15.56	0.68
300Hz	0.05s	15.28	0.70
400Hz	**0.05s**	**15.01**	**0.66**
500Hz	0.05s	14.87	0.73

Figure 11: Tuning results - Pareto Front

MPC control results

The settings of the MPC controller are now fixed. We can compare its results with one control law coming from the state of the art, a PI controller (). The PI settings roughly correspond to those which are implemented in the standard autopilots. To obtain these data, we make 4 simulations based on navigations log:

PI control simulation
MPC control simulation $(Q_1 > Q_2)$
MPC control simulation $(Q_1 \approx Q_2)$
MPC control simulation $(Q_1 < Q_2)$

The $(Q_1 > Q_2)$ simulation shows a better precision level than the PI control. However, a great precision leads to an increase of the energy consumption. In fact, it moves the rudder in a more reactive way. For about the same power consumption than the PI, the $(Q_1 \approx Q_2)$ simulation still gives a better precision (+40%). Finally, in the $(Q_1 < Q_2)$ simulation, the MPC control enables a better precision than the PI control with a reduced power consumption (15%). It moves the rudder into the lowest possible power mode.

Based on its knowledge of the system, the MPC control does more than driving the rudder from a point A to a point B, it uses the best way with the most appropriate speed and acceleration which reduce the energy consumption. In the case of a lack of power on board, the skipper (or the embedded system) can change online the load factor that means moving a cursor corresponding to the power he wants to allocate to the steering. The limits of the cursor are defined in advance in order to prevent a system divergence. To do so, the feasibility of the cost function minimization is checked at each bound.

All the conventional feedback design techniques are not compared with the MPC control in this study for lack of time. Some comparisons can be found in the state of the art (ex: "McGookin, et al., (2008)"). This study approach aims to test a new control method which is not used in the sailing area but largely known in aviation, ever closer to our activities. In fact, the aviation area concerns the same actuators (rudder, flap, elevator, etc.) and the same attitude control issues than the hydrofoiling control. These results are encouraging. The MPC control puts a physical problem in mathematical form thanks to one equation. Other conventional methods would address this issue in many parts as velocity gain scheduling, integrator anti windup, dead bands, etc.

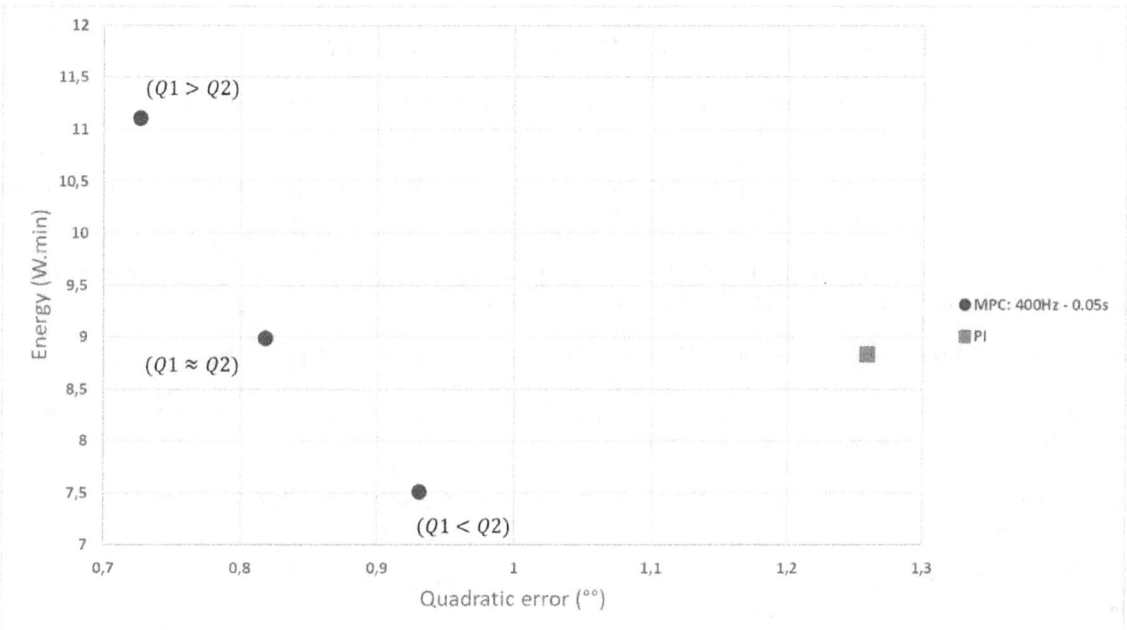

Figure 12: MPC / PI results

CONCLUSION

Our main goal is to design an energy aware autopilot for a heading and foil dynamic control. To answer this problem, in this paper, we propose a new control method which is aware of the energy consumption. This control method is based on a MPC controller which minimizes a cost function taking into account two criterions: the square error and the energy consumption. It makes a prediction of the system behavior from an electric actuator model and a load model. This control method is then applied on a rudder control. In fact, this control loop is the first step to completely manage the autopilot system and its energy. For the same energy consumption than a PI controller, we increase by 40% the precision level and we are able to reduce the consumption by at least 20%. This controller optimizes the use of the energy onboard to steer the boat. It can be tuned online to reduce the power consumption according to the battery level. Now, we have to develop a hydrofoil control law which uses this method to manage the actuators. The next step, in the medium term, will be the development of a flight stabilization system which is aware of the available energy on board. It will be tested in simulation as this first control method before being implemented in a real hydrofoil sailboat.

REFERENCES

Alfi A., Shokrzadeh A., Asadi M., "Reliability analysis of H infinity control for a container ship in way point tracking", Applied Ocean Research, Vol. 52, 2015.

Ang K. H., Chong G., Li Y., "PID control system analysis, design, and technology", IEEE Transactions on Control Systems Technology, Vol. 13, Issue 4, pp. 559 576, 2005.

Aung W. P., "Analysis on Modeling and Simulink of DC Motor and its Driving System Used for Wheeled Mobile Robot", World Academy of Science, Engineering and Technology, Vol. 32, pp. 299 306, 2007.

Briere Y., "Sailing robot performance: maximum speed tracking vs energy efficiency", Field Robotics 14th International Conference on Climbing and Walking Robots and the Support Technologies for Mobile Machines, Paris, France, 2011.

Douguet R., "A New Real Time Method for Sailboat Performance Estimation based on Leeway Modeling", SNAME 21st CSYS, Annapolis, MD, 2013.

Lewis F.L., Vrabie D., Syrmos V.L., "Optimal control", Hoboken, New Jersey : John Wiley & Sons, Inc., 2012.

Li Z., Sun J., "Disturbance Compensating Model Predictive Control With Application to Ship Heading Control", IEEE Transactions on Control Systems Technology, Vol. 20, Issue 1, pp. 257 265, 2012.

McGookin E. W., Murray Smith D. J., Li Y. Fossen, "Ship steering control system optimisation using genetic algorithms", Control Engineering Practice, Vol. 8, Issue 4, pp. 429 443, 2000.

McGookin M., Anderson D., McGookin E., "Application of MPC and sliding mode control to IFAC benchmark models", UKACC International Conference on Control., Manchester, UK, 2008.

Roberts G. N., "Trends in marine control systems", Annual Reviews in Control, Vol. 32, Issue 2, pp.263 269, 2008.

Shi Y., Shen C., Fang H., Li H., "Advanced Control in Marine Mechatronic Systems: A survey", IEEE/ASME Transactions on Mechatronics, Vol. 22, Issue 3, pp. 1121 1131, 2017.

Tiano A., Zirilli A., Yang, C., Xiao C., "A neural autopilot for sailing yachts", 9th MED '01 Proceedings, pp. 27 29, Dubrovnik, Croatia, 2001.

Tomera M., "Nonlinear Controller Design of a Ship Autopilot", International Journal of Applied Mathematics and Computer Science, Vol. 20, Issue 2, pp. 271 280, 2010.

Velagic J., Vukic Z. et Omerdic E., "Adaptive fuzzy ship autopilot for track keeping", Control Engineering Practice, Vol. 11, Issue 4, pp. 433 443, 2003.

Zimmermann H J., "Fuzzy Set Theory and Its Applications", New York : Springer Science+Business Media, 1996.

THE 23RD CHESAPEAKE SAILING YACHT SYMPOSIUM
ANNAPOLIS, MARYLAND, MARCH 2019

Sailboat Routing with Multiple Objectives for Sailing Races

Goulven Guillou, Lab-STICC, University of Brest, France
Laurent Lemarchand, Lab-STICC, University of Brest, France
Jean-Philippe Babau, Lab-STICC, University of Brest, France

ABSTRACT

Sailboat routing consists in computing the best route for a sailboat taking into account the characteristics of the vessel and environmental data such as weather forecast. In the context of sailing races, the best route computation is usually based on the isochrone algorithm, a sub-optimal solution to optimize the time to destination (TtD) criterion by computing a route as a sequence of waypoints. In this paper, we propose to compute a set of possible routes by considering two criteria : the time to destination and the *stress*. The time to destination is evaluated according to weather forecast and boat polar diagrams. The stress function is a combination of human and environmental factors. The set of possible routes is then obtained by using an iterative multiple objective optimization algorithm. Isochrone algorithm is used for initializing the set of routes. Then mutation operators are used to explore alternative solutions. Applied to realistic test cases, our search strategy allows to obtain routes with very different characteristics in terms of time to destination and stress values, asserted by experimented sailors. Concerning the main objective of minimizing time to destination, we obtain very similar results by comparison with commercial softwares such as MaxSea or Adrena.

NOTATION

TWA	True Wind Angle ($°$)
TWS	True Wind Speed (kt)
TWD	True Wind Direction ($°$)
AWS	Apparent Wind Speed (kt)
VMG	Velocity Made Good (kt)
OSA	Optimal Sailing Angle ($°$)
TtD	Time to Destination (h)

INTRODUCTION

Sailboat routing considering weather forecast and the characteristics of the boat, such as polar diagrams, is common in a racing context (Philpott and Mason, 2015). The objective is to find an optimal route in terms of time to destination. Nowadays most commercial navigation softwares provide a routing tool based on the isochrone algorithm (Hagiwara, 1989) or by finding the shortest path in dynamic graphs (Walther *et al.*, 2016). These approaches suffer from two limitations. They consider static models of boat and environment and they consider only one objective (minimizing the time to destination).

Tools integrate variability of the system by characterizing the boat (configuration of water ballasts, configuration of the appendages, etc), the environment (sea state, weather forecast) through the setting of parameters and lookup tables. Actually, the setting may be far from actual conditions, leading to non optimal routes. Moreover, routing optimization requires to consider different objectives. For instance, in the context of maritime transport and motor vessels, the literature proposes to analyze trade-offs between time to destination and fuel consumption, one crucial economic aspect. For sailing races, the degree of fatigue, and consequently the morale are very important criteria for the capacity of the skipper or the crew to manage the boat. So, minimizing crew fatigue and boat stress while maintaining a speed goal helps to maximize the final performance of the boat.

In this paper, for offshore sailing races, we propose to add a new optimization objective called *stress*. The idea is to integrate mechanical and human factors. These aspects are related to the weather and the sea conditions as well as the type of boat and the duration of the race. In our approach, the stress objective function computes a weighted combination of the mechanical and human factors like the type of maneuver, the wind speed and the point of sail. Considering these two criteria (time to destination and stress), Multiple Objective Optimization (MOO) frameworks allow to obtain a set of solutions proposing different trade-offs between the two optimization criteria.

First section introduces the necessary concepts considered in this paper to solve the sailboat routing problem. Related work is then presented. The following section details our approach, by presenting the stress criterion computation and the search process mechanism, based on the PAES algorithm. Solutions for relevant test cases are computed and analyzed in the experiments section. At last, conclusion and future works are discussed.

BACKGROUND AND RELATED WORK

We present in this section the notions related to multiple objective optimization and the related work.

Problem formalization

The goal of this work is to compute a *best* route or a set of *best routes* from a starting point w_s to an arrival point w_a. A route is defined by a sequence w of n waypoints $\{w_s, w_1, w_2, ..., w_{n-1}, w_a\}$. We do not consider obstacles in this paper, but the approach can be adapted to take such constraints into account. The routing starts at time t_0.

The input data of the problem are the weather forecast F and the polar diagram P of the boat. F is defined for a sufficient period of time, over the time interval $[t_0, t_0 + l]$ where l is a bound of the duration of the race.

The evaluation of a route is computed over an objective vector $f = < f_1, ..., f_k >$, of size k. We consider here $k = 2$ objectives. The time to destination is one of them (f_t), computed from (w, t_0, P, F), without possible waiting time at waypoints. A *stress* function f_s is also considered, as a combination of different factors (see the Approach Section).

The solution space to explore is defined by all feasible routes with a time to destination less than l, excluding the routes containing segments w_i, w_j starting at time t at w_i with a null speed (going against the wind, or entering into a no-wind area).

Multiple Objective Optimization

When considering a MOO problem, a single solution optimizing all of the objectives values $< f_t, f_s >$ simultaneously rarely exists. Let $f : S \rightarrow Z$ be an objective function mapping solutions $s \in S$, the search space, to the objective space Z, with $Z = \mathbb{R}^m$. MOO algorithms look for solutions $s \in S$ such that $z = f(s)$ is optimized (in the sequel, we consider minimization). $z \in \mathbb{R}^m$ is a point of the objective space Z, with each of the $z_i = f_i(s), i \in [1..m]$, being one of the objective function values to be minimized. Many approaches rely on the *dominance* concept to choose, among a set of solutions S, those ones that represent the best trade-offs of objectives within the search space. We say that a solution $s \in S$ *dominates* another one $s' \in S$ iff $\forall i \in [1..m], f_i(s) \leq f_i(s')$ and $\exists i \in [1..m], f_i(s) < f_i(s')$. It is denoted as $s \succ s'$. Namely, s is as good as s' on all objectives, and better than s' for at least one of them. Solutions that are not dominated by any member of S are *efficient* solutions, and constitute the *Pareto set* $\mathcal{PS}$. The set of points $z \in Z$ corresponding to the efficient solutions is the *Pareto front* $\mathcal{PF} = \{z \in \mathbb{R}^m | \exists s \in S, f(s) = z \text{ and } \nexists s' \in S, s' \succ s\}$. In the sequel, we say for short that a point $z \in Z$ dominates $z' \in Z$ if $z = f(s), z' = f(s'), s \succ s'$.

The goal of MOO algorithms is generally to determine or approximate points of $\mathcal{PF}$ and associated solutions in $\mathcal{PS}$. Basic MOO algorithms combine different objectives into a single one by an aggregation method (weighted sum of objectives). Finally, these approaches provide one solution, and so one specific trade-off, among those included in the Pareto front. A very common class of MOO algorithms are EMOA (Evolutionary MO algorithms), that mimic natural process such as genetic selection for explorating the search space. Popular EMOA frameworks are NSGA2 (Deb *et al.*, 2000), MOGA, SPEA2 (Zitzler *et al.*, 2002), PAES (Knowles and Corne, 2000), etc. These methods are adapted to a particular problem by defining objective functions and exploration operators, that transform a solution into another within its neighborhood. They maintain a population of non dominated solutions, while trying to improve accuracy and diversity of this population over the generations.

Related Work

Work in the field of multiple objectives routing concerns mainly motor vessels. The considered optimization objectives are the time to destination (TtD) and economic objectives like fuel consumption. They are based on graph theory, specific methods such as the isochrone method (for the TtD objective), or heuristics such as EMOA (see survey by Walther *et al.*, 2016) for multiple objectives.

Many methods are based on graph traversal. They used techniques for finding the shortest path considering dynamic graphs. They eventually combine them with methods for multiple criteria (Gandibleux, 2006). These methods require the building of a graph covering the routing area and one or more labeling functions for the edges of the graph, depending on time and environmental data. By discretizing the search space and using the results of the graph theory, the approach can eventually lead to exact methods to compute the Pareto set (with some restrictions about the dynamic graph properties). For instance, (Veneti *et al.*, 2015) proposes a graph labeling algorithm to optimize both fuel consumption and risk. However, these approaches are based on a global evaluation of a route, computed as the sum of the values of the edges. This is not suitable for sailboat routes where the evaluation depends on relationships between the edges (e.g. heading changes).

A second class of methods uses a sequence of possible waypoints to represent a route. Waypoints are not restricted to nodes of a graph, leading to large solution space explored by heuristics. In the mono objective case, the isochrone algorithm is the most popular method. It is widely used in routing commercial softwares dedicated to sailboats. The isochrone algorithm was originally proposed by R.W. James in (James, 1957) for motor vessels. In this approach, the polar diagram depends of the sea state forecasts (the direction and the height of waves) and remains convex. In the convex case, Bijlsma in (Bijlsma, 1975) proved that the problem of minimal-time routing has a solution. Unfortunately, the sailboat polar diagrams are not convex and Hagiwara proposed in his PhD (Hagiwara, 1989) an isochrone-based approach to face this problem. The approach proposes different versions for distance or fuel objectives but remains mono objective. Furthermore, the method can be easily implemented on a computer.

In the case of multiple objectives, EMOA approaches are mainly used. Hinnenthal in (Hinnenthal, 2007), and Marie & Courteille in (Marie and Courteille, 2007) used a MOGA EMOA framework for routing motor vessels with time and fuel consumption objectives. Szlapczynska (Szlapczynska, 2007) based his approach on SPEA (Zitzler *et al.*, 2002) for the same objectives. An improved version was developed, with risk evaluation according to the maximal wind value encountered on a route. A decision phase is then performed for electing a route among the Pareto set (Szlapczynska and Smierzchalski, 2009). In (Veneti *et al.*, 2015), risk and fuel objectives are taken into account, with a search based algorithm onto the NSGA2 framework. Waypoints are here restricted to nodes of a grid. This choice allows to elaborate a cross-over, which mixes routes that intersect at some point of the grid. The approach is based on the exploration of the search space starting with time-oriented solutions obtained with an effective isochrone algorithm. All of these EMOA methods show the effectiveness of MOO for the motor vessels routing domain. As compared to previous EMOA approaches for multiple criteria routing, e.g. MEWRA algorithm (Szlapczynska and Smierzchalski, 2009), our method does not require a crossover operator. A crossover operator, which mixes partial routes, is not well suited to our routing problem because of the existence of no-go zones due to wind direction or strength. Our search is realized using local transformations (mutation operators) applied to the waypoints.

Other approaches mix previously mentioned methods, or target other objectives than TtD or fuel consumption. (Böttner, 2007) transforms the MOO problem into a single objective one by aggregation (weighting objectives within the fitness function). He uses a second technique based onto multi-weight shortest path algorithm for graphs for near coast routing of motor vessels. In (Hinnenthal and Clauss, 2010), a set of forecasts is given as input (with a main one), and the robustness is measured as the number of weather forecast variants when the route remains feasible according to load and other vessel constraints. In (Philpott and Mason, 2015), robustness is also considered, using a set of weather scenarios and associated probabilities, with a route built according to a shortest path search into a stochastic graph.

In other domains, a lot of research works show the interest of using MOO to integrate impact of human and machine fatigue. We give here an overview of some representative ones. In design domain, (Brintrup, 2008) considers qualitative (designer perception) and quantitative criteria (posture quality) as optimization objectives to explore and evaluate different solutions. The use of optimization genetic algorithms is here interesting for an iterative evaluation of several strategies, calculated from a "good" strategy. This work illustrates the necessity of considering qualitative and quantitative criteria as optimization objectives when taking a complex design decision. In automation domain, physical fatigue of workers may be used as a specific optimization objective (Ma, 2009). This work considers human fatigue when optimizing posture to save performance along the time for a stressing work. Recent work in automotive domain (Fang, 2015) considers vehicle fatigue to improve performance in presence of uncertainty. The use of multi-objective optimization technique helps the designer on determining the best compromise, improving both performance and robustness.

In the domain of inshore racing and match racing (Tagliaferri *et al.*, 2017, 2014) propose to integrate a notion of risk to improve the probability to arrive before the competing boat. This work shows that the space discretization and the consideration of other criteria than TtD are relevant to improve the routing strategy. As an example, the race leader prefers to stand between the trailers and the finish line even if another route seems better, because of the risk that the wind turns unfavorably is too important. In this specific domain, the race area and the duration of the race are limited. These conditions make that the material and human fatigue factors can be ignored. In the domain of offshore racing, the race conditions make that the integration of risk of damages is a key factor to improve the chance of finishing and win. For instance, in the recent Route du Rhum 2018, five boats out of six, engaged in the Ultime class, had major mechanical problems related to weather conditions. Two had to retire and the one who did not have a mechanical problem won the race. For offshore racing, the skipper has, especially if he is alone on board, to find a good trade-off between speed and mechanical and human fatigue.

In the domain of offshore racing, even if some of the above presented methods can be used to avoid bad weather phenomenons, there is no work, to our knowledge, that considers the mechanical and human factors as an optimization objective. In commercial softwares some mechanical or human aspects can be integrated in a certain manner in the computation of the best route. For example the level of performance of the boat can be defined through a percentage of its polar or the loss of time related to a maneuver can be given in terms of minutes. However these settings are directly used in the evaluation of the isochrones and are not an objective by themselves.

APPROACH

In this paper, we propose an approach to compute different routes representing different trade-offs between time to destination and stress. A route is viewed as a sequence of waypoints. The PAES EMOA framework (Knowles and Corne, 2000) is used as a search engine for the exploration. We present first the implementation of isochrone algorithm, used to initialize the search algorithm. Then our original stress function is detailed. To finish, we present the mutation and fixing operators and the overall algorithm, embedding PAES algorithmic structure.

Objective functions

Isochrone implementation In order to obtain good candidates in terms of minimal TtD, the isochrone method is used for the MOO algorithm initialization.

The isochrone algorithm assumes that the knowledge of the polar diagram of the boat and the weather forecast for the whole duration of the trip. From the starting point, at a given time, the maximum reachable distances in any direction over the chosen time step are determined thanks to the polar diagrams and the weather forecast. The set of the associated reachable points defines the first isochrone and the process is repeated from each point of the current isochrone until the last one oversteps the arrival point of the route. To avoid the problem of the combinatorial explosion, the computed trajectories need to be pruned. Several techniques have been proposed but all are not convenient for sailboats. For example, deleting the routes for which the last point is at the same time in the wake of the last point of another route may be not suitable. Sometimes, a sailing boat has to make a large detour to avoid some weather phenomenons like a high pressure center. In our implementation, we adapt the Hagiwara's idea of using sectors by decreasing their width as one moves away from the starting point in order to keep distances possibly reachable by a sailboat between two neighboring sectors. In a given sector, all the points of the last computed isochrone standing behind the nearest of the arrival point are deleted with the associated routes.

Finally, the last segment of the routes can be problematic. The route can bring the boat very close to the final arrival point but the last meters can be unfeasible. To avoid to discard these routes, the TtD evaluation is done by considering that the skipper has to tack. The remaining distance is the sum of the length of each tack:

$$\frac{d \sin \alpha}{\sin(2\gamma)} + \frac{d \sin \theta}{\sin(2\gamma)}$$

where d is the distance between the two last points, $\gamma = |OSA|$ is the Optimal Sailing Angle (the true wind angle which maximizes the VMG which is itself the projection of the boat velocity in the upwind direction), $\theta = \gamma - |TWD - bearing|$ (*bearing* is the bearing between the two last points) and $\alpha = 2\gamma - \theta$.

Our algorithm has been tested for several starting and arrival points and several weather forecasts and compared to two commercial navigation softwares: MaxSea Time Zero (MaxSea, 2018) and Adrena (Adrena, 2018). Figure 1 shows the set of the generated routes and isochrones with our algorithm for a sailing off the Bay of Biscay starting on Monday, October 5th 2015 at 18h55mn44s UTC. The starting (resp. arrival) point is located up right (resp. down left) in the corner of the figure. The boat is a *Grand Surprise* and the weather forecast file is generated thanks to the Global Forecast System (GFS) model and provides three hourly forecast for 10 days with an approximate horizontal resolution of 60 nautical miles. The least-time route is displayed in bold.

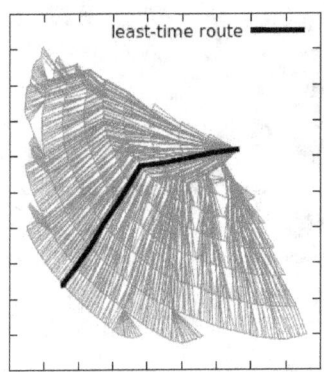
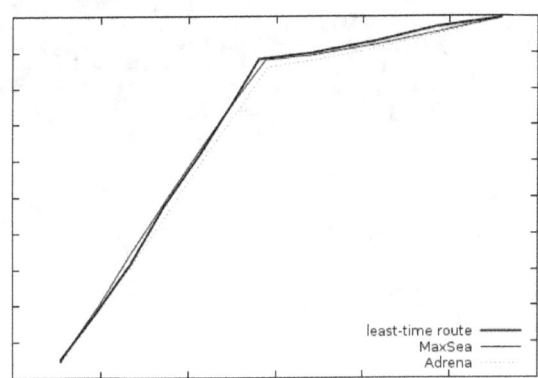

Figure 1: Routes and isochrones (left) and the least-time route compared to those found by MaxSea and Adrena (right)

The same routing with the same time step between two consecutive isochrones has been performed with MaxSea Time Zero and Adrena by paying attention to disabling the options allowing to take into account the currents or the waves. Figure 1 (right) shows that the least-time routes are very similar. The duration time is 1 day 6 h 45 mn with MaxSea, 1 day 6 h 12 mn with Adrena and 1 day 6 h 6 mn with our algorithm. The differences may be explained by the interpretation of weather files and polar diagrams. For our implementation, we choose to perform a trilinear interpolation for the weather forecast (interpolation in time, in direction and in speed) and a bilinear one for the polar diagram (interpolation in TWA and in TWS). Commercial implementations are not public. However as the three tracks look very similar and the durations are close, we consider that our isochrone implementation is relevant for the routing purpose.

Stress Offshore sailboats have become much more difficult to manage. For example the multiplication of mechanical parts like canting keels, canting and turning masts, daggerboards or foils increases the risk of breakage and makes the maneuvers longer and more difficult. In particular sailing on a extreme racing boat for a single-handed sailor has become a true challenge. Actually, an IMOCA 60 has about 300 m^2 and upwind sail area of 300 m^2 and a downwind one of 500 m^2 whereas an ultime multihull like MACIF has an upwind sail area of 430 m^2 and a downwind one of 650 m^2. As a consequence the boats have become more and more hard to sail and the stress, and consequently the morale, take a more and more important part in the final result of a race.

The beginning of the Route du Rhum 2010 is very illustrative about that concern. In the Ultime class (maxi multihulls) Francis Joyon (Idec) chose a south route, longer (see Figure 2 the bold track) but faster and more comfortable because downwind whereas Thomas Coville (Sodebo) chose a north route, shorter but more difficult because against a strong apparent wind. Thereafter Francis Joyon unfortunately had to jibe several times to catch a stronger wind but despite this he finally finished ahead Thomas Coville. Sodebo and Idec were quasi sisterships. Conversely the outcome of the race was the opposite for IMOCA 60, the boats in the south have finished behind those in the north. To deal with variability, the competitors use different tool settings and different sources of weather forecasts. When two different routes seem comparable in terms of TtD, they choose one rather than the other according to criteria such as comfort, risk, fatigue or number of maneuvers. This is the purpose of the stress function. The stress function is used to model the trade-off between TdD and other criteria. It is based on a linear combination of terms related to mechanical and human fatigue. The main interest of a linear combination is to explicit the relative weight of the different elements. The weights can be adapted to a specific boat or to abilities of a given skipper. The approach facilitates the consideration of new aspect of the problem (for example the spinnaker hoist is not considered in this paper).

The stress evaluation between two considered points of a route is as follows:

$$S = \alpha_t * tack * (1 + 1/\Delta T) + \alpha_j * jibe(1 + 1/\Delta T) + (\alpha_{ld} * ldw + \alpha_{su} * suw + \alpha_{ld} * lw + \alpha_{su} * sw) * \Delta t$$

whereas the stress function is evaluated by summing the different terms along the route. Since the wind changes both in space and time, the evaluations are performed at each waypoint of the route and at each instant corresponding to the time step of isochrones if the travel time between two consecutive waypoints exceeds this time step.

Repetitive maneuvers imply fatigue due to the sailor's operations (move all the mobile weights to the other side of the boat, ...). Tacking and jibing may lead to damages and takes a lot of the sailor energy. To model the impact of maneuvers,

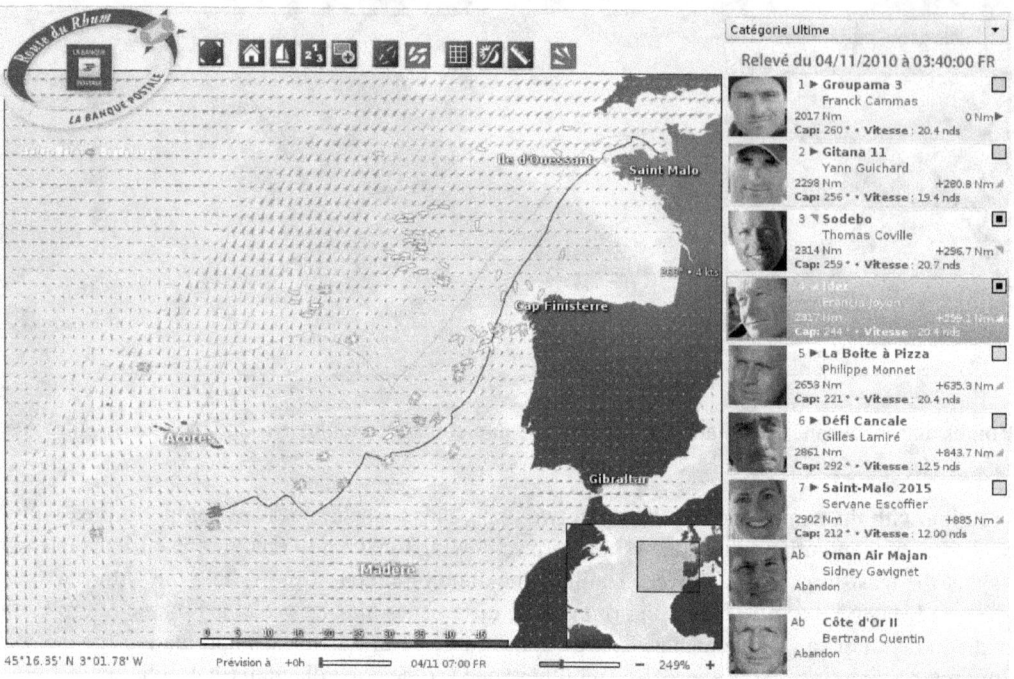

Figure 2: Route du Rhum 2010, Idec (bold path) vs Sodebo (light path). Geovoile screenshot.

tack and *jibe* indicate if respectively a tack or a jibe was performed since the previous segment, the values is 0 or 1. The weight of a maneuver includes a constant part and a part inversely proportional to the time ΔT since the previous maneuver. Indeed, maneuvering at short time intervals is considered as tiring. α_t and α_j allow to distinguish the costs of a tack and a jibe respectively. Generally a jibe costs more energy than a tack. More generally the α coefficients represent the weight of each term of the sum S.

ldw and *suw* take into account that sailing downwind in a light wind (*ldw*) or upwind in a strong one (*suw*) is generally risked and penalizing. Actually, the weather forecast models trend to sub evaluate the speed of the wind in strong wind and over evaluate it in light wind and sailing in these weather conditions requires more efforts for the skipper. *ldw* and *suw* (like for *sw* and *lw*) depend on AWS and thresholds. For example for *suw*, if AWS is less than the associated threshold, *suw* is zero else *suw* is proportional to the difference between AWS and the threshold. Δt is the travel time required for the considered portion of the route.

Although storms and centers of high pressures are theoretically navigable in terms of polar diagrams (the target speed is greater than zero), in practice these two cases require a boat in perfect condition and a good physical health. Then *sw* (for strong wind) and *lw* (for light wind) represent this aspect. The different coefficients α and the thresholds depend on the type of the boat and of the ability of the skipper and must be set by an expert (often by the sailor himself). As previously mentioned, the weight of a jibe is much more important than the weight of a tack just because of the risk of breakage due to the trajectory of the beam.

While in commercial softwares the loss of performances related to maneuvers and boat behavior is directly included in the computation of isochrones, in our approach, the stress function remains an objective function in its own right. The stress objective is then used to find interesting trade-offs between performance and stress.

Search Operators

The search operators transform a solution, represented as a sequence of waypoints $\{w_s, w_1, w_2, ..., w_{n-1}, w_a\}$ (see Section Related Work) into a slightly different one (in the neighborhood of the search space). We propose two operators. The mutation operator modifies (mutates) one waypoint into another one. However, the obtained route can be an unfeasible solution (null speed between 2 waypoints). Therefore, a fixing operator is proposed to make the solution feasible again.

Mutation the mutation operator proceeds as follows: it first decides randomly the type of mutation to apply: (a) delete a waypoint, (b) add a waypoint or (c) modify a waypoint. Deleting a waypoint is straightforward. Adding a waypoint consists in choosing randomly a waypoint w_i of the route (except the final one w_d), computing the time t spent going from w_i to w_{i+1}, choosing randomly a deviation angle a between $-30°$ and $30°$ and adding a waypoint w_k between w_i and w_{i+1}, going for a time period of $t/2$ to the direction $(w_i \to w_{i+1}) + a$. Modifying a waypoint is made in the same way, replacing w_{i+1} by the location reached after a time t going to the direction $(w_i \to w_{i+1}) + a$. These operations are illustrated in Figure 3 (a). After mutation, the modified route is checked for feasibility and quality.

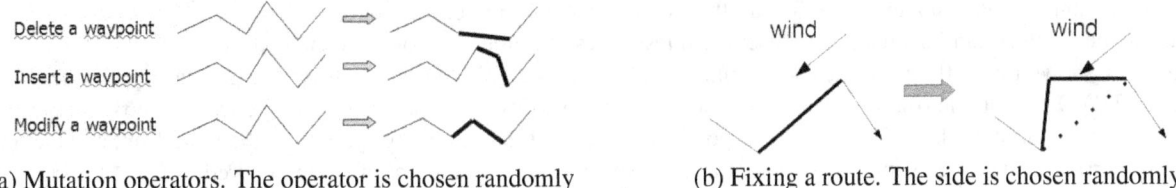

(a) Mutation operators. The operator is chosen randomly   (b) Fixing a route. The side is chosen randomly

Figure 3: Route operators

Fixing in some cases, 90% of the mutation attempts can lead to a zero-speed segment (w_i, w_{i+1}). To face this problem, the fixing operator works as follows, as shown in Figure 3 (b): a new point w_k, located at $45°$ left or right of direction $(w_i \to w_{i+1})$ and distance $dst(w_i, w_{i+1})/\sqrt{2}$ of w_i is inserted between w_i and w_{i+1}. If the segment (w_i, w_k) remains unfeasible (in the sense that the speed is still 0), w_k is computed at $-45°$ opposite left or right of direction $(w_i \to w_{i+1})$. If it is still unfeasible, the repair procedure is abandoned. If successful, the new segment (w_k, w_{i+1}) is also checked in the same way, eventually leading to the insertion of a supplementary waypoint. The repair procedure ends when w_d is reached.

Overall MOO Algorithm

The MOO algorithm implementation is based on the PAES framework (Knowles and Corne, 2000). The generic algorithm principles are the following:

- It starts with an initial solution s generated randomly or computed by an ad hoc algorithm, and an empty archive A (of limited size). s is added to the archive. Eventually A can be filled with non dominated solutions obtained by ad-hoc heuristics.

- At each iteration, the current solution s is mutated into s'. s' is evaluated. If s' is dominated by s ($s \succ s'$), s' is discarded. If $s' \succ s$, s and elements of A dominated by s' are discarded. s' becomes the current solution s. Last, if neither s or s' dominates the other, s' is added to A, replacing eventually a solution in A if A is full. Replacement policy of PAES aims at providing solutions well spread over the objective space, i.e., in our case, representing various trade-offs between TtD and stress. The next current solution is either s or s', depending on the replacement policy.

- The algorithm stops after a fixed number of iterations.

Contrary to standard EMOA methods as NSGAII, PAES does not require a crossover operator. Crossover usually mixes 2 partial solutions into a single one. In our case, it should join partial routes, with the drawback that connecting waypoints from very different routes can lead to coming backward, or going through a no-wind area. We have customized PAES as follows.

Initialization The initial solution s_0 is the isochrone route. In order to avoid a biased search to TtD objective function, to favor least-time route search only, we fill the archive A (of size m) during this initial step, while checking dominance properties. We generate m random solutions according to 3 techniques: (a) we compute the TtD-oriented isochrone solution, and mutate it repetitively to produce a new solution (20% of the archive); (b) we generate randomly a solution with approximately the same number of waypoints as s (60% of the archive); (c) we mutate the direct route $(w_s \to w_d)$ (20% of the archive). In all cases, unfeasible and dominated routes are not included into A. The current solution s is set to s_0.

Mutation We attempt to mutate the current solution 20 times per iteration, trying to fix it if it is unfeasible after mutation. If the 20 trials fail to produce a feasible solution, the current solution is mutated again at next generation.

EXPERIMENTS

For illustrating our approach, a North Atlantic race on an IMOCA 60 has been chosen with a classical wind scenario. The boat sails upwind to the southwest when a warm front passes over. On Sunday, April 9th 2016, over the considered area, the wind is blowing from WNW 20 knots, then it is backing 12 hours later to SSW by decreasing up to 14 knots for finally slowly veering to W during the next 24 hours while increasing up to 28 knots. This case is very different of that presented in Figure 1: another boat, another route, another day and year, thus another weather forecast.

With the same weather and boat data, we present two test cases, differentiated only by the precise location of the arrival point. Figure 4 shows two different routes with some wind indications plotted with barbs, one for each test case, both optimized for TtD. The first one (on the left) is much longer but more downwind and the second one (on the right) is more direct and shorter but close-hauled. The latter corresponds to a classical racing strategy: going to the wind front and tacking when the front arrives. One interesting aspect is that a slight shift of the arrival point of some hundred of meters brings out one route rather than the other with a less than 10 minutes time difference. In the following, the first test case is called "west" whereas the second test case is called "direct" and is the quickest in term of time.

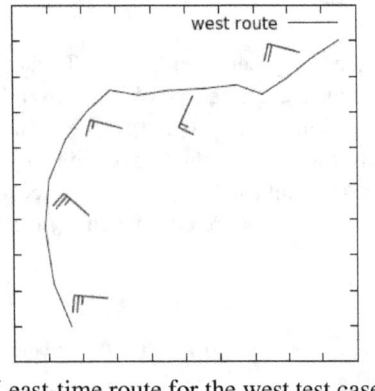
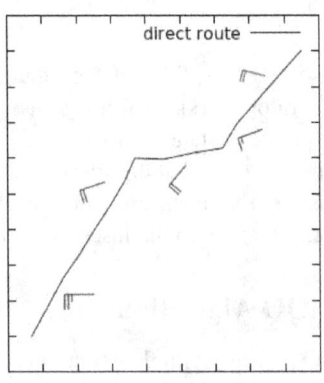

Least-time route for the west test case   Least-time route for the direct test case

Figure 4: West and direct routes, the starting point is up right and the arrival one is bottom left

The corresponding Pareto fronts are depicted in Figure 5 and give the best trade-offs between performance and stress. The annotations indicate the correspondences with the routes of Figure 6. The points close to the vertical axis correspond to fast routes whereas those close to the horizontal axis correspond to stress-less routes. Therefore, an ideal route, if it exists, stands near the origin of coordinates. These results have been obtained using the following parameters for our EMOA algorithm: it ran for 4,000 iterations, with an archive size of 100. The time step for isochrone computation was of 3 hours, and the angle step of 5°. The execution time is around 3 seconds on a Linux virtual machine (with one i7-5000 Intel CPU core at 2.7GHz and 2GB of RAM of dedicated resources).

In both cases, as expected, the least-time route is or is almost the most stressful. However the least-time route computed by isochrone algorithm does not belong to the front, it is dominated by faster routes for a handful of seconds ! Indeed the mutation which consists of deleting a waypoint replaces two segments by a single. When this mutation is valid, a faster route, compared to the route computed by the isochrone method, can be found. In the same way, the mutation which consists of modifying a waypoint may lead to angles not reachable by the isochrone method. In both cases, the algorithm is able to get a slightly better route in terms of time to destination optimization and also some routes very close in terms of stress and time. The obtained solutions present some interesting trade-offs.

For the west test case, the Pareto front presents three groups of routes, one around 42 h, another around 47 h, and last around 56 h. The last group proposes too slow routes for a race. The second group proposes less stressful solutions (the stress is divided by almost 3). The fastest route of this group is the westerly route (the route number 1 on the left side of Figure 6). The other routes (like the route number 2 in the same figure) are a variant of the least-time route. Analyzing the context, the most westerly route appears less stressful because the boat sails always more downwind. For the fastest routes of the first group, the route number 2 is the stress-less with almost identical duration than the least-time route. According to the morale of the crew and the state of the boat, the skipper can then choose the adapted route.

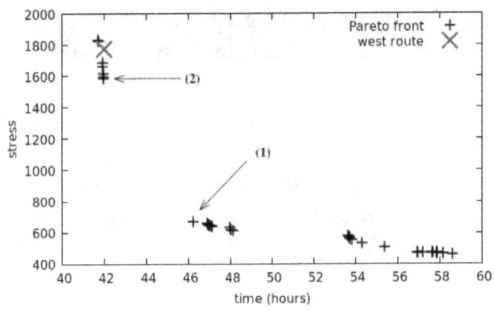

Pareto front for the west test case

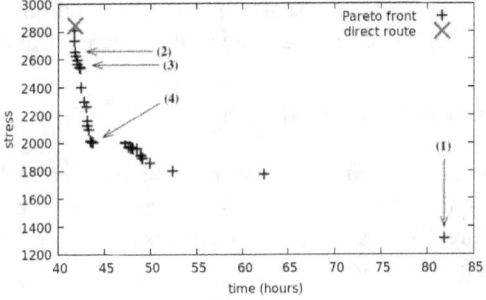

Pareto front for the direct test case

Figure 5: Pareto fronts

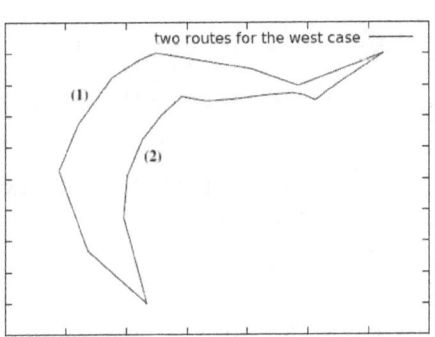

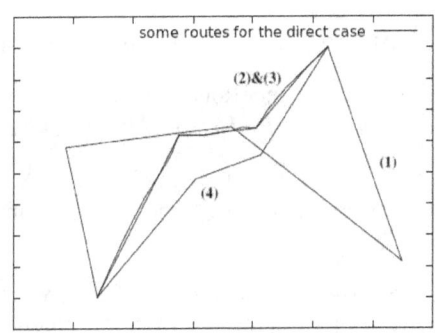

Figure 6: Some good trade-off routes for the west (left) and direct (right) test cases

For the direct test case (see Figure 5 on the right), the Pareto front is more compact. Four routes are displayed on the right side of Figure 6. The zigzag track (numbered 1) corresponds to the slowest route (81.84h) and is sailed downwind. The two close routes (2 and 3) correspond to the least-time and show the kind of variants proposed by the optimization algorithm. The fourth route is a compromise between a level of stress of 2000 for a time of 43.16 h. It starts more downwind, therefore with less stress, and seems keeping ongoing more upwind along the two last stages but the wind direction is slightly more favorable than on the least-time route (because both the location and the time are different) and allows to stand longer out of the upwind range.

Furthermore, the algorithm compares favorably to the isochrone algorithm in terms of time to destination optimization, thanks to the discretized search process and the efficiency of the exploration operator. As we have seen, the execution time required on legacy hardware is in few seconds for a quick exploration (and around 5 minutes for a deep exploration of the solution space with 400,000 iterations). This allows an on-board application of the algorithm.

CONCLUSION

This paper proposes an Evolutionary Multiple Objective Optimization Algorithm to determine Pareto fronts in the domain of sailboat routing. The algorithm computes a large set of routes providing the best trade-offs between time to destination and stress. The stress function models the crew fatigue due to tack and jibe maneuvers, the wind strength and the wind direction. Stress parameters are tunable to consider different boat/skipper pairs.

Application to two realistic test cases shows the ability of the algorithm to detect small strategic changes leading to routes with different combinations of time/stress characteristics. For instance, a small degradation in time to destination may save the crew a lot of effort.

For future works, we plan to extend the stress function by adding other maneuvers like hoisting a spinnaker, changing a sail or heading changes. We also plan to improve robustness by exploring different arrival point definitions, stress functions and weights, sailing configurations and weather forecasts. We are also thinking about other mutation operators in order to be able for example to avoid storm area by slowing down the boat. The idea is to propose alternative routes (risky, balanced, easy) to the skipper considering possible weather evolutions and stress functions. Having different routes with different associated trade-offs is very convenient for the skipper if he can easily exploit them. For this purpose a graphical user interface in which the sailor could select a route and visualize the associated road-book with the associated maneuvers should be proposed.

ACKNOWLEDGEMENTS

Thanks to Nicolas Lunven double winner of the Solitaire du Figaro and navigator during the Volvo Ocean Race 2017-18 on the boat Turn The Tide of Plastic for his enlightenments on how to use weather forecast and navigation software for routing. Thanks to Emilien Lavigne, a Mer Concept team member (Ultime boat MACIF), the François Gabart's team, for the routings with Adrena.

REFERENCES

ADRENA http://www.adrena.fr. Visited nov. 2018

S.J. BIJLSMA *On minimal-time ship routing*. The Netherlands, Technical University Deft, Dissertation, 1975

C.U. BOTTNER Weather Routing for Ships in Degraded Condition. In: *International Symposium on Maritime Safety, Security and Environmental Protection.* Athens, Greece, 2007

A. BRINTRUP, J. RAMSDEN, H. TAKAGI, A. TIWARI Ergonomic Chair Design by Fusing Qualitative and Quantitative Criteria Using Interactive Genetic Algorithms In: *IEEE transactions on evolutionary computation.* 12 (2008), Nr. 3, S. 343–354

K. DEB, S.AGRAWAL, A.PRATAP, T. MEYARIVAN A Fast Elitist Non-Dominated Sorting Genetic Algorithm for Multi-Objective Optimization: NSGA-II. *Parallel Problem Solving from Nature PPSN VI.* Berlin : Springer, 2000, S. 849–858

J. FANG, Y. GAO, G. SUN, C. XU, Q. LI Multiobjective robust design optimization of fatigue life for a truck cab In: *Reliability Engineering and System Safety.* 135 (2008), S. 1–8

X. GANDIBLEUX *Multiple criteria optimization: state of the art annotated bibliographic surveys.* Bd. 52. Springer Science & Business Media, 2006

H. HAGIWARA *Weather routing of (sail-assisted) motor vessels.* The Netherlands, Technical University Deft, Dissertation, 1989

J. HINNENTHAL *Robust Pareto - Optimum Routing of Ships Utilizing Deterministic and Ensemble Weather Forecasts.* Germany, Technical University Berlin, Dissertation, 2007

J. HINNENTHAL, G. CLAUSS Robust Pareto-optimum routing of ships utilising deterministic and ensemble weather forecasts. In: *Ships and Offshore Structures* 5 (2010), Nr. 2, S. 105–114

R.W. JAMES Application of wave forecast to marine navigation. In: *US Navy Hydrographic Office* (1957)

J.D. KNOWLES, D.W. CORNE, Approximating the nondominated front using the pareto archived evolution strategy. In: *Evolutionary computation* 8 (2000), Nr. 2, S. 149–172

L. MA, W. ZHANG, D. CHABLAT, F. BENNIS, F. GUILLAUME Multi-Objective Optimisation Method for Posture Prediction and Analysis with Consideration of Fatigue Effect and its Application Case In: *Computers and Industriel Engineering.* 57 (2009), S. 1235–1246

S. MARIE, E. COURTEILLE Multi-Objective Optimization of Motor Vessel Route. In: *TransNav - International Journal on Marine Navigation and Safety of Sea Transportation* 3 (2007), Nr. 2, S. 133–141

MAXSEA https://mytimezero.com. Visited nov. 2018

A. PHILPOTT, A. MASON Optimising Yacht Routes under Uncertainty. In: *15th Chesapeake Yacht Symposium,* 2015

J. SZLAPCZYNSKA Multiobjective approach to Weather Routing. In: *TransNav - International Journal on Marine Navigation and Safety of Sea Transportation* 1 (2007), Nr. 3, S. 273–278

J. SZLAPCZYNSKA, R. SMIERZCHALSKI Multicriteria Optimisation in Weather Routing. In: *TransNav - International Journal on Marine Navigation and Safety of Sea Transportation* 3 (2009), Nr. 4, S. 393–400

A. VENETI, C. KONSTANTOPOULOS, G. PANTZIOU An evolutionary approach to multi-objective ship weather routing. In: *2015 6th International Conference on Information, Intelligence, Systems and Applications (IISA)*, July 2015, S. 1–6

F. TAGLIAFERRI, I.M. VIOLA, R.G.J. FLAY A real-time strategy-decision program for sailing yacht races. In: *Ocean Engineering* 134 (2017), S. 129–139

F. TAGLIAFERRI, A.B. PHILPOTT, I.M. VIOLA, R.G.J. FLAY On risk attitude and optimal yacht racing tactics. In: *Ocean Engineering* 90 (2014), S. 149–154

L. WALTHER, A. RIZVANOLLI, M. WENDEBOURG, C. JAHN Modeling and Optimization Algorithms in Ship Weather Routing. In: *International Journal of e-Navigation and Maritime Economy* 4 (2016), S. 31 – 45. – ISSN 2405-5352

E. ZITZLER, M. LAUMANNS, L. THIELE SPEA2: Improving the Strength Pareto Evolutionary Algorithm for Multiobjective Optimization. In *Evolutionary Methods for Design, Optimisation and Control with Application to Industrial Problems, proceedings of the EUROGEN2001 Conference* Athens, Greece, September 19-21, 2001.

Scoring Methods for Handicap Yacht Racing
Jim Teeters, Newport, Rhode Island

ABSTRACT

Two boats sailing in close proximity may inevitably compare their boat speeds, perhaps even "race" each other. Over the last few centuries there have been numerous handicap systems designed to estimate the performance of different boats relative to one another. Corollary to handicapping is another arcane art, that of scoring races. Scoring methods have both technical options, in part determined by handicap rules, as well as "human engineering" options in the sense that different solutions can work best for different constituencies, be they race organizers or sailors. Options may include single vs. multiple ratings, time on distance vs. time on time, pre/during/post race handicapping, attempts to predict the environmental conditions on the race course, constructed courses, pursuit vs. staggered vs. fleet start racing, and performance curve scoring. The underlying assumptions and motivations for these choices are presented along with the consequences of adopting them. The expectations of the competitors, and indeed their ability to intuitively grasp the fundamentals of how elapsed times are transformed into race rankings, are discussed with a view towards finding solutions that achieve a successful balance of fairness, transparency and acceptance.

NOTATION

CSYS	Chesapeake Sailing Yacht Symposium
Fn	Froude Number
SLR	Speed Length Ratio
TOD	Time on Distance
TOT	Time on Time
TCF	Time Correction Factor
SPM	Seconds per Mile
ET	Elapsed Time
CT	Corrected Time
PHRF	Performance Handicap Racing Fleet
PCS	Performance Curve Scoring
H0	Zero Hour GRIB Files and an Associated Scoring Method

INTRODUCTION

The methods available to scoring sailboat races are entirely dependent on the methods with which sailboats themselves are handicapped or rated. The two terms "handicap" and "rating" are often used interchangeably and may imply different concepts to different folks. In this paper both refer to the generation of time allowances that are applied to boats elapsed times to "allow" a slower to boat to compete against a faster one, hopefully with something close to competitive parity. This paper refers to handicap/rating systems only to the extent that they influence scoring options.

Early Approaches to Scoring

Length, the apparent size of a boat, was a reasonable approach to assigning time allowances in that there is some validity to the concept that "bigger is faster." In fact an effective boat length was used to generate handicaps. There is a technical basis for using length to determine boat speed.

William Froude established the concept of a formulation that bears his name (Froude Number) defined as:

$$Fn = Vel \div \sqrt{(Gravity \times Length)}$$

in which Vel represents the velocity or speed of a boat (feet per second in Standard units), Gravity is Newton's gravitational constant (32.17 feet/sec/sec) and Length is a linear fore and aft dimension that characterizes how long a boat is. Sailing waterline length is a primitive definition of length. As rating rules progressed many other features of a sail boat were expressed as modifiers to length, including sail area and stability(!), so that length ultimately represented the speed potential of a sailing boat.

The above equation can be manipulated to express potential speed (Vel):

$$Vel = Fn * \sqrt{(Gravity \times Length)}$$

For general use the Fn was re-cast as the more familiar speed length ratio SLR. Gravity is a constant, so is the factor that converts feet/per second to knots.

$$SLR = Vknots \div \sqrt{Length}$$

When a boat sails at a speed such that SLR = 1.34 the wave generated at the bow has a length equal to value of "length" in the SLR formula. This means that the second crest of that wave is aft of the first crest by the distance Length and there is a long trough in between. For the heavy displacement boats of past eras this SLR marked a speed "limit" that was difficult to exceed. As the second crest moves aft of the boat, the boat squats down in the trough, trimming down by the stern. The speed a boat has when its SLR equals 1.34 is termed the "hull speed" which, correctly or not, was used to define the maximum speed of a yacht.

So, if length became the rating parameter for a sailboat, how was that used for scoring? The following mathematics illustrates one such procedure:

$$Velocity\ (V) = Distance\ Sailed\ (D)/Elapsed\ Time\ (ET)$$

Therefore $ET = D/V$

And the elapsed time difference between two boats with respective speeds V1 and V2:

$$ET2 - ET1 = D * (\frac{1}{V2} - \frac{1}{V1})$$

If V = some fraction of hull speed f, then the elapsed time difference in sec/mile is:

$$\delta ET spm = \frac{3600}{f \times 1.34} \times (\frac{1}{\sqrt{L2}} - \frac{1}{\sqrt{L1}})$$

According to the Handbook on American Yacht Racing Rules (Parsons et al 1934):

$$SPM\ allowance = 2160 \times (\frac{1}{\sqrt{R2}} - \frac{1}{\sqrt{R1}})$$

Where L has been replaced with a rated length R.

This formula gives a time allowance between two boats of different speed potential as determined by their rated lengths. The time allowance is expressed as seconds per mile (SPM), a method known as "Time on Distance" scoring.

SINGLE NUMBER SCORING

As in the example above a single number, or rating, can used to characterize the performance of a sailboat. There are two general approaches to scoring with a single rating, time on distance and time on time.

Time on Distance Scoring (TOD)
Time allowances between boats are expressed as seconds owed per mile. To score a race it is necessary to pick one boat

as a common reference boat from which time allowances are calculated. Each boat is given its time allowance, a deduction to its elapsed time, to achieve a "corrected" time (CT). The boat with the lowest corrected time is the winner.

$$CT = ET - SPMallowance \times Distance$$

For the International Offshore Rule (IOR) the SPM allowance was calculated using the square root of rated length as explained above. A common contemporary rule, PHRF, typically also uses time-on-distance handicapping but with ratings expressed as SPM. If one boat rates 40 SPM and a slower one 60 SPM then the faster boat owes the slower one 20 SPM:

$$CT = ET - (60 - 40) \times Distance$$

TOD scoring is relatively easy to calculate after or during a race. All that is required is a common mark or finish line, the distance from the start to that mark, and the elapsed time of each boat at that mark. The choice of the reference boat, termed the "scratch" boat, makes no difference in the rankings of the boats. It simply shifts the corrected times to larger or smaller numbers depending on the choice of scratch boat. Common practice has been to use the fastest boat in a fleet as the scratch. It should be noted that the corrected time of a boat has been transformed into the "world" of the scratch boat. In the case of PHRF, the scratch boat has a rating, by definition, of zero.

Time on Time Scoring (TOT)

TOD scoring assumes that boat A always owes boat B a set time allowance per mile, say 10 SPM, for all races. An alternative concept, time on time (TOT), assumes that boat A is faster than boat B by a certain percentage of speed. If 3% then boat A is assumed to have a speed that is 1.03 times the speed of boat B. Ratings are then ratios of boat speeds. A scratch boat with some rated speed Vs is chosen. Any other boat with a speed Vb has a rating of Vb/Vs. If our boat B above is the scratch, then boat A has a rating of 1.03, often termed a Time Correction Factor (TCF.) The TCF of the scratch boat is, by definition, equal to 1.00 (Vb/Vb.)

It is also possible to establish TCFs without ever having a specific boat speed for the scratch boat or any other boat. One boat could be assigned a rating of 1.000 and other boats given ratings faster or slower by some logical method (such as converting PHRF TOD ratings to TOT with a few assumptions.) PHRF New England uses the following:

$$TCF = A/(B + PHRF)$$

PHRF is the PHRF TOD rating of a boat and A and B are constants. B is chosen such that the sum with PHRF gives a reasonable average sec/mile for the boats. B is course specific. A is used to scale the ratings to desirable values but has no effect on the rankings of the boats when scored. Note that if A is set to be equal to B + PHRF for a specific boat, then that boat has a TCF of 1.00 and can be considered a scratch boat.

TOT scoring is then calculated as:

$$CT = ET \times TCF$$

All that is required to score boats using TOT is their elapsed times at a mark (or finish.) The distance to that race is not required, certainly a simplification for race management.

TOD vs. TOT

Figure 1 shows a graph of SPM ratings, at wind speeds of 6, 8, 10, 12 and 16 knots, for each of two boats, the Kernan 65 PELIGROSO and the SNC70 HOLUA. These ratings, unlike as in PHRF, are actual predicted boat speeds produced by a velocity prediction program (VPP). PELIGROSO takes about 527 seconds to sail one mile in 10 knots. The course content may be Windward/Leeward (equal distances sailed upwind and downwind) or any other course of interest. (The course content is irrelevant for this discussion.) HOLUA takes about 580 seconds to sail one mile in 10 knots. So in 10 knots the faster boat PELIGROSO owes the slower HOLUA 53 SPM. Taking PELIGROSO as the scratch boat and the 10 knots ratings we can generate rating curves for HOLUA relative to PELIGROSO. Using the TOD assumption the dashed magenta curve is created by adding 53 SPM to the PELIGROSO curve at all wind speeds. To generate the TOT curve we use PELIGROSO as the scratch boat, calculate the ratio of boat speeds (3600/580)/(3600/527) to get a TCF for HOLUA of .909. Dividing 3600, the number of seconds in an hour, by the SPM speed prediction converts SPM to knots. An equivalent calculation is to simply take the ratio of 527/580 (scratch SPM divided by target boat SPM.) Since figure 1 is showing SPM, not knots, we generate the TOT curve by multiplying PELIGROSO's values by the inverse of .909, 1.101. The magenta dotted line is the result. The two magenta curves represent HOLUA's performance across the full range of wind speed, as assumed by TOD and TOD single number scoring.

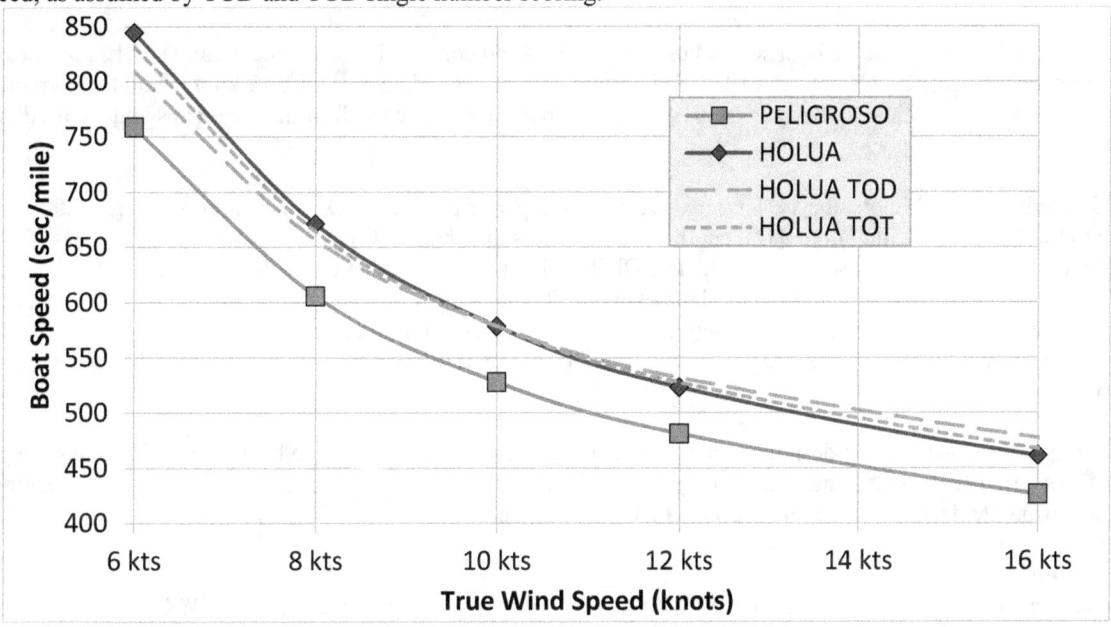

Figure 1 – TOD vs TOT

To compare TOD versus TOT we need to see how well the curves replicate the blue curve of "actual" speeds for HOLUA. It is visually obvious that the dotted TOT curve more closely follows that curve. To state the obvious: in a 10 knot wind either method is a completely accurate replication of the actual relative speeds of the boats because we used the 10 knot solutions to derive the ratings. For this particular set of predicted ratings for two boats TOT does a better job than TOD when the wind is lighter or heavier than the wind used to generate the ratings. In short, the assumption of being X% slower than the scratch boat is more accurate than the assumption of being Y SPM slower. This is borne out by the growing use of TOT scoring for numerous events. Is TOT always better? Certainly not if the wind dies away for a period in the middle of a race. The time allowances in TOT are based on elapsed time so while all the boats are sitting around not moving relative allowances are growing.

Figure 2 illustrates scoring for a J125 and a C&C35 in a two mile windward/leeward race.

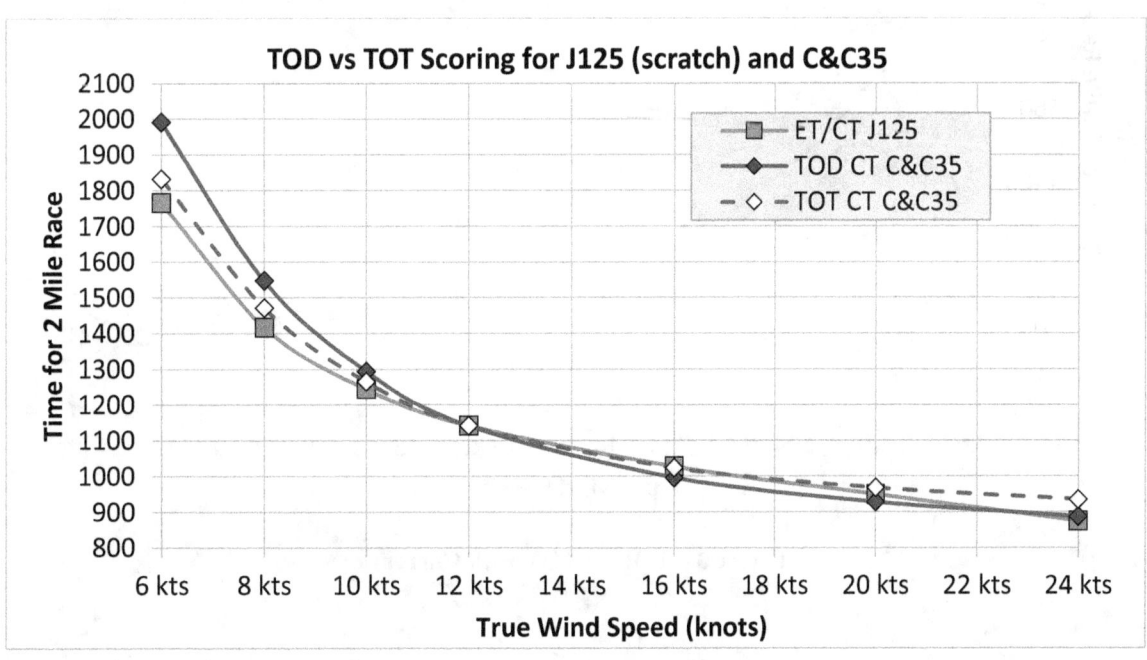

Figure 2 – TOD vs TOT for WW/LW

The J125 is chosen as the scratch boat so its elapsed time (ET) and corrected time are identical. The 12 knot boat speeds were used to create TOD and TOT ratings of 131 SPM and a TCF of .813. In a race with 12 knots wind the corrected times (CT) of the C&C35 are identical to the J125 because the same polar tables are used for the handicaps and for predicting elapsed times.

In wind speeds below 12 knots, the TOT ratings are clearly superior to TOD. TOD does not give sufficient time allowance to the C&C35. Neither does TOT but the "error" is less than half. TOT is also better at 16 knots but TOD scoring is equal to TOT at 20 and superior at 24. The ORR VPP has the J125 planing in 20 and 24 knots wind. The C&C35 is a relatively heavy boat that will not plane offwind. To be competitive with the J125 planing performance in strong winds the C&C35 is better served by using the larger SPM TOD rating established at 12 knots of wind. In planing conditions, at least for these two boats, the TOD assumption of constant differences is more accurate than the TOT assumption of constant ratios.

This scoring comparison of two designs with fundamentally different performance in heavy wind illustrates the difficulties of scoring such boats against each other with a single number rating. There are more sophisticated methods of scoring such disparate designs. These are presented later in this paper.

Current Effects

An interesting challenge for scoring is dealing with tides and currents. If the same J125 and C&C35 are raced on a windward leeward course, with one knot of current aligned with the wind against the boats when sailing upwind, the time allowances in Figure 3 result. Note that at 12 knots the C&C35 is no longer scored even with the J125. Because the J125 is faster upwind she is less affected by the current than is the C&C35. The race has been slowed down as well since the boats are slower sailing upwind than down and are therefore spending more time sailing against the current than sailing with the current.

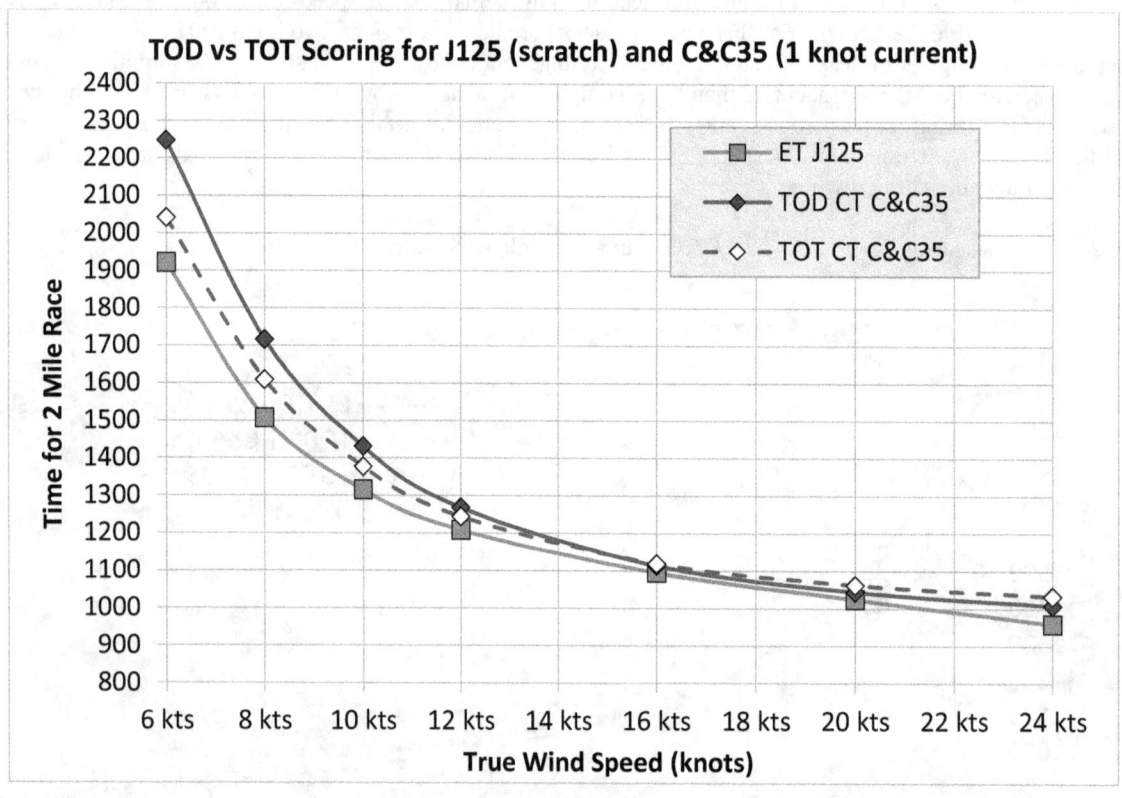

Figure 3 – TOD vs TOT with Current

TOT gives much more accurate time allowances below 16 knots with only a small loss of accuracy compared to TOD at larger wind speeds. It have become very common for racing locales with significant tidal currents, such as San Francisco Bay or the Solent in the UK, to use TOT scoring.

WIND AVERAGING

How much wind speed variation occurs during a race? Figure 4 shows wind data recorded by the Farr 43 PAX NZL during an American Yacht Club series. Diamonds show the true wind speed in knots, crosses the magnetic wind direction in degrees. The solid black curve is a running average of the wind speed. The record covers about 2 hours of racing. The average wind speed starts at 8, dips to 7 and ends around 8.5 knots. Of particular interest is the variation in wind speed. A statistical index of the variation of the wind speed is the standard deviation. By definition, 95.5% of the data in a record lies within +/- 2 standard deviations of the mean (average.) The standard deviation (as defined here as the deviation from the running average) is .76 knots. The graph shows curves above and below the running average, separated by 2 standard deviations, 1.52 knots. This analysis indicates, for at least one race, that the short term wind variation can have a standard deviation of about .75 knots. Because the mean wind speed itself varies, there is a greater spread of wind speeds throughout the duration of the race.

The wind direction has its own pattern of variation.

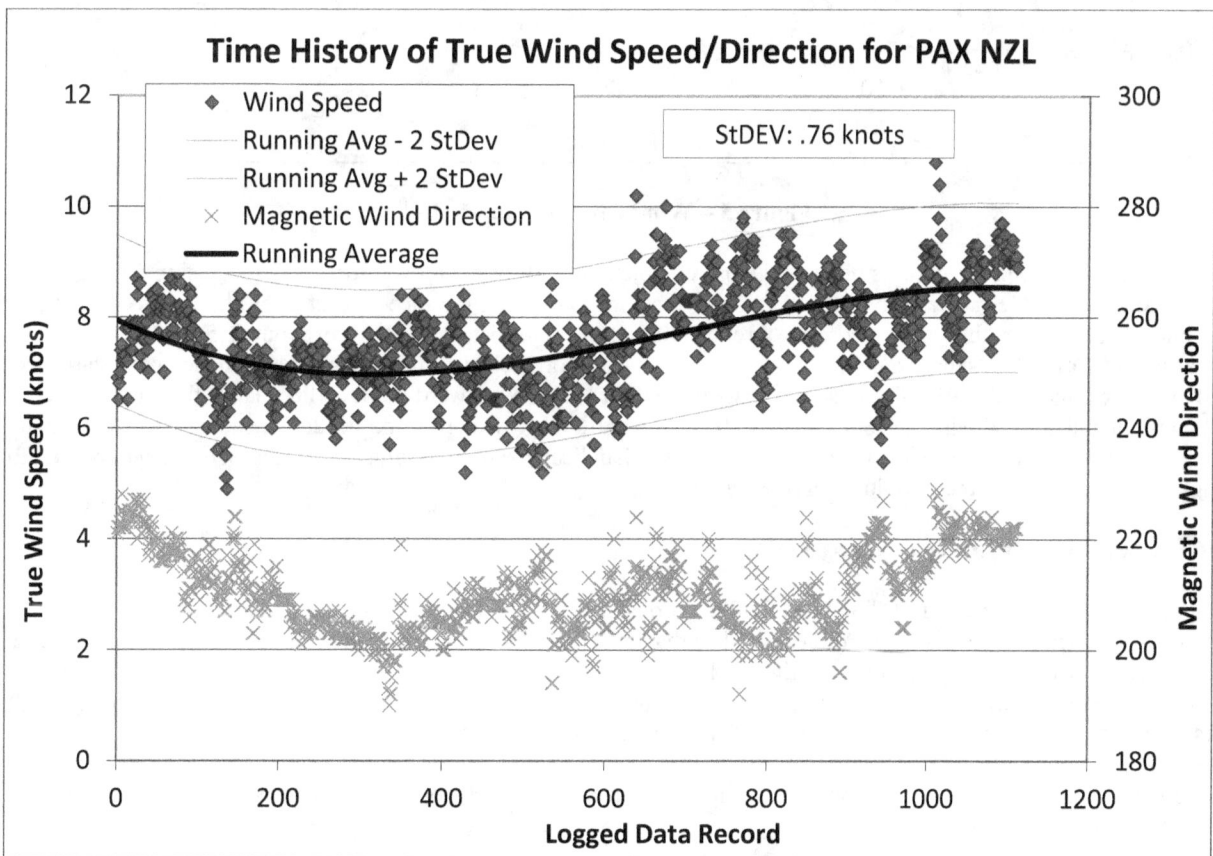

Figure 4 – Wind Averaging

So what are the implications of the wind variations on handicaps and scoring? A more accurate representation of what happens on the race course would include a distribution of VPP derived boat speed predictions centered around the nominal wind speed. How broad should the distribution be?

The two hour record presented above for Pax NZL would suggest the distribution in Figure 5 for a nominal wind speed of 12 knots and a standard deviation of .75 knots.

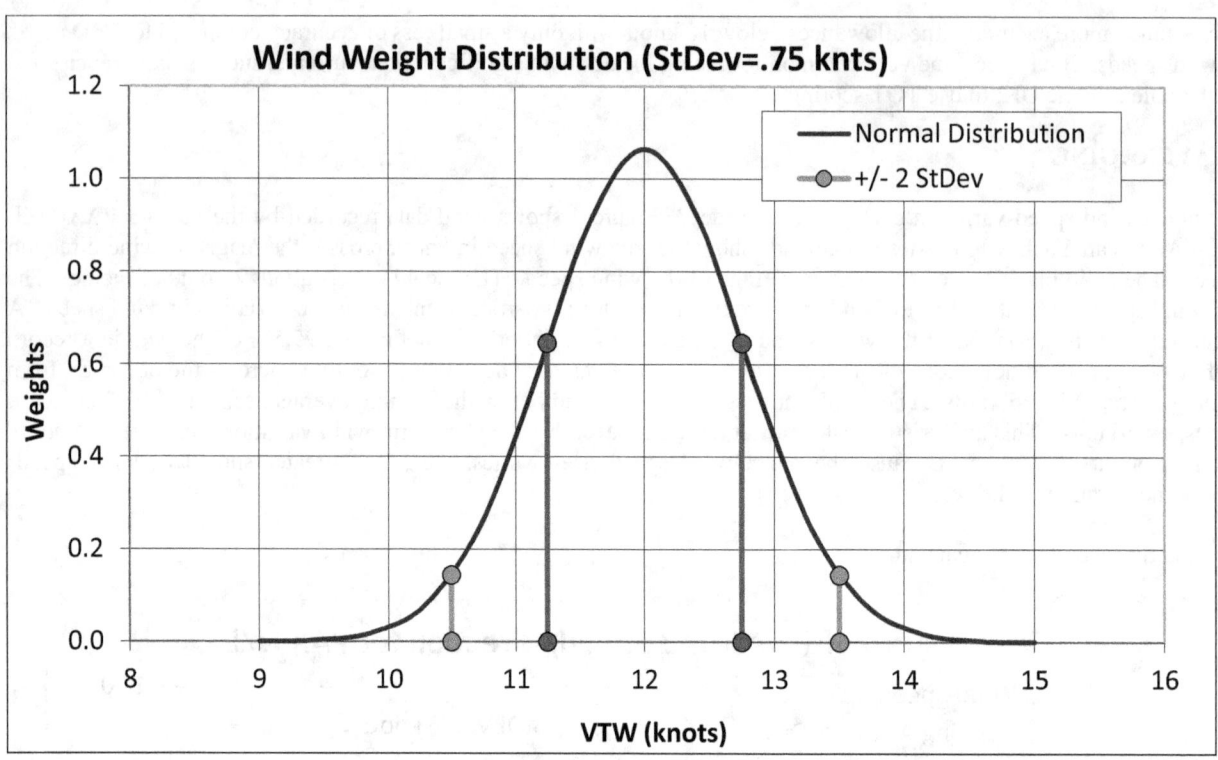

Figure 5 – Wind Distribution Model

The variation in wind direction is perhaps a more complex issue. For a windward/leeward race using dropped marks it can be assumed that the race committee will relocate those marks during the race to best align the course with the wind. The first 30 minutes of this race, records 1 through 300, show a persistent shift from about 225 to 200 degrees. The tactically astute, or lucky, sailor would leave the starting area on port tack, sail into this header then tack onto starboard and lifted up to the windward mark. The upshot is that boat will get to the windward mark earlier than if the wind direction had been constant and aligned with the course marks. If wind variations provide lifts and headers then boats will get around the race course faster than if the wind were steady. Perhaps handicaps should assume some degree of wind shifts and use ratings that are faster than they would otherwise be.

SINGLE VS. MULTIPLE RATING SYSTEMS

The comparison of TOD and TOT showed the limitations of single number scoring. While TOT generally does a better job than TOD when the wind is lighter or heavier than that wind for which the ratings are tuned, a better option is to use light and heavy wind ratings. This means a light or heavy wind race can be scored with a set of ratings specifically designed for light or heavy wind. Not only can the wind speed vary from what is built into the ratings, but so can the course content. Ratings can be focused on upwind, reaching, downwind or specific combinations.

While the justification for this, greater fairness in scoring races, seems obvious there are some issues raised. So what the pros and cons for using single or multiple ratings?

- In a single rating system all the boats know their rating, there is no confusion. It is simple.
- Multiple ratings provide greater fairness. The ratings are tuned to expected wind or course content.
- Single ratings can deter boats from showing up at the start. If the wind is very strong the time allowances may be too generous for the slow boats. Performance differences between fast and slow boats tend to get closer with the exception of boats that can really take off downwind. The same problem occurs when the wind is lighter. The slow boats are not getting as much time as they need from the faster boats. Why go racing if you can't win?
- OK, for single rating systems the luck of the weather favors and disfavors different boats. Another way of saying this is that they promote the syndrome "every dog has his day." There is an upside to this: more dogs win races during the season, the joy of winning is spread around. A boat that is not as skilled as another might continue to

participate because they know in certain weather conditions they can beat the better boat. Unfairness due to Mother Nature might actually lead to greater participation.

- If using multiple ratings, someone needs to make the decision before the race as to which rating will be used. Although the decision may be based on weather forecasting services, the decision maker is vulnerable to second guessing by the competitors, particularly those that would have been helped by a different choice.
- A really important issue is the expectations of the competitors. A class of professional sailors may want absolute fairness so that the best sailed boat wins. A group of less committed amateurs who are just having fun may appreciate having a bit of luck in the results.

Some conclusions may be drawn from the above:

- Professional programs sailing under a single number rule would be best served if the boats were similar to each other. They would all be effected somewhat the same as the wind and course content vary. I.e., a highly type-formed fleet should have good racing no matter what happens. One design racing is the ultimate expression of type-forming.
- A diverse fleet is likely better served with multiple rating scoring. The rating appropriate to the race conditions can be chosen ahead of the start and better represent boat speed differences. You would not want to use a rating that works for windward/leeward courses on a point to point distance racing that likely will have more reaching.
- If a single number rule is used for a single race then the expectation for close finishes is diminished.
- Conversely, a single number rule can be quite effective for a diverse fleet over the course of a season. More boats win races during the season, the best sailed boats should come out on top if enough races are run. Again, the key is to understand the expectations of the sailors.

Multiple Rating Systems: Custom Ratings
A multiple rating system, in particular one that uses a VPP to predict performance at all combinations of wind speed and wind angle, can create multiple single number solutions each customize to a specific race.

Let's look at some specific rating examples. Figure 6 shows a table of single number ratings from a 2019 ORR certificate. Each rating represents the expectation of what occurs during that race. The Chicago Mac Race and the fleet in Acapulco have various options they can choose before the start which gives them greater flexibility.

Custom Single Number Ratings					
Chicago Mac Upwind:	0.826	TCF	Puerto Vallarta:	0.812	TCF
Chicago Mac AP:	0.816	TCF	Cabo San Lucas:	0.811	TCF
Chicago Mac Offwind:	0.812	TCF	Acapulco WW/LW:	0.773	TCF
Bayview Mac:	0.814	TCF	Acapulco Random:	0.771	TCF
Marion Bermuda:	545.6	SpM	Acapulco WW/LW5:	0.769	TCF
Marion Bermuda:	0.829	TCF	Acapulco Stat Fit:	0.776	TCF
California Offshore:	0.813	TCF	San Francisco Bay:	0.834	TCF
Pacific Cup:	0.8107	TCF	Transpac:	0.7156	TCF

Figure 6 – Ratings Customized To Specific Events

The Transpac Yacht club has, through years of experience and analysis, developed a table of what wind speeds and wind angles they expect their fleet to experience sailing from Long Beach to Honolulu. Figure 7 shows the table used for the 2017 race. Wind speeds range from 6 to 24 knots, wind angles from best upwind, through reaching, to best downwind. Each entry is the probability of occurrence for each combination of speed and angle. Cells are shaded according to the value of the probability. The combination of 16 knots wind at 135 degree reaching has the highest value, .0899, or about 9%. The assumption is that 9% of the race distance will be sailed in that condition.

Heading	6	8	10	12	14	16	20	24	
00d	0.00030	0.00020	0.00050	0.00065	0.00120	0.00155	0.00070	0.00000	0.5%
52d	0.00020	0.00020	0.00050	0.00065	0.00120	0.00155	0.00070	0.00000	0.5%
60d	0.00000	0.00080	0.00200	0.00260	0.00480	0.00620	0.00280	0.00000	1.9%
75d	0.00000	0.00160	0.00400	0.00520	0.00960	0.01240	0.00560	0.00000	3.8%
90d	0.00000	0.00320	0.00800	0.01040	0.01920	0.02480	0.01120	0.00000	7.7%
110d	0.00000	0.00480	0.01200	0.01560	0.02880	0.03720	0.01680	0.00000	11.5%
120d	0.00000	0.01040	0.02000	0.02340	0.03360	0.03100	0.01400	0.00000	13.2%
135d	0.00450	0.00600	0.02100	0.02990	0.06000	0.08990	0.04060	0.00600	25.8%
150d	0.00250	0.00640	0.01600	0.02080	0.04080	0.05270	0.02380	0.01200	17.5%
180d	0.00250	0.00640	0.01600	0.02080	0.04080	0.05270	0.02380	0.01200	17.5%
	1.0%	4.0%	10.0%	13.0%	24.0%	31.0%	14.0%	3.0%	100.0%

Figure 7 – TRANSPAC Probability Density Matrix

The next step is to create a polar table for each competitor such as the following from the ORR VPP. These are speed predictions expressed as sec/mile for a variety of combinations of wind speed and sailing angle.

Sec/Mile Speed Predictions

True Wind Speed	6 knots	8 knots	10 knots	12 knots	16 knots	20 knots	24 knots
Opt Up Angle	44.6	43.5	39.9	37.3	34.8	34.3	34.6
Best VMG Up	904.6	729.8	646.0	603.6	564.5	549.6	542.4
52d	580.4	478.9	443.2	425.4	406.8	396.3	389.9
60d	535.4	455.6	425.6	409.5	391.6	380.8	373.7
75d	497.6	435.3	404.7	388.7	369.0	355.4	344.4
90d	478.6	423.5	397.0	374.4	349.2	330.0	313.2
110d	482.5	421.3	384.6	362.6	327.9	302.5	282.2
120d	506.2	435.5	394.9	362.0	316.6	288.3	270.5
135d	597.9	482.4	433.4	399.4	337.7	272.9	247.4
150d	731.0	583.1	503.1	449.7	384.2	326.4	264.3
Best VMG Down	844.0	673.3	580.9	517.8	439.0	376.8	305.1
Opt Down Angle	136.6	138.8	146.2	153.1	158.1	143.3	142.7

Figure 8 – Boat Speed Table Predicted by VPP

The Transpac rating is generated by multiplying each cell in the wind table by the corresponding number in the speed table. The value of .0899 for sailing in 16 knots at 135 degrees is multiplied by the 337.7 spm. The sum of all such multiples represents the expected average speed for the boat. This is converted to time on time using a Transpac specific scratch boat.

CONSTRUCTED COURSE

The Transpac rating approach uses a probability distribution of wind based on observations of many races or analysis of historical wind records. It represents an averaged likelihood of what will occur on the race course. In a sense it is a constructed course. A similar approach can be used for races of shorter duration in which known marks, whether permanent or dropped, are used to find the course layout. A wind speed and direction are selected. (In fact each leg of the course may have its own wind speed and direction.) Figure 9 illustrates a constructed course.

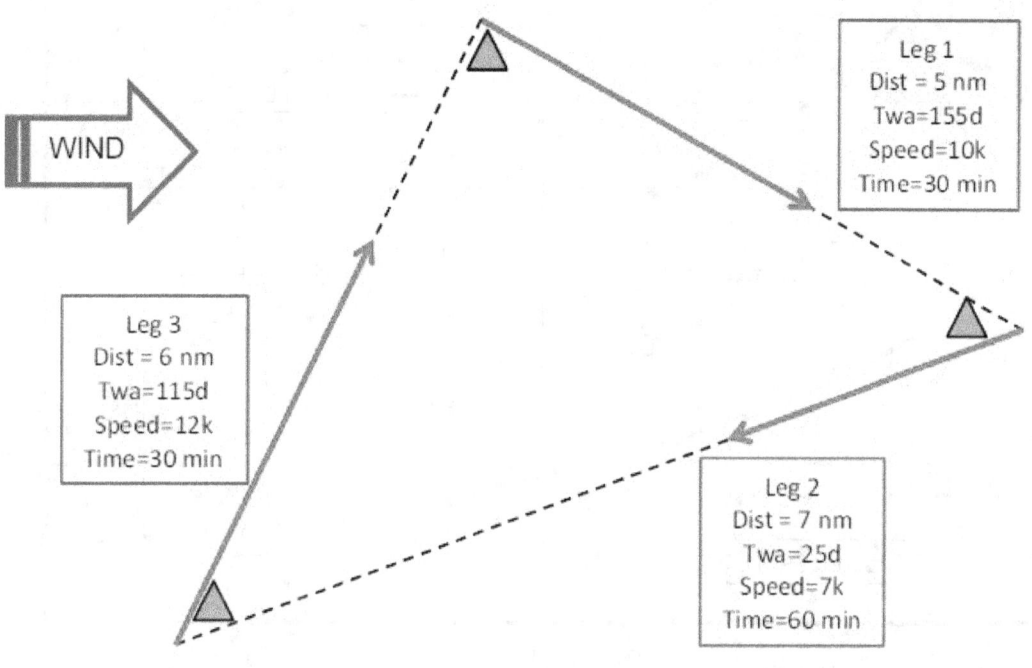

Figure 9 – Constructed Course

This is a triangular race in which the wind is constant but each leg has its own true wind angle (Twa) and distance (Dist). For each boat and for each leg there is a predicted boat speed that is used to calculate the time to sail each leg. The total elapsed time can then be used as a handicap for that boat. Allowances between boats are just the difference in predicted total elapsed times. (It is of course trivial to convert total elapsed time to sec/mile and then we have the same units of rating in the previous examples. And the choice of a scratch boat can lead to time on time ratings.)

PERFORMANCE CURVE SCORING

One of the more elegant (and fairest) methods of scoring is to use Performance Curve Scoring (PCS). PCS uses a continuous curve of ratings across wind speed, built from a series of ratings at specific wind speeds. Figure 2 showed some examples of ratings at 7 different wind speeds with an interpolating curve. The concept of PCS is to use the rating that fits the wind speed. This is especially desirable in the case where one boat is faster than another in one wind but slower in another. But it is also applicable to any fleet of boats whose relative ratings vary with wind speed in ways that neither single number TOT or TOT can accurately model. Which wind would you use for a single rating handicap? You could pick the wind ahead of time, and many regattas do that, but in a long distance race with variable conditions that becomes really problematic. The PCS method automatically slides along the wind speed scale picking the right handicap. How? Figure 10 shows ratings and interpolating curves for three boats zoomed in to the wind range of 6 to 16 knots.

Let's start with Boat A. AT the end of the race her elapsed time is divided by the rated course length to get an average sec/mile. She could have experienced a wide variation of conditions of wind strength, direction, storms, currents, etc. but for the purposes of scoring only her elapsed time matters. Her elapsed sec/mile is 594 which is marked off on the vertical axis of the graph. If we run across horizontally to her curve (blue) and then down vertically to the horizontal wind speed axis we get a wind of 11.9 knots. Her finish time, coupled with her predicted ratings, imply that she sailed in that wind despite the variety of conditions she may have actually experienced. This "implied wind" is an index of how well she sailed.

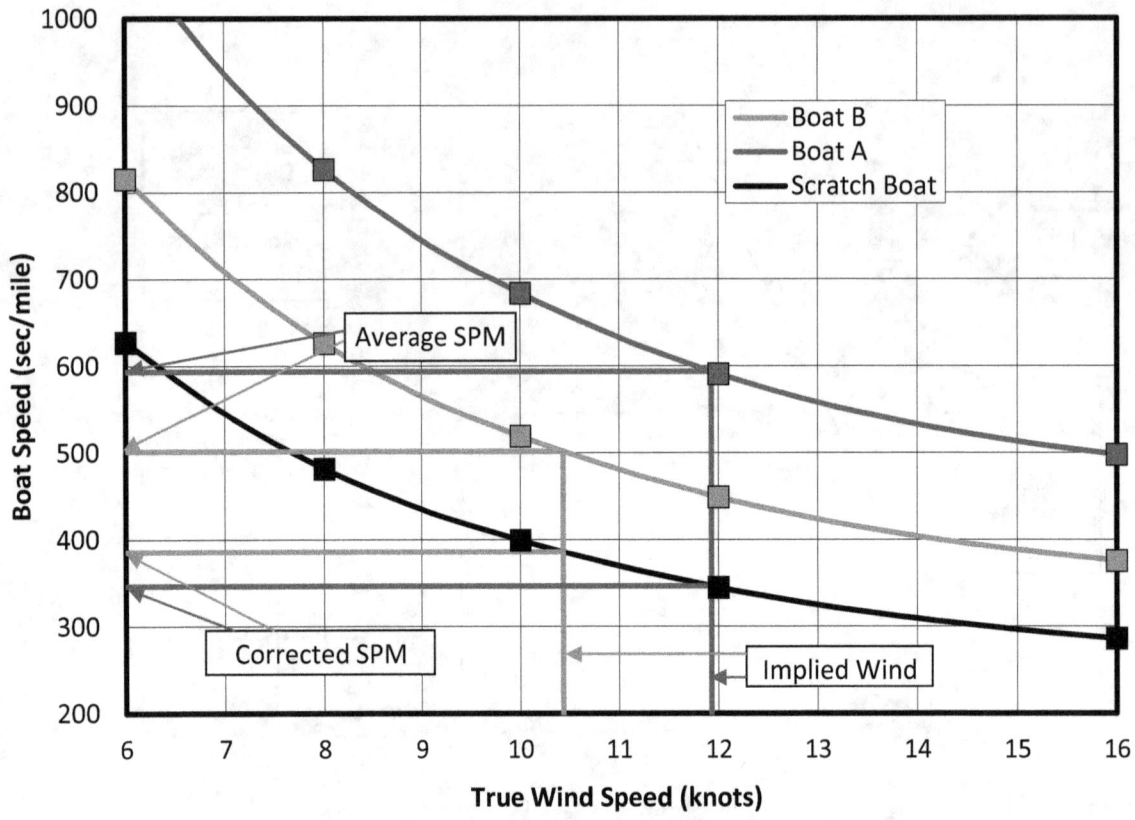

Figure 10 – Performance Curve Scoring

Following the same procedure with Boat B she has an implied wind of 10.4 knots. Although she had a faster sec/mile of 502 her elapsed time makes it appear that she had less wind. Boat A beats boat B. No one chose the wind speed. Each boat's elapsed sec/mile determined her implied wind which ranked the boats. Because the implied wind may not be, and for a long distance race likely will not be, the same as what was experienced onboard it is useful to convert the implied wind results into corrected time results. For that a scratch boat is required. For each of the boats in the fleet, we find the point on the scratch boat curve for the implied wind of each boat. For each boat we run across to the vertical axis and we can get the elapsed sec/mile of the scratch boat if it had the same implied wind. Multiply by the rated distance and we have a corrected time. Each boat's performance has been translated into the performance of the scratch boat, which is the essence of scratch boat scoring.

An especially attractive feature of PCS is that course content can be different for each of the wind speed ratings. The Newport to Bermuda Race uses PCS with this feature. If there is a very slow race, determined by the elapsed time, there are likely two reasons: light wind and a low point of sailing (sailing dead upwind and downwind.) So the 6 knot handicaps can include quite a bit of windward/leeward best VMG sailing. A fast race can occur for two reasons: heavy wind and a fast point of sailing (broad reaching.) The handicaps used for the high wind speeds can include a lot of broad reaching. The handicaps for the intermediate wind speeds are a sliding blend between the two. The race authority, the Bermuda Race Organizing Committee, does not know ahead of time that a fast race has a lot of reaching and a slow race a lot of beating and running, but making those assumptions is better than not making them.

The 2012 Newport to Bermuda Race was run in record setting weather conditions, reaching in very strong Northeasterly winds. Course records were set. A single number scoring system would use the only ratings it has available, which to be useful for all races would represent average conditions, certainly not reaching in high winds. Figure 11 shows fleet scoring under PCS with ORR handicaps. Newport Bermuda typically uses different scratch boats for each division to promote the concept of races within races by precluding cross divisional comparisons. Figure 11 was generated using a common scratch boat so that the corrected times are compatible. There is broad representation in the top 20 boats across the fleet. Slow, heavy boats in the St. David's Division, rating over 500 sec/mile, did quite well because the scoring recognized that they were slow broad reaching when compared to the super0-fast offwind boats in the Gibbs Hill Division.

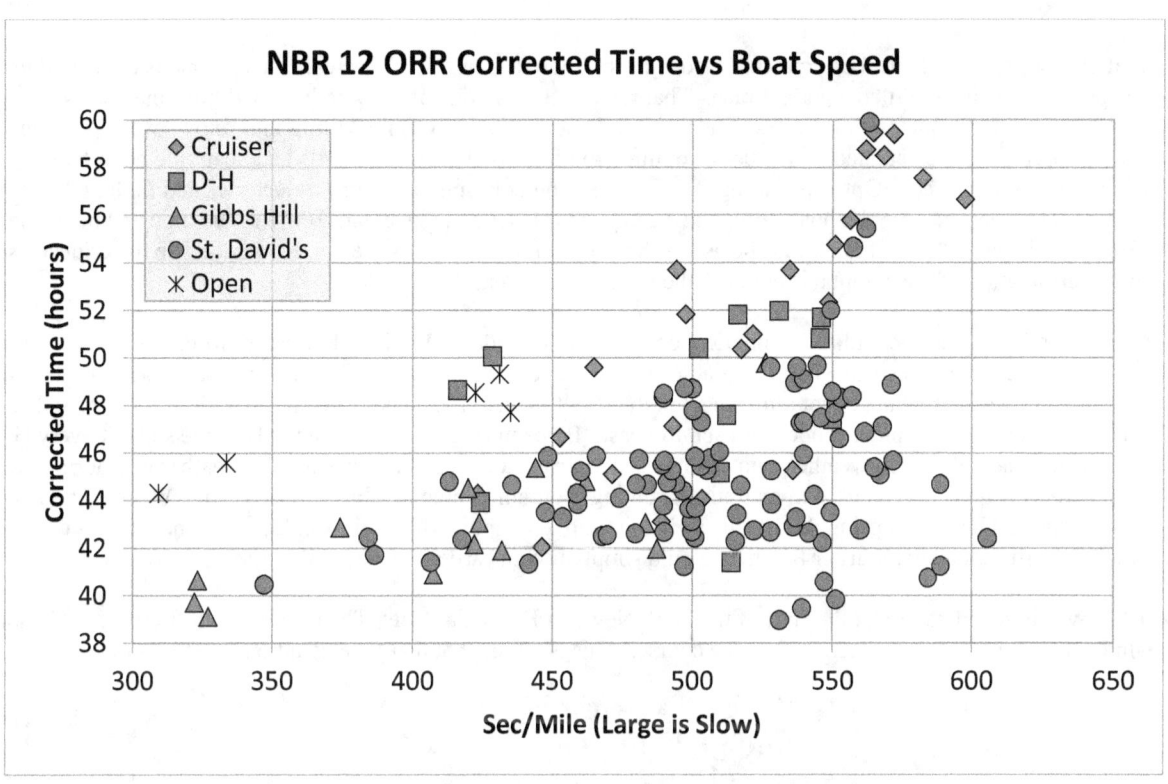

Figure 11 – PCS for the 2012 Newport to Bermuda Race

The assumptions underlying the model of PCS used for the Newport to Bermuda Race did an excellent job in 2012 at scoring the race in a highly unusual weather pattern. Again, the scoring is automatic, no human decisions were made other than any handicap decisions made before the Notice of Race was published.

WE HAVE A PERFECT RULE, NOW WHAT?

Let us assume we have a rating methodology such as a "perfect VPP" or perfect table of boat speed polars. Can that provide perfectly fair scoring? Perfect: never, acceptable: yes. At some point the race organizers need to pick or create a handicap from that table of perfect polars. If they can totally control the environmental conditions on the race course, then the handicaps would achieve perfectly fair scoring. But the weather never cooperates to that extent. There has been any number of offshore races where a building breeze favored the slow and a dying breeze favoring the fast. The uncertainty in weather can far overshadow any rating rule biases, and even sailing skill, in determining who wins and loses a race.

The most common choice is to create a handicap, or choice of handicaps, before the race based on expected conditions. It is common to have light/moderate/heavy wind solutions at hand for buoy racing and make the choice before the start. The Chicago Mac race has 3 choices for mostly upwind, mostly downwind, and a mix of everything. They make their choice the night before. It takes a bit of courage to make that decision. Someone might accuse them of making the wrong choice but with modern weather forecasting it is generally better to make a choice than not make one.

And, there are really brave souls who will make the decision after the racing is finished. Hindsight is great for customizing the ratings to what occurred on the race course but opens the door even further to accusations regarding who is favored or disfavored by the choice. This requires trust in the integrity and impartiality of the event's handicapper.

It has been suggested that each boat record the weather pattern it sailed in and that each boat be scored against how well they sailed in that pattern. This of course would reward those sailors who made incorrect tactical choices and sailed into the slowest conditions. Sailboat racing usually rewards those we make the better choices. The concept of post race scoring is however a precursor to a new method that has gained interest lately: H0 scoring.

Whatever the scoring method, the uncertainty of the weather makes a compelling argument to consider class results as a better indicator of how well boats performed. Within a class the boats are likely to have experienced the same pattern.

H0 Scoring

The underlying premise of H0 scoring is that a boat can be scored against itself: actual elapsed time is compared to the optimal elapsed time if it had sailed a perfect race. That requires knowledge of the weather conditions the boat encountered, but unlike the method of the previous paragraph, no credit is given for sailing into "slow" weather. The perfect race elapsed time is calculated by the use of optimal routing: combining the boat's handicap rule polar table with a computer model of the weather. Optimal routing means discarding non-optimal course trajectories and finding the fastest one. This yields the elapsed time the boat would have achieved if it sailed to its polars 100% of the time and made the correct course decisions 100% of the time. The use of the handicap rule polars is the correct choice. So the critical issue then is the determination of the computer model of the wind (and current.)

NOAA, the National Oceanic and Atmospheric Administration, produces worldwide weather forecasts every 6 hours using its Global Forecast System (GFS) and available as a GRIB (Gridded Binary) file. Each file starts with an "H0" or "hour zero" state of weather which is not a forecast but the result of analyzing data measured worldwide. The H0 data is used to initialize NOAA's computer model for each forecast. By capturing and retaining the H0 time slots of every GFS GRIB file during a race, those H0 weather data can be used as an accurate estimate of what the weather was during the race. For scoring purposes the H0 file is used in an optimal routing program to predict a best elapsed time. Whenever a boat finishes a race she can be scored using the H0 file as it exists up to the point of her finish. Later finishers might use a file that has more records added on. Earlier finishers would simply use an earlier portion of that file.

Figure 12 shows the wind pattern at the start of the 2018 Newport Bermuda Race. The rhumb line to Bermuda is in solid black, wind vectors are distributed across the race course. This is a snapshot that is relevant only to the start.

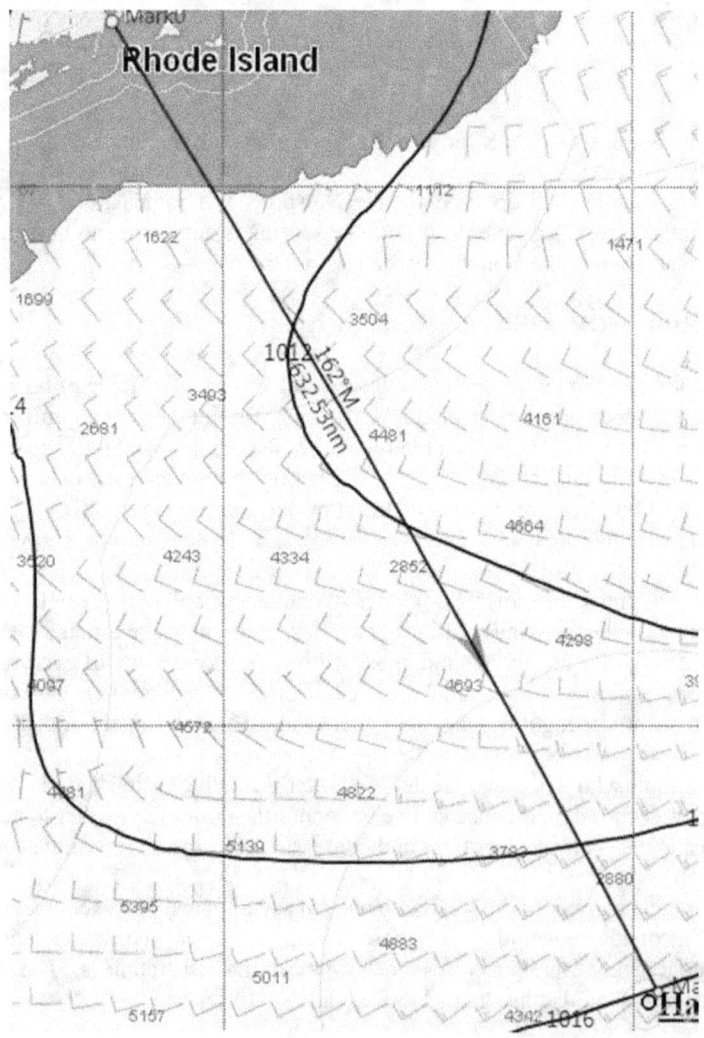

Figure 12 – Wind at Start of 2018 Newport to Bermuda Race

Figure 13 shows the results of optimal routing using the program Expedition. The blue curves are isochrones, lines that show where the boat could be at equal times. The heavy blue line shows the optimal route to make best use of the wind and the boat's performance characteristics (polars) to get to Bermuda in the least amount of time. Along that path, at every isochrone is a display of the wind vector that would exist at that location at that time, according to the GRIB file. The wind vectors displayed across the entire graph are those that would exist at one point in time. This particular illustration shows them when the boat is close to half way to Bermuda as indicated by the blue arrow on the optimal route and the heavy blue isochrone. (The contraction of the isochrones spacing to East and West of the rhumb line half way through the race shows just how slow it would have been to deviate to left or right.)

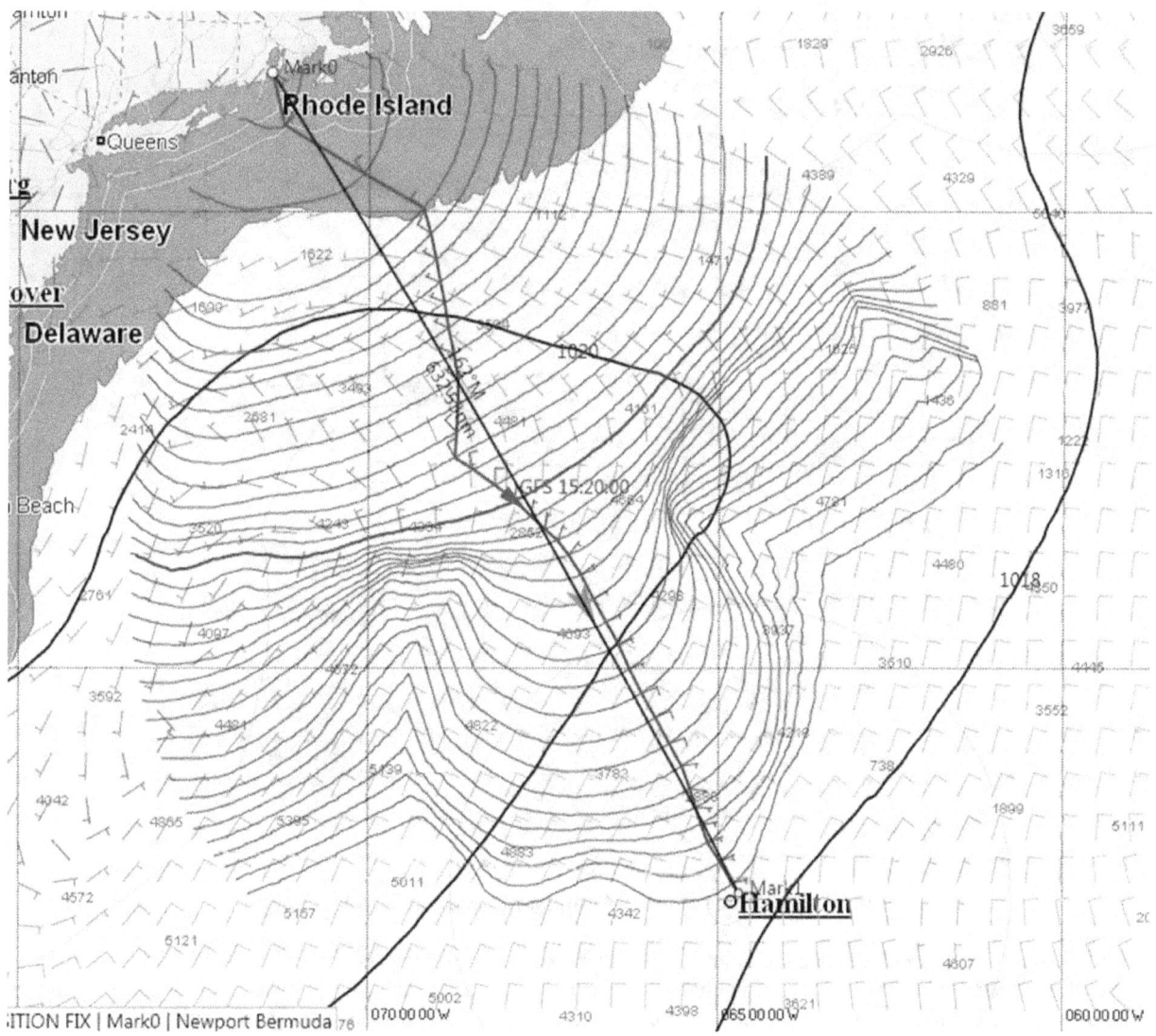

Figure 13 – Track Derived from Optimal Routing

Figure 14 shows yet another optimal routing solution, this one including the actual route (red track) taken by a boat during the 2017 Marion Bermuda Race. There is some similarity between actual and optimal routes.

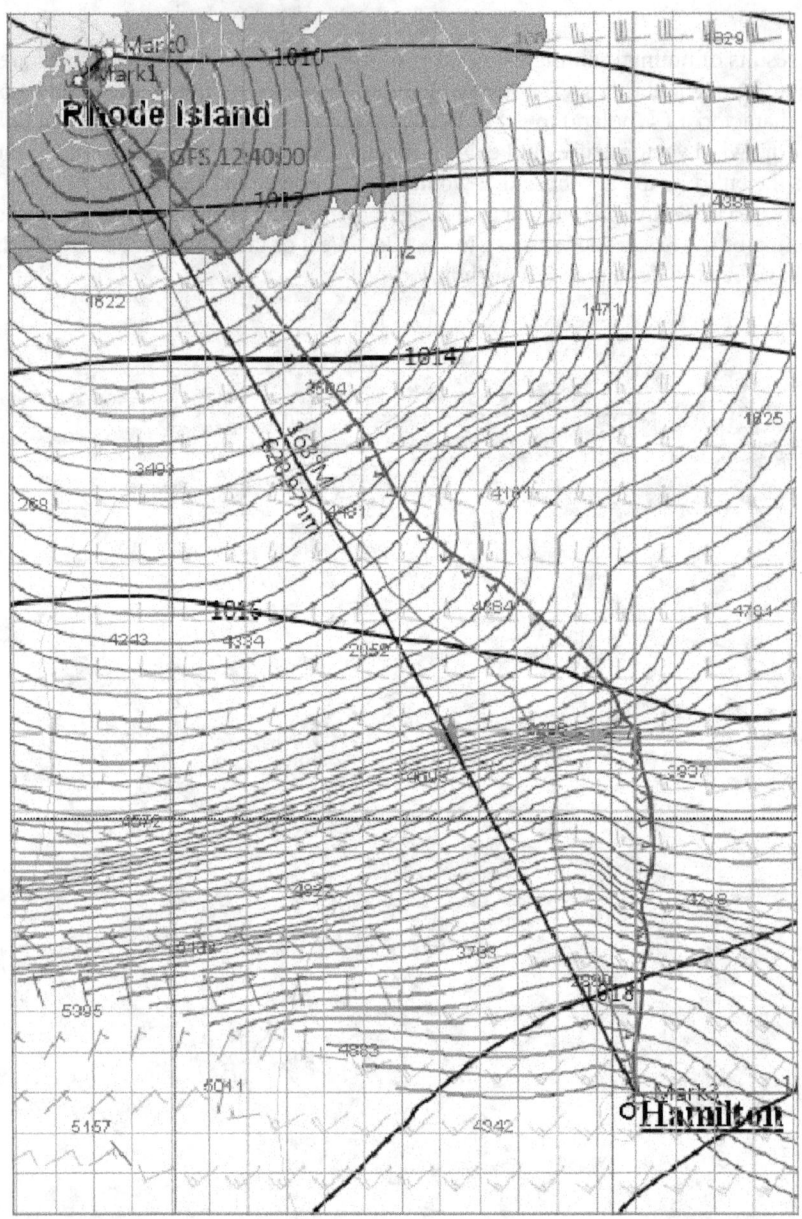

Figure 14 – Comparison of Actual and Optimal Tracks

H0 scoring would be achieved by taking the ratio of actual elapsed time to the elapsed time predicted by optimal routing. It is the ratio of your performance vs. the theoretically best performance you could have achieved by sailing perfectly in the weather conditions that existed while you were on the race course: the lower the ratio the better your ranking. (If desired, this can be converted to a corrected time with the choice of a scratch boat.) The attraction is that some degree of luck has been removed from race results. Boats racing in different environmental conditions can be fairly scored against each other.

H0 scoring also raises many of the issues presented earlier. What is acceptable, or suitable, for the different types of sailors and sailing programs that participate in racing? For those that want ultimate "fairness" an H0 GRIB optimal routing solution has promise, assuming the accuracy of the GRIB files and the optimal routing scheme. But it means that no one knows what their ratings will be in advance of the race, or even during the race. It would be more difficult to know how you are doing against the competition. This effects strategy. There is the potential that online technology might be able to continuously compute and broadcast mid-race and projected finish results. H0 scoring can also be difficult to explain. It introduces complexity when many sailors would prefer simplicity. Ultimate technical precision may not be the answer for some sailors. There is also the attraction that: "if the weather gods shine on me, I just might beat that hot racing boat with the hotshot crew." These are difficult questions that must be answered by race organizers to best serve their events.

CONCLUSIONS

The scoring of handicap sailboat races can be very challenging, the more complex the handicap rule the more varied are the scoring options. Some options are arguably fairer than others. But there are pros and cons associated with each.

If the metric of success for a race event is the number of participants then it must be a priority for the organizers to consider composition of their fleet and the expectations and sophistication or experience of their competitors. They can then make informed choices, including how to score their event, that will best grow participation. This would include the balance of on one hand the "pursuit of scoring perfection" and on the other the "magic of uncertainty" resulting from A thought to keep in mind was expressed by Greg Stewart of Nelson Marek Yacht Design: "we are in the entertainment business." That applies to rule developers, event organizers, designers, sailmakers, boatyards and everyone else that supports the racing of sailboats.

And perhaps a different perspective should be kept in mind. I quote the wizard Bill Lee: "More sailboat races are lost by not reading the NOR than anything else."

ACKNOWLEDGEMENTS

The author would like to acknowledge just a few of the countless people and organizations who have spent considerable effort as volunteers in developing and promoting handicap race scoring: Jim McCurdy, Karl Kirkman, Dan Nowlan and Pete Reichelsdorfer from the US IMS Committee of the 1990s; numerous members of the International Technical Committee; Terry Kohler for his commitment to science based handicapping; and the years of effort and support by the Offshore Racing Association and Bermuda Race Organizing Committee. Specifically, Bruce Nelson, Alan Andrews and Stan Honey have provided invaluable support into the exploration of H0 scoring.

Lastly, the late Alan McIlhenny of the Cruising Club of America originated the concept of Performance Curve Scoring to provide greater fairness but also, in a fascinating turn in later life, expressed the belief that perfect handicapping and scoring is probably not the ultimate goal, that a combination of simplicity and a bit of "every dog has his day" may provide a better racing experience for the vast majority of sailors that enter our regattas.

REFERENCES

Froude, W., "The Resistance of Ships," 5966-No. 23; Bureau of Navigation, Navy Department, Washington: Government Printing Office, NJ, 1888.

A Case Study on the Effect of Sweep and Variations in Free-Surface Cross Section Geometry on the Lift and Drag of Transom-Hung Sailboat Rudders

Paul H. Miller, United States Coast Guard Academy

ABSTRACT

Conventional transom-hung rudders are often used on small sailboats because of their simplicity compared to rudders mounted under the hull; however, they present substantial performance penalties, including (1) the rudder is more likely to ventilate by drawing air down from the free surface, (2) the effective aspect ratio, and therefore the lift-to-drag ratio, is not increased by the mirror-plane of the hull bottom and (3) there is additional spray and wavemaking resistance that arises as a result of the rudder passing through the free surface. This case study focuses on a means to mitigate the last of these penalties, the increased spray and wavemaking resistance. While many transom-hung rudders are essentially parallel, or tapered with the maximum chord at the top where it meets the tiller handle, the reader will recognize that having the largest cross section of rudder at the free surface will generate significant spray and wavemaking resistance, especially when the rudder is turned. This study investigated the use of minimizing the rudder chord length where it passes through the free surface, demonstrating the findings by full-scale towing tests of a series of rudders designed for a *Fireball*-class dinghy. Running the tests at full-scale, therefore matching Reynolds number and Froude number, eliminated questions on scaling. Experimentation on the effects of sweep angle, section shape and chord length at varying angles of attack and velocities showed a noticeable increase in lift-to-drag ratio of foils with reduced chord length at the free surface and by sweeping the rudder forward. To complete the case study, a velocity prediction program was used to estimate the change in speed around a notional race course.

INTRODUCTION

Rudders mounted on a vessel's transom, such as that of the *Gjoa*, the 1872 sailing vessel shown in the frontispiece that was the first to transit the Northwest Passage, and the modern racing dinghy shown in Figure 1, are often used due to their low cost, high reliability and accessibility. Most sailboats over eight meters long however, locate the rudder underneath the canoe body to take advantage of the hull's end plate effect that improves the effective aspect ratio of the rudder, and increases the lift-to-drag ratio. Indeed, apart from a few one-design classes like the J/24, it is uncommon to see a keelboat designed for racing with a transom-hung rudder. The popularity is reversed in the centerboard classes however, where apart from developmental boats like the International Canoe, it is rare to see rudders anywhere but attached to the transom.

Practical reasons such as cost, weight and issues related to launching and sailing in shallow or weedy water are the reasons, but the desire to increase performance is still strong in the development classes and evolutionary developments, such as deeper, less sweep and higher aspect ratios have improved transom-hung rudder performance.

Figure 1 - Transom-Hung Rudder of Fireball Class Dinghy (Photo credit from Fireball International, Robin Inns Photographer, Fireball South Australian Championship, March 5, 2016)

Relevant Previous Work

Hoerner (1965, page 10-13) provided a discussion of the drag of surface-piercing struts, including wave and spray drag. It was shown that the theoretical maximum wave resistance occurs at a chord-length Froude number of approximately 0.5, and steadily drops to zero as speed increases.

$$F_C = \frac{V}{\sqrt{gc}}$$

Where,

F_C is the chord-length Froude number

V is the velocity (m/s)

c is the chord length (m)

g is the gravitational acceleration (m/s²)

For a typical rudder with a 0.25m chord length, $F_C = 0.5$ corresponds to roughly 0.8 m/s or 1.5 knots, which is a fairly low speed for sailing. At high speeds, the pressure on the strut causes a jet of water that forms spray. Hoerner provides the following empirical equation for the spray drag of a strut, based on tests of an ogival section. The ogivial shape is often used for high-speed surface piercing struts because it lacks the blunt leading edge of a hydrofoil section, reducing the spray height.

$$D_{SPRAY} = 0.24 \frac{1}{2} \rho V^2 t^2$$

Where,

ρ is the water density (kg/m³)

t is the thickness of the strut (m)

This equation has been used for many years to justify the choice of surface-piercing struts that have as small a thickness as possible, resulting in large chord lengths to provide structural rigidity. This equation is only applicable for chord length Froude number $F_C \geq 3$ (or 9 knots for the notional 0.25m chord length rudder), however. The choice of an ogival section is based on technology for high-speed hydrofoil boats rather than low-speed transom hung sailboat rudders. For low speed boats, an ogival section may not be ideal, and so this study explores the effect of section shape. Hoerner's equation is also intended only for zero angle of attack, which is not the case for a rudder.

Figure 2 shows a sketch of the effect of angle of attack on the projected thickness of two struts. The top strut is 50% thinner than the bottom one. By Hoerner's equation (and most reasoning) this strut would be expected to have less spray resistance than the shorter, thicker one. However, the bottom half of the figure shows both struts at a 10-degree angle of attack, where t' represents the projected thickness perpendicular to the direction of the flow. In this case, the thinner strut has a much greater projected thickness t' and would be expected to have substantially more resistance. Thus, the guidance for thin struts with large chord lengths may not be applicable for the case of a rudder.

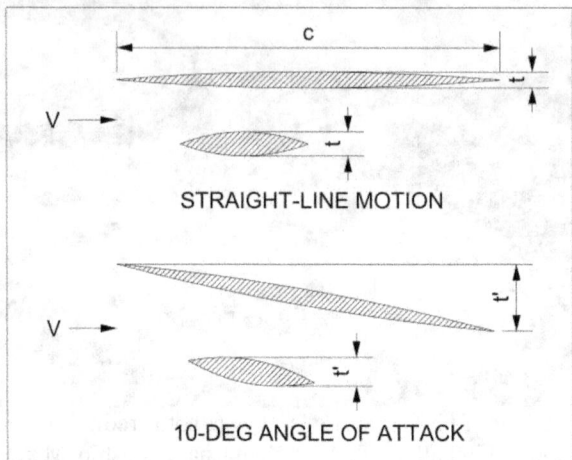

Figure 2: Comparison of thin and thick struts in straight-line motion and at angle of attack

Larsson, Eliasson and Orych (2014) explain the advantage of tucking a foil under the hull is the increase in effective aspect ratio and hence, efficiency. When a lifting foil (such as rudder, keel or wing) is ended against a nearly flat plate such as the hull, its effective aspect ratio is increased due to the elimination in spanwise flow (vorticity) at the flat plate. The result is that a reflected image is created effectively doubling the aspect ratio of a rudder that does not have an endplate, significantly improving the lift-to-drag ratio. In practice however, a hull may not be a perfect end plate as the vessel moves through waves and heels and therefore the gain in aspect ratio is usually not doubled.

At very low speeds the free surface itself may act as an endplate, but the mirrored surface effect diminishes as wavemaking and induced drag is formed on the rudder at the surface. In Molland and Turnock (2007), it was stated the flow conditions around rudders near and at the free surface are "confused and detailed conclusions are not possible." Nonetheless, the book references a study by Millward (1967) that concluded that by the time a rudder reaches a Chord Froude Number F_c of 0.8 the mirrored effect of the surface is essentially gone. The study spanned the F_c range of 0.5 to 1.1 while modern small boat rudders are generally above one. The rudders discussed later in this study ranged from 0.6 to 3.4, indicating the mirror effect is not present.

More recent rudder research for sail craft has focused mainly on the larger and faster boats associated with the America's Cup, where the rudders are mostly submerged under the hull. One notable exception was the full-scale testing of a 3.5 meter International Moth. Beaver and Zseleczky (2009) tested three surface piercing rudders (of nearly identical chord length) at 11.8 knots (F_c = 5.7). Their focus was on T-foil hydrofoil rudders partially supporting a foil-borne craft. They concluded that

- Wave and spray drag was 30% higher than Hoerner predicted
- The advantage of using laminar section foils was more significant at higher Rn than previously thought
- Qualitatively, thinner foils (<13% thickness-to-chord ratio) performed better in surface piercing conditions.

Other papers have explored the effect of adding end plates to rudders at the tip to further increase their effective aspect ratio or at the surface to reduce ventilation. Miller (2007) showed peak lift and hence bending moment values could be higher than American Bureau of Shipping design values. Keuning and Verwerft (2009) built on the earlier work by Keuning, Katgert, and Vermeulen (2007) and developed an improved method of calculating lift for the Delft Velocity Prediction Program (VPP), but the method assumed no free surface effects. None of the published work dealt with surface piercing rudders in the laminar transition region however.

Present Study

While there has been a wealth of published data on rudders in general, the literature review showed very few papers addressing the design of rudders for smaller sailing dinghies, and methods of mitigating the performance disadvantages associated with transom-hung rudders. While there are many factors to consider, the present study focuses on three factors associated with the surface-piercing portion of the rudder design:

- Chord length at free surface

- Section shape
- Sweep angle

The rudders on sailing dinghies typically operate across the laminar and turbulent flow regimes and generate significant spray at higher speeds. As a result, they are very difficult to model experimentally. To avoid the problems associated with the scaling of surface effects which require matching both Reynolds and Froude numbers, the tests were run at full-scale in the 36 meter long tank at the United States Naval Academy, which has a maximum carriage speed of seven knots.

Choice of Fireball Rudder

Although this study has general applicability, the researchers decided to start with a known design with the goal of improving its characteristics, to provide essentially a case-study on the effectiveness of the various proposed designs. A review of the boat designs with International status accorded by World Sailing identified more than a dozen potential candidates. While strict one-design classes such as the Laser have the widest popularity, their class rules restrict the ability to modify their rudder designs, limiting the results' applicability. Each candidate design was then evaluated to meet both the towing tank requirements and the ability to allow development of the rudder design. Finalists included the 505, Fireball, International 14 and Moth. The Fireball was ultimately selected due to its lower speed, favorable class rules that would allow complete freedom to modify the rudder and easy-to-duplicate stern design. The last was a factor as the stern was recreated for the test setup. Figure 3 shows a Fireball, a 4.9-meter dinghy designed by Peter Milne. To-date, over 15,000 Fireballs have been built and they are raced in more than a dozen countries. To ensure that the results were relevant to current rudder designs, a duplicate of the rudder used to win a recent Championship in the United Kingdom was purchased from the builder (Winder Boats of West Yorkshire, UK) as a baseline. The baseline rudder was then modified after the initial tests to incorporate different shapes at the free surface.

Figure 3. A Fireball with its transom-hung rudder.
(photo credit: Helena Richards, UK Fireball Association)

EXPERIMENTAL DESIGN

The main test parameters were:
- Velocity
- Angle of attack
- Sweep Angle
- Chord length at the free surface
- Section shape at the free surface

Two section shapes were compared, both with a thickness to chord ratio of 13%. The section shapes selected for comparison to the baseline design were a Selig/Donovan SD8020 and a tangent ogive. The SD8020 was selected due to its high lift-to-drag and gentle stall characteristics at Reynolds Numbers typically seen in this application, and the ogival was in response to existing design guidance for surface piercing struts of high-speed craft. Figure 4 shows the two foil sections.

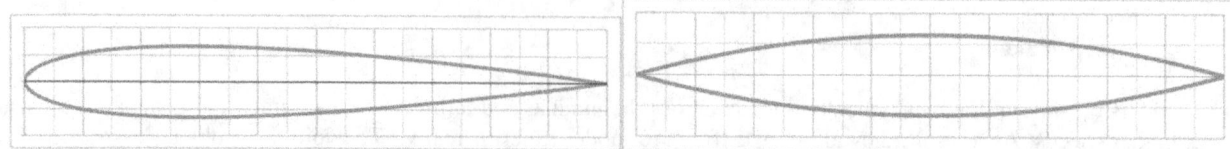

Figure 4. 13% SD8020 (left) and tangent ogival (right) sections.

To minimize variations in spanwise flow from 3D effects, the span was kept constant for all tests at 773 mm. Varying the chord length at the free surface necessarily changed the total projected area, and so the results were normalized to an equivalent surface area. Chord length selection for the shortest chord was based on structural capacity with the goal of making the smallest chord length possible while maintaining the 13% thickness to chord ratio. This was driven by the bending moment at the point of maximum lift at the maximum speed. To gain the highest structural capability the two rudder sections with the smallest chords were designed to a yield factor of safety of 1.5 and machined from H1150 heat-treated 17-4ph stainless steel which had a hardness-derived yield strength of 141 ksi (972 MPa). The larger sections were machined from Renwood polyurethane as sleeves that slipped over the smaller stainless sections.

Sweeping the rudder forward puts it partially under the hull, potentially creating a partial end-plate effect, increasing the effective aspect ratio and therefore the lift-to-drag ratio. To determine the effect of sweep on the lift-to-drag ratio, it was necessary to include a hull to provide the partial end-plate effect. It was not possible to tow a full-sized boat in the towing tank, so a small portion of hull (LWL = 610 mm, Beam = 762 mm) representative of the centerline portion just ahead of the rudder was tested (Figure 5). In each test the rudder was set 50 mm aft of the hull at the free surface and the hull immersion was fixed at 38 mm.

Test speeds of 2, 4, and 6 knots are representative of typical sailing conditions for the Fireball class. These speeds correspond to speed-length ratios $V_{KNOTS}/\sqrt{L_{FT}}$ of approximately 0.5, 1 and 1.5, or hull Froude Numbers of 0.15, 0.3 and 0.45.

Figure 5. Baseline rudder with simulated Fireball "hull" attached.

An angle of attack range of 0 to 16 degrees by increments of 2 degrees was chosen, with the upper limit close to the estimated stall angle of the SD8020 section.

The test matrix is as follows:

For Each Rudder and Sweep Combination:
      Speed = 2, 4, 6 knots
      Angle of Attack = 0 to 16 by 2 degrees
      Runs = 27
Baseline Rudder Sweep Tests
      Sweep = -15 to +15 by 5 degrees
      Total Conditions: 7
      Total Runs: 189
Rudder Geometry Tests
      Section Shapes = Ogival, SD8020, Winder (baseline)
      Chord Lengths = 84mm, 163mm, 241mm
      Total Conditions: 6
      Runs: 162

UNCERTAINTY ANALYSIS

Uncertainties were evaluated using the ITTC approach. Geometric modeling uncertainties were not needed as the testing was done at full scale with the actual rudder and proposed designs. The SD8020 and ogival sections were NC-machined and ground to final thickness. The blades were checked for fairness and thickness and were within 0.25 mm of the design values.

With a tank width of 2.44 m and a depth of 1.83 m, the hull's and rudder blades' maximum cross section was just under the 1% value that suggests blockage correction. Similarly, by restricting and fixing the span as a constant value of 42% of the tank depth, end plate effects were not a significant factor in the results.

Initial alignment was performed using a laser plane on the rudder quadrant and was accurate within 0.2 degrees. Due to the large lift forces that would be produced by the rudder at six knots in the maximum lift condition; a two-strut system similar to those used in submarine tests was adopted to minimize rig deflection. This provided a fairly rigid yaw condition. The rig and the rudders were tested independently using the expected load range to develop a correlation curve that accounted for rig deflection and rudder twist and deflection so that the indicated angle of attack at the head could be correlated to the actual angle of attack at the rudder's geometric center. At zero angle of attack setting the blade sometimes twisted to a positive angle of attack and sometimes to a negative. This can be seen in Figures 10-12 and 20-25 and was accounted for in the results.

The lift force block was calibrated to 0.23% accuracy (0.115 pounds for the 50 pound lift gage) and the drag block was 0.2% (0.01 pounds for the 5 pound drag gage) and were checked four times using certified weights. The consistent bias was noted and included as a correlation adjustment. The water temperature was checked at the beginning of each day.

A direct measurement uncertainty analysis was carried out each day during the study by repeating a shorter span baseline rudder test at four knots with zero sweep and four degrees angle of attack. Five runs were performed each time to determine repeatability. The combined 20 results gave relative uncertainties and were checked using a Student T-Distribution. The coefficients of variation were 1.8% in side force and 1.2% in drag. The data reduction equation uncertainty for the lift/drag calculation was a COV of 1.9%. A Student T-Distribution confirmed the repeatability within a confidence interval of 90%.

Baseline Tests

The baseline Winder rudder (241 mm chord at the root) was tested at zero degrees sweep and speeds of two, four and six knots, corresponding to Reynolds numbers of 220,000, 440,000 and 660,000. The angles of attack varied from 0 to 16 degrees. Figure 6 shows the results. Note that in Figures 6-25 (except 13) the angle of attack is corrected for the jig deflection. The lift coefficient, C_L decreased with increasing speed as the free surface became more disturbed by wavemaking and spray. This finding agrees with the study by Millward (1967) that indicated that the mirror plane effect of the free surface diminishes above a chord-length Froude number of 0.8. This would account for the reduction in lift, brought on by a reduction in effective aspect ratio. The curves peak between 13 and 16 degrees. The rudder began to intermittently ventilate at the extreme condition of 6-knots and 16-degree angle of attack, and this unsteady point was omitted from the plots.

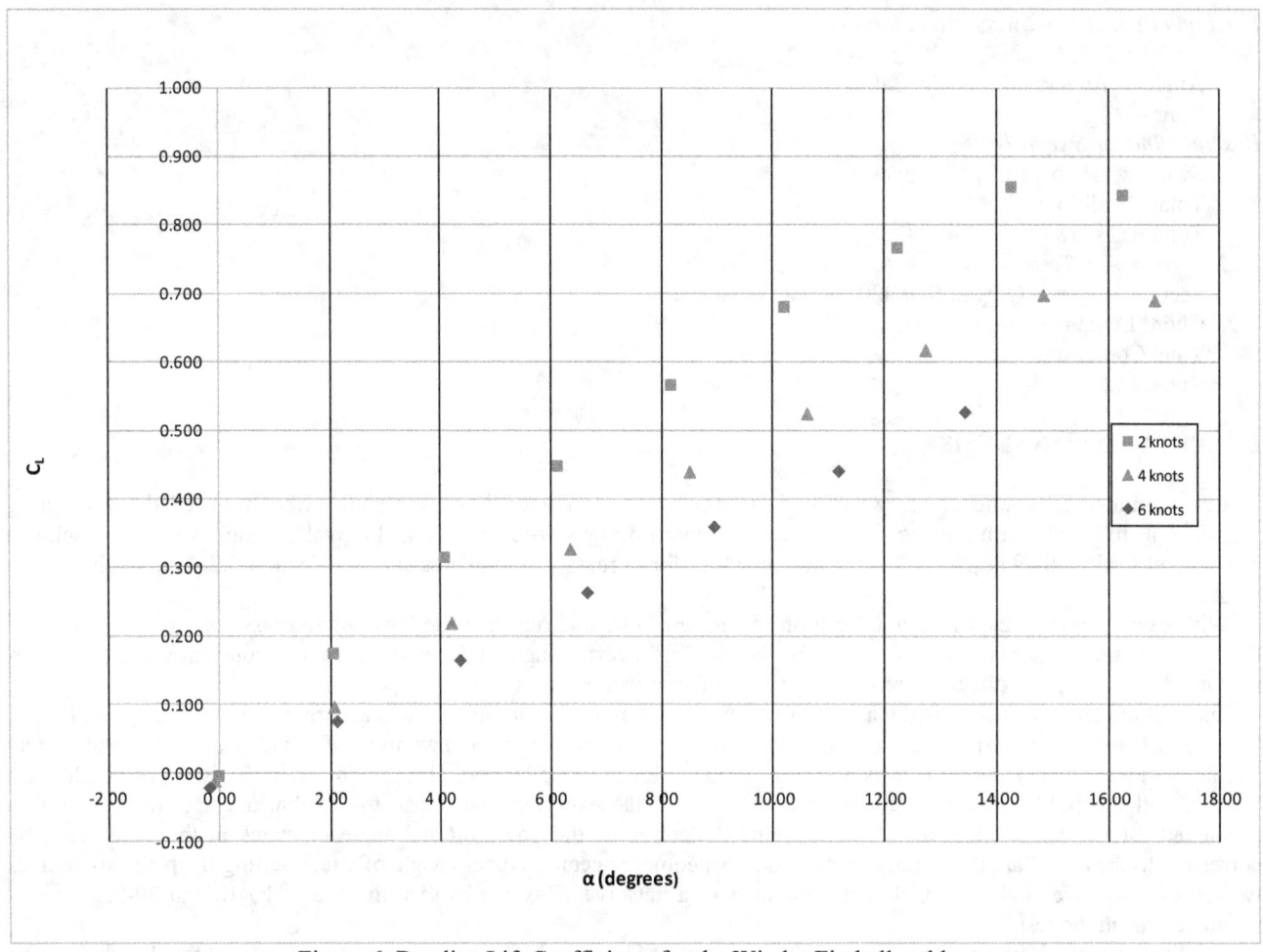

Figure 6. Baseline Lift Coefficients for the Winder Fireball rudder.

Sweep Experiments

Marchaj (1979, pg 452, Figs. 2.143 and 2.147) concluded that sweep angles of less than five degrees produced the lowest drag and the highest lift slope. He also found that for some taper ratios the lowest induced drag may occur with a forward sweep. Practically however, it is also likely that a forward sweep may not produce a comfortable helm feel if using a vertical axis rudder mounting system as the rudder may have too much area in front of the axis of rotation to be well-balanced. Nonetheless, three reasons were behind the decision to experimentally investigate the effects of sweep on transom-hung rudders.

1)   The first was due to practical considerations based on the nature of the boats. In many places where the boats sail weed is present and aft sweep can aid in weed shedding.

2)   Secondly, due to downwash from lift, aft sweep with constant span has the effect of moving the center of pressure lower on the foil and away from the free surface (Kroo, 2011), while the opposite is true of forward sweep. This means that aft sweep might reduce drag due to the lower disturbance on the free surface at the cost of potentially higher heeling moment.

3)   The third was a question whether forward sweep that would put the rudder under the hull might be beneficial due to the end-plate effect of the hull and the potential for the near-surface pressure field to decrease the wave field. To answer these questions sweep angles ranging from –15° (forward sweep) to 15° (aft sweep) at five-degree increments were tested.

Figures 7-9 show the effect of sweep on the lift coefficient for the baseline rudder at speeds of 2, 4 and 6 knots. At six knots, it is evident that that increasing forward sweep (negative sweep angle) increases the lift coefficient. This indicates that the hull may have some beneficial end plate effect. The results at the lower speeds of two and four knots are more mixed, likely because of the reduced magnitude of the lift and the circulation. It is not presently known if the effect of forward sweep seen here is dependent on the hull form, or if it would have also occurred with no hull ahead of the rudder. The findings shown, and the questions that are raised, warrant further study on this interesting phenomena.

Figures 10-12 show plots of lift-to-drag ratio C_L/C_D versus C_L. There is substantial scatter at 2-knots. There are two key

causes of this scatter.
1)   At this speed, Reynolds number is 2.2E5, based on the mean chord length, which likely caused transitional flow. Ordinarily, turbulence stimulation would be applied to a model test rudder to ensure turbulent flow; however in this case, the results are full-scale. The transitional flow observed in these tests would also be observed in operation of the boat, and would depend on the initial turbulence intensity and water temperature.
2)   An additional cause of the scatter is the extremely small magnitude of drag forces measured at this speed, relative to the forces seen at other conditions. The lift-to-drag ratios for these rudders is approximately 20, and the force increases with the square of the velocity. Thus, for a given rudder and angle of attack, the 6-knot case will have 9 times the force as the 2-knot case. Taking into account the lift-to-drag ratio, the lift measured at 6 knots will be 180 times the drag measured at two knots. The disparity increases with differences in angle of attack. While the instrumentation is built to withstand the large forces at the extreme conditions, it is less sensitive at the lower forces.

At the higher speeds the results are more consistent. From these it is seen that while the maximum forward sweep of 15 degrees generated the greatest lift, it is not the most efficient due to a greater amount of drag. A slight advantage is seen at four and six knots with a forward sweep of five degrees, which placed about one quarter of the foil forward of the transom.

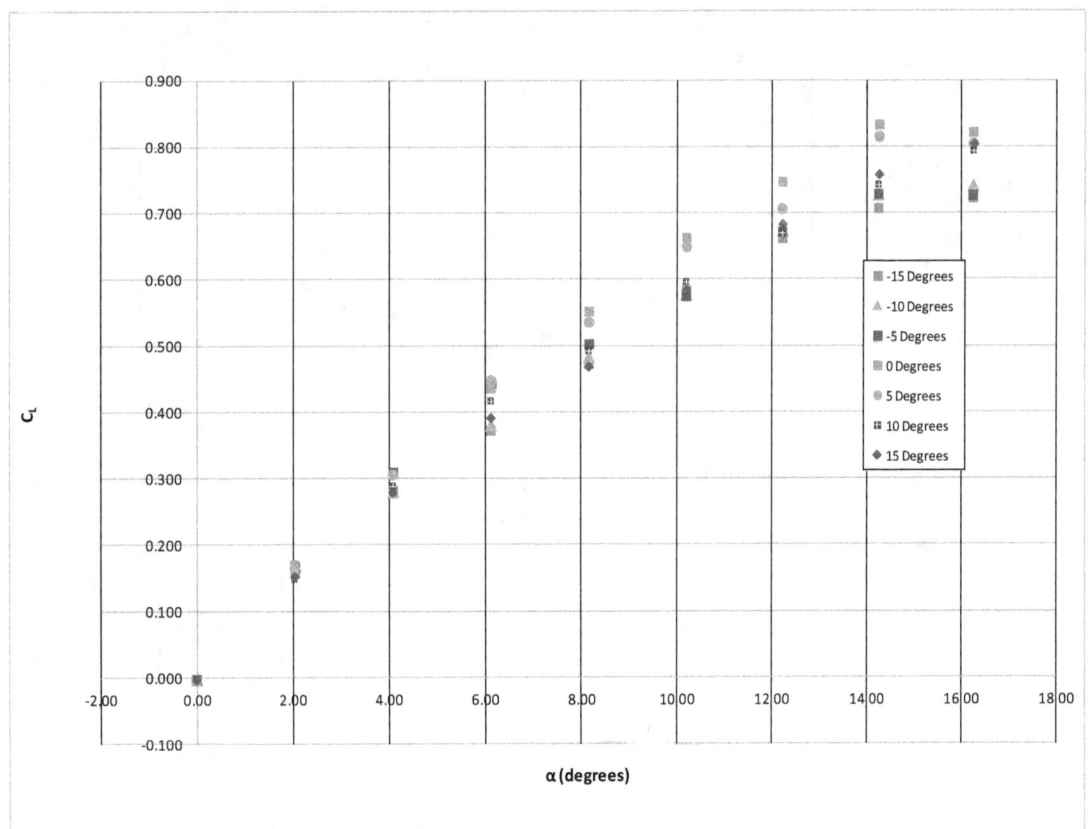

Figure 7. Sweep versus the coefficient of lift at two knots. (negative sweep is forward)

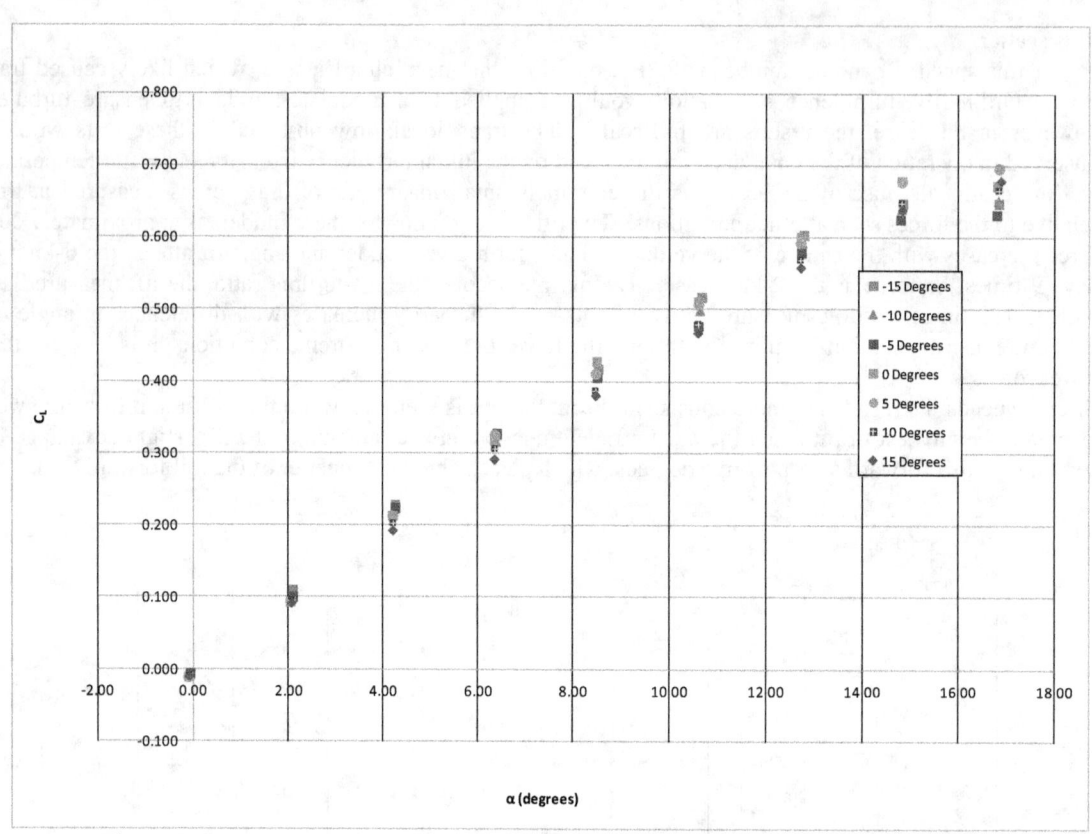

Figure 8. Sweep versus the coefficient of lift at four knots. (negative sweep is forward)

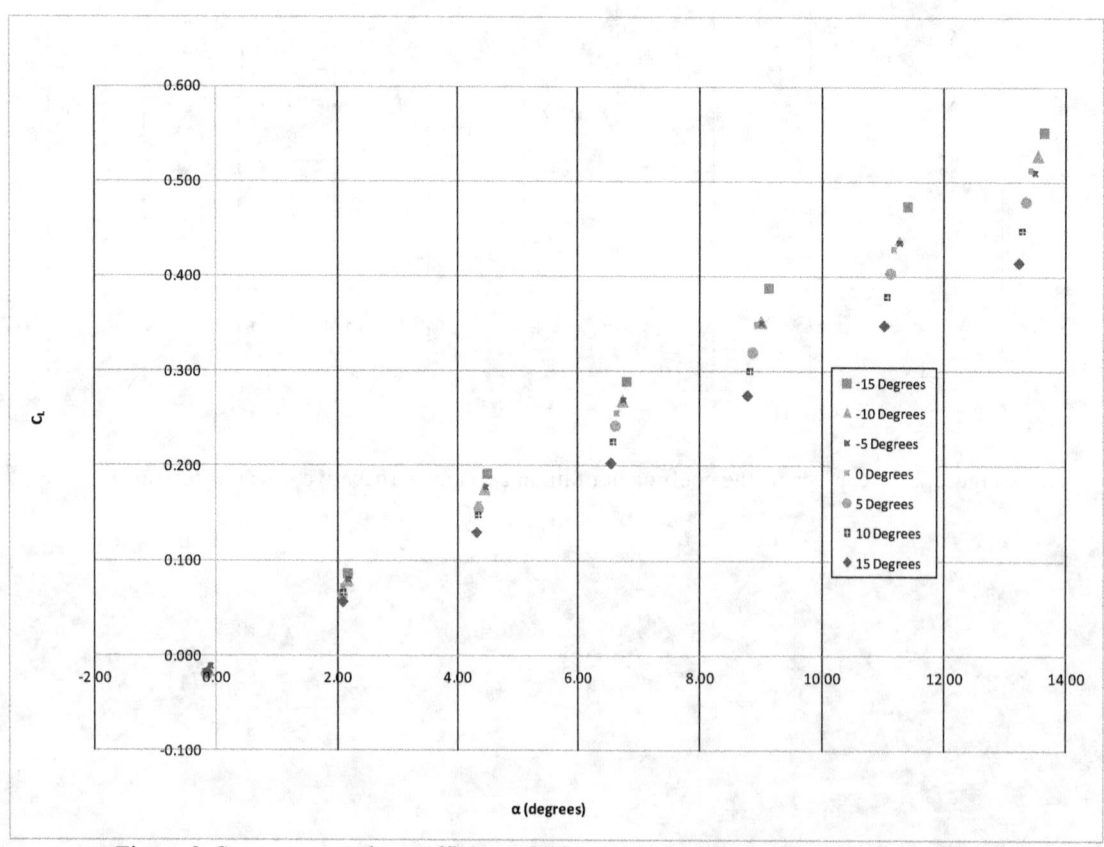

Figure 9. Sweep versus the coefficient of lift at six knots (negative sweep is forward).

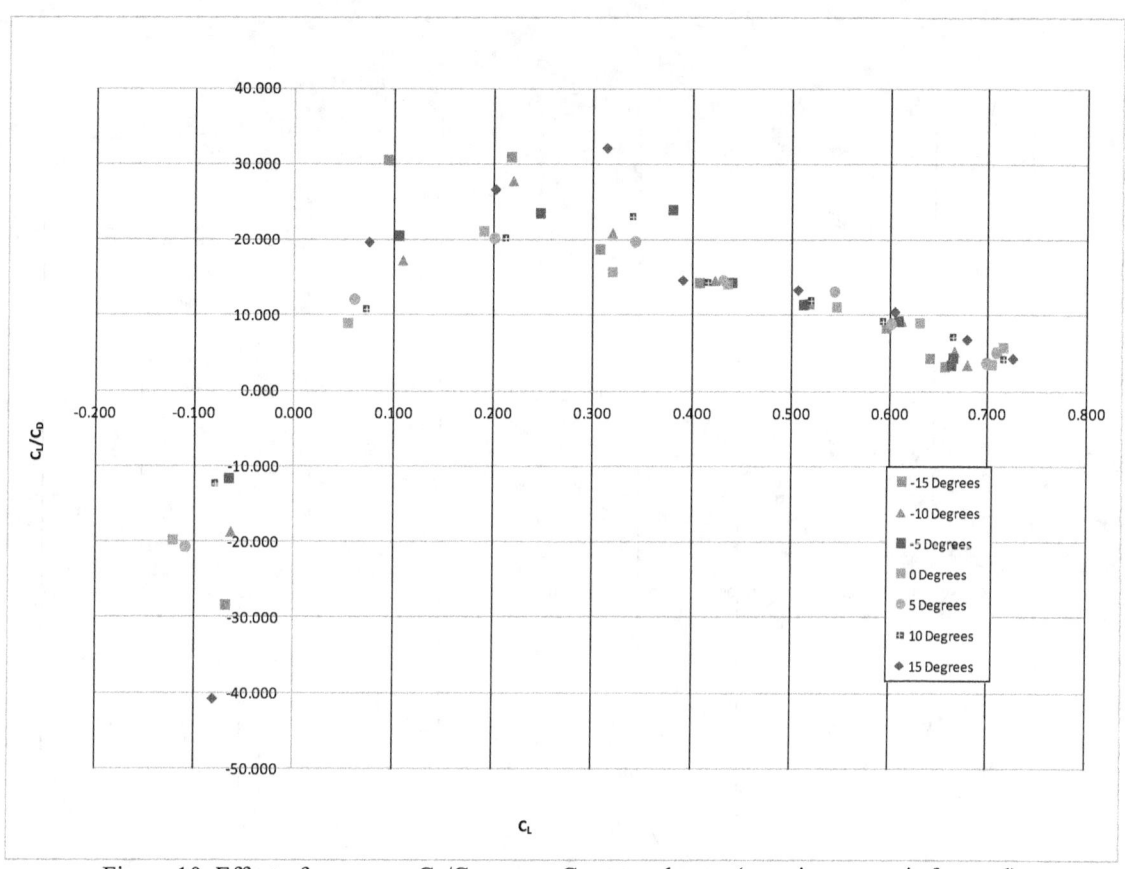

Figure 10. Effect of sweep on C_L/C_D versus C_L at two knots. (negative sweep is forward)

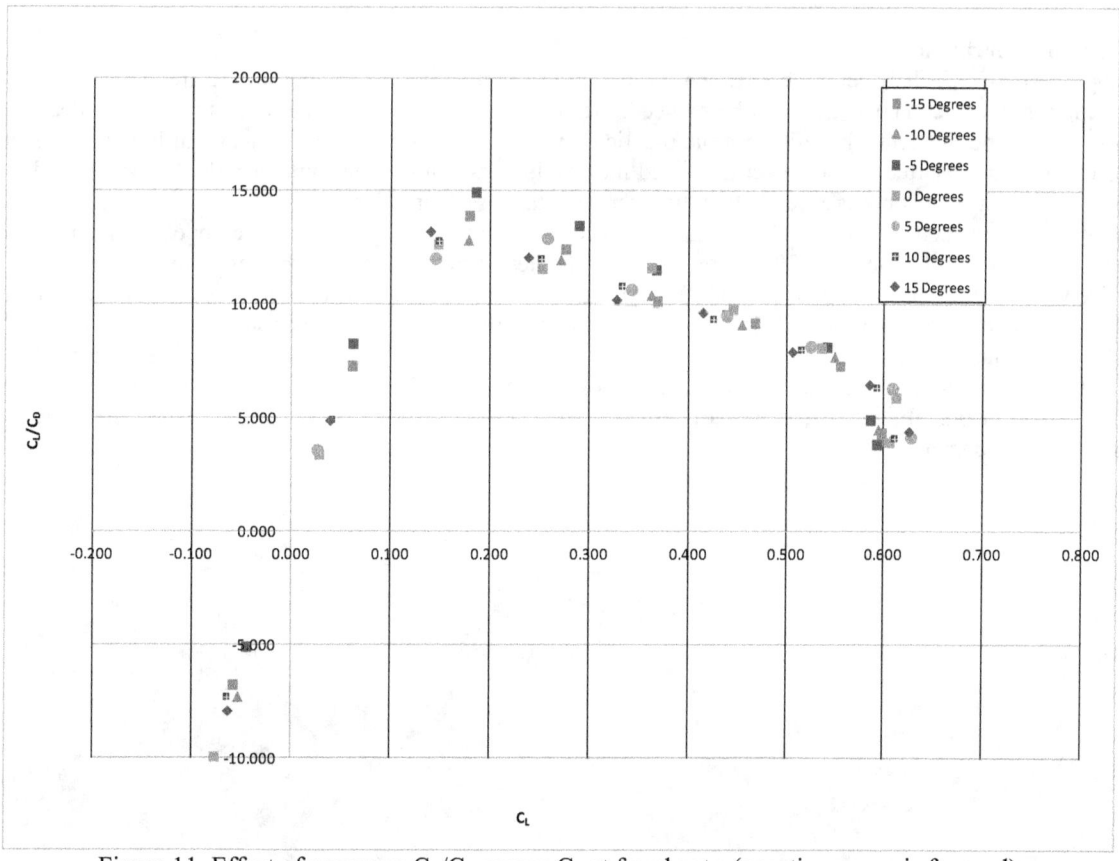

Figure 11. Effect of sweep on C_L/C_D versus C_L at four knots. (negative sweep is forward)

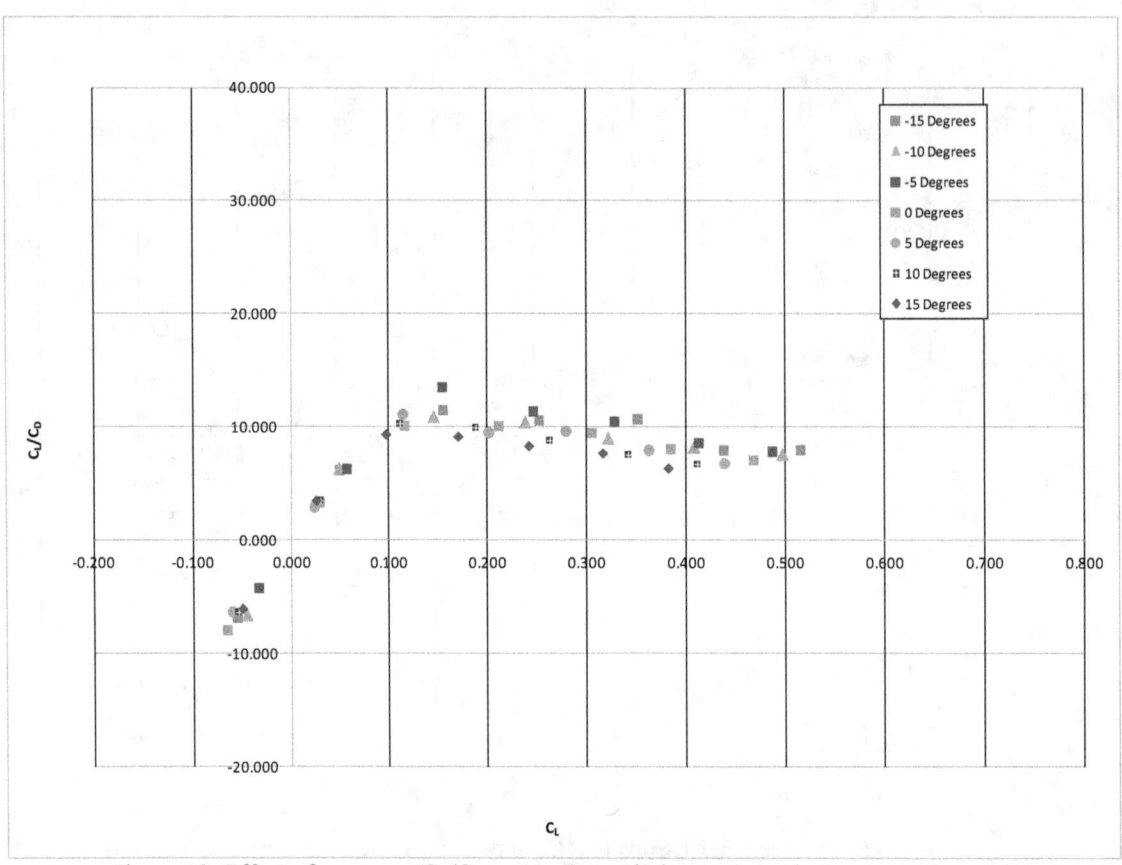

Figure 12. Effect of sweep on C_L/C_D versus C_L at six knots. (negative sweep is forward)

Chord Length Experiments

The hypothesis of the chord length experiments was that by minimizing the volume at the free surface the wave and spray-making resistance components would be reduced, increasing the efficiency. While this is not a new idea, the goal of these tests was to find the actual benefit from the practical limit that is driven by current material limits. The metal struts described earlier were inserted into a pocket machined into the lower section of the Winder rudder. The larger foil sections were machined from Renwood as sleeves that fit over the stainless struts. To minimize twist the shear center of the machined strut was located over the zero yaw moment location of the foil. Figure 13 is a composite picture showing the three test configurations for the SD8020 section with the stainless strut showing on the left. The ogival sections looked similar. The white lower part of the rudder is the Winder rudder cut 200 mm below the waterline and the sleeves are seen as the pinkish Renwood. The yellow stripes are thin tape used to seal the gaps. The three chord lengths were 84, 163 and 241 mm, representing the smallest ("min"), middle ("50%") and the baseline ("100%") chord lengths. The baseline sleeve closely matched the Winder rudder profile and the section shapes were slowly transitioned from the Winder section to either the SD8020 or ogival strut. The constant chord section at the free surface ended 89 mm below the waterline where the transition region began.

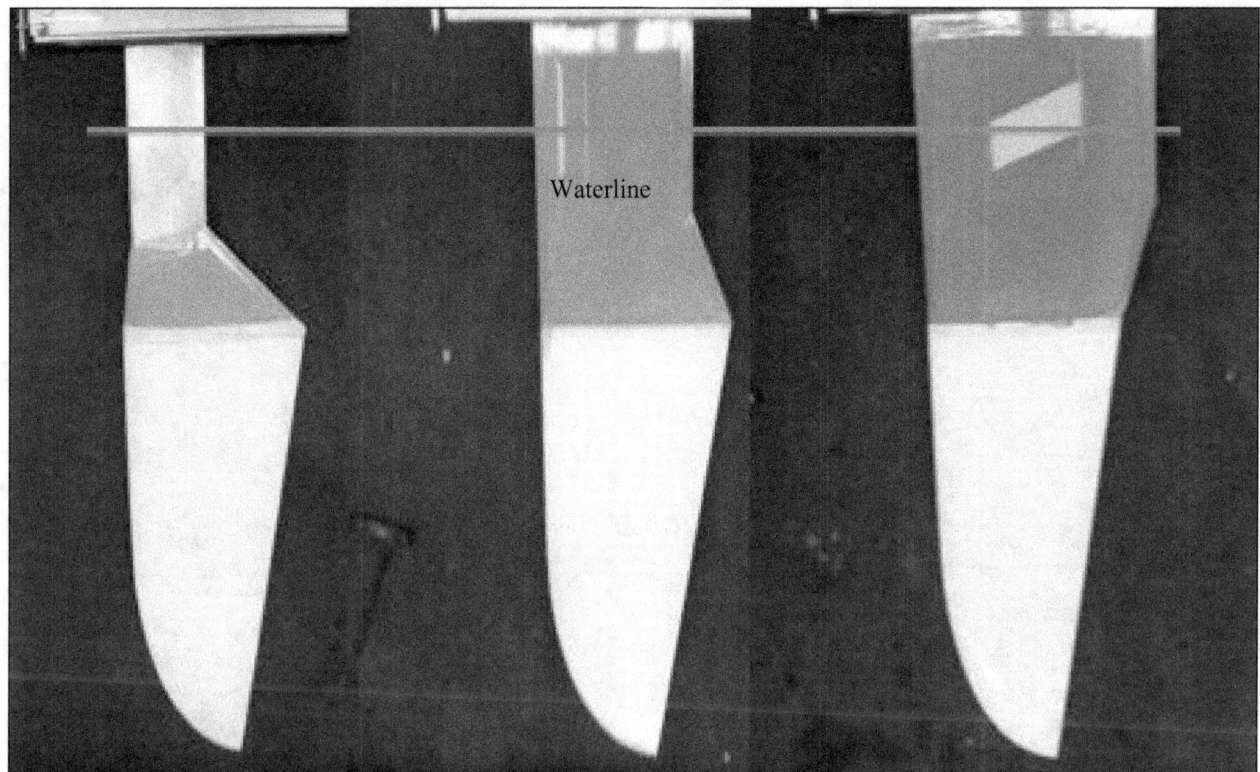

Waterline

Figure 13. Profiles of the chord length experiment rudders. The blue line indicates the designed waterline.

As with the sweep test the rudders were aligned 50 mm aft of the hull and the hull had 38 mm of transom immersion. Figures 13-18 show the results of the chord length at the free surface on the coefficient of lift. At two knots the longest chords produced the highest lift coefficients. This is likely due to the free surface still acting as a reflecting plane. At the two higher speeds however decreasing the chord length increased the coefficient of lift. There was little difference between the two section shapes.

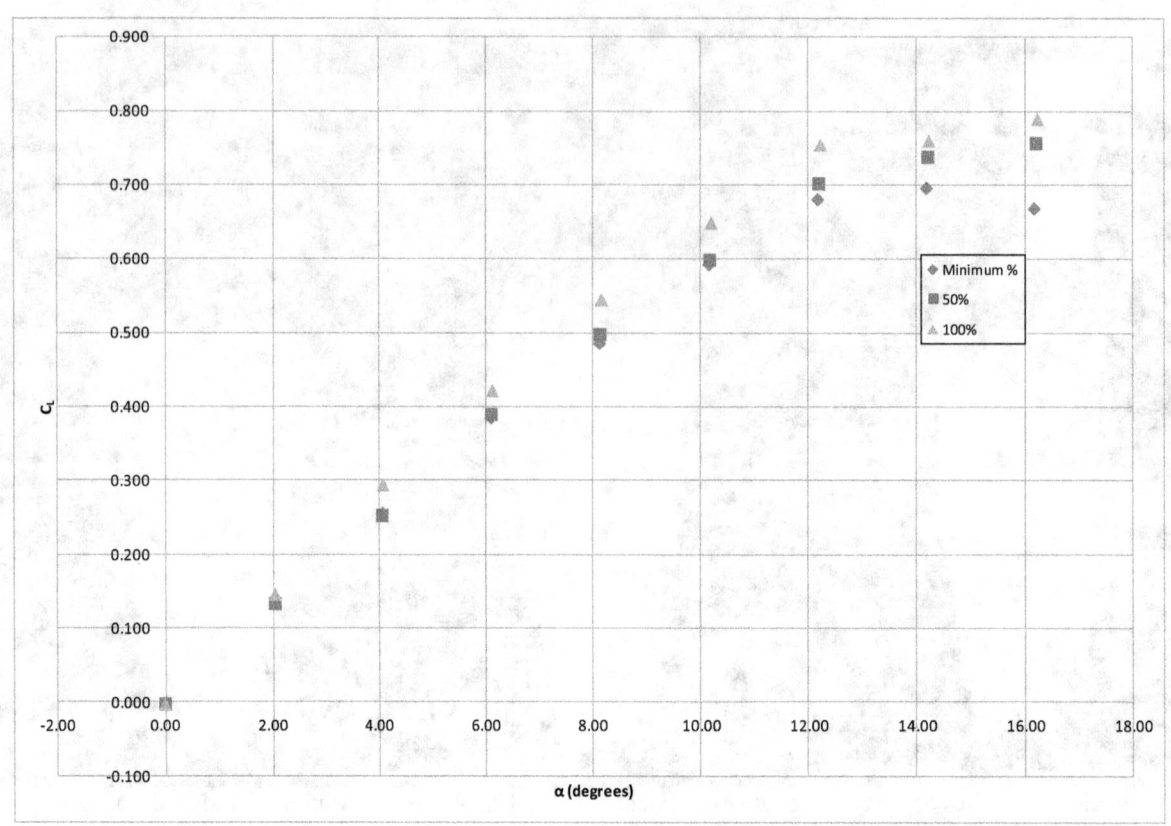

Figure 14. Effect of chord length variation on SD8020 lift coefficient at two knots.

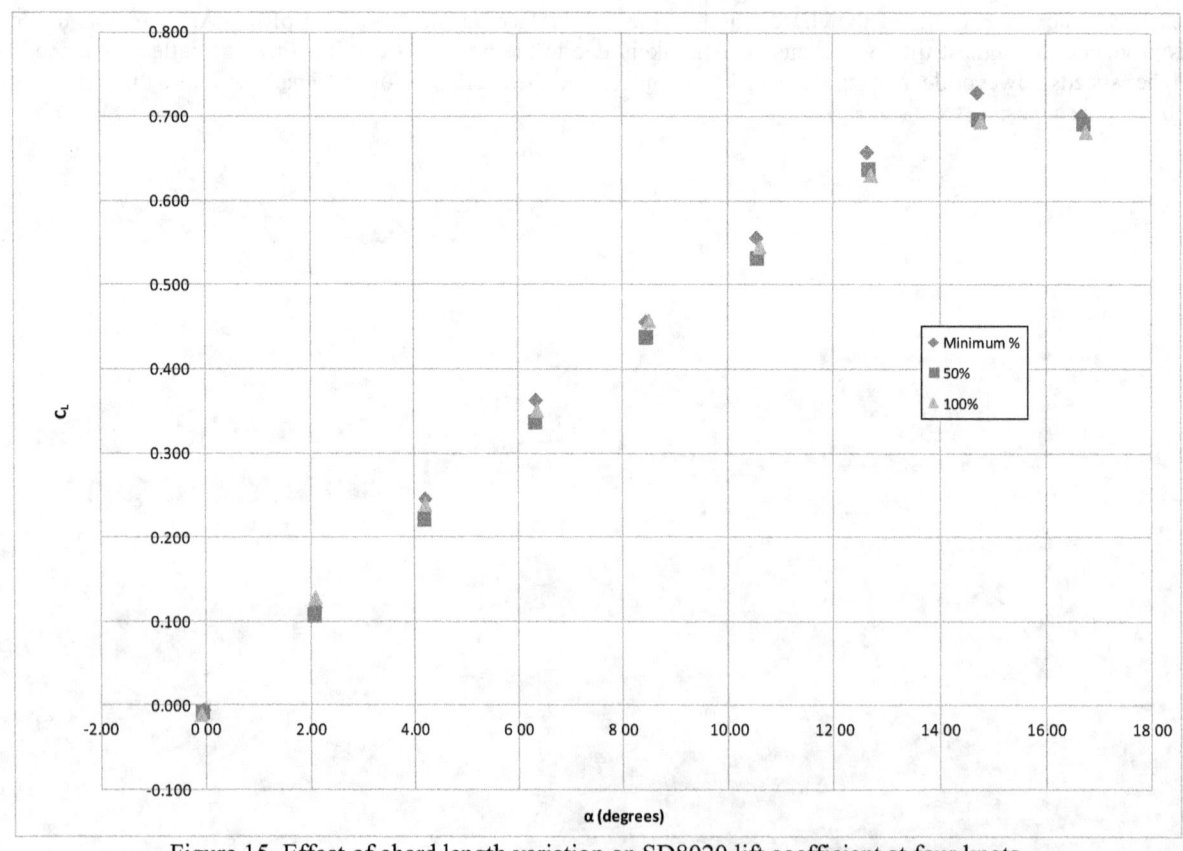

Figure 15. Effect of chord length variation on SD8020 lift coefficient at four knots.

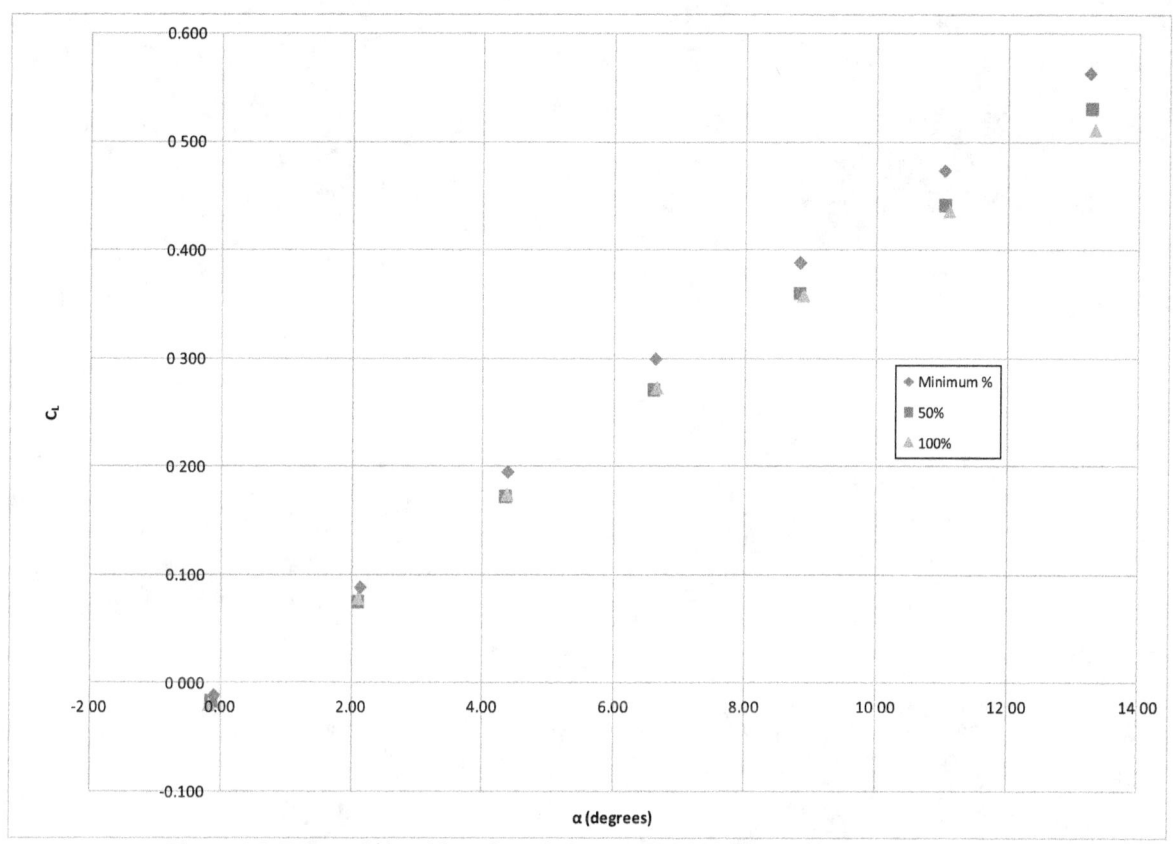

Figure 16. Effect of chord length variation on SD8020 lift coefficient at six knots.

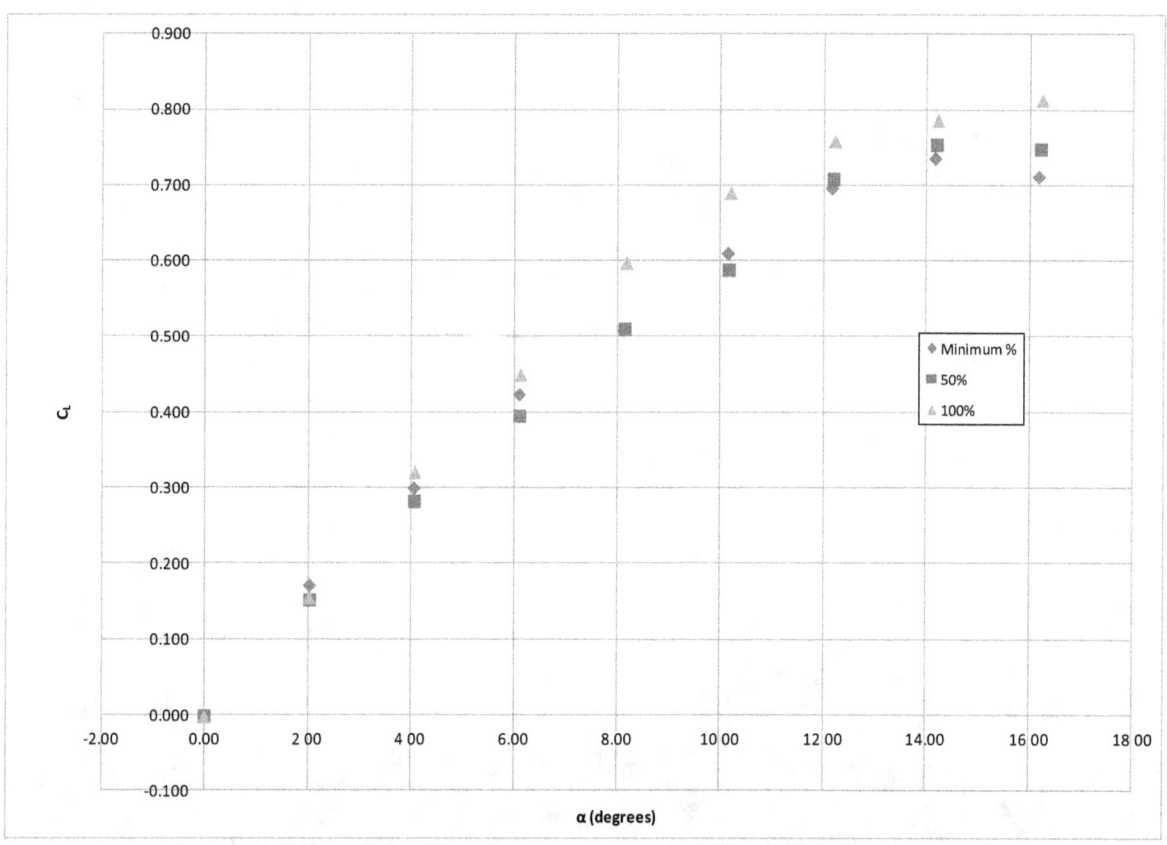

Figure 17. Effect of chord length variation on ogival lift coefficient at two knots.

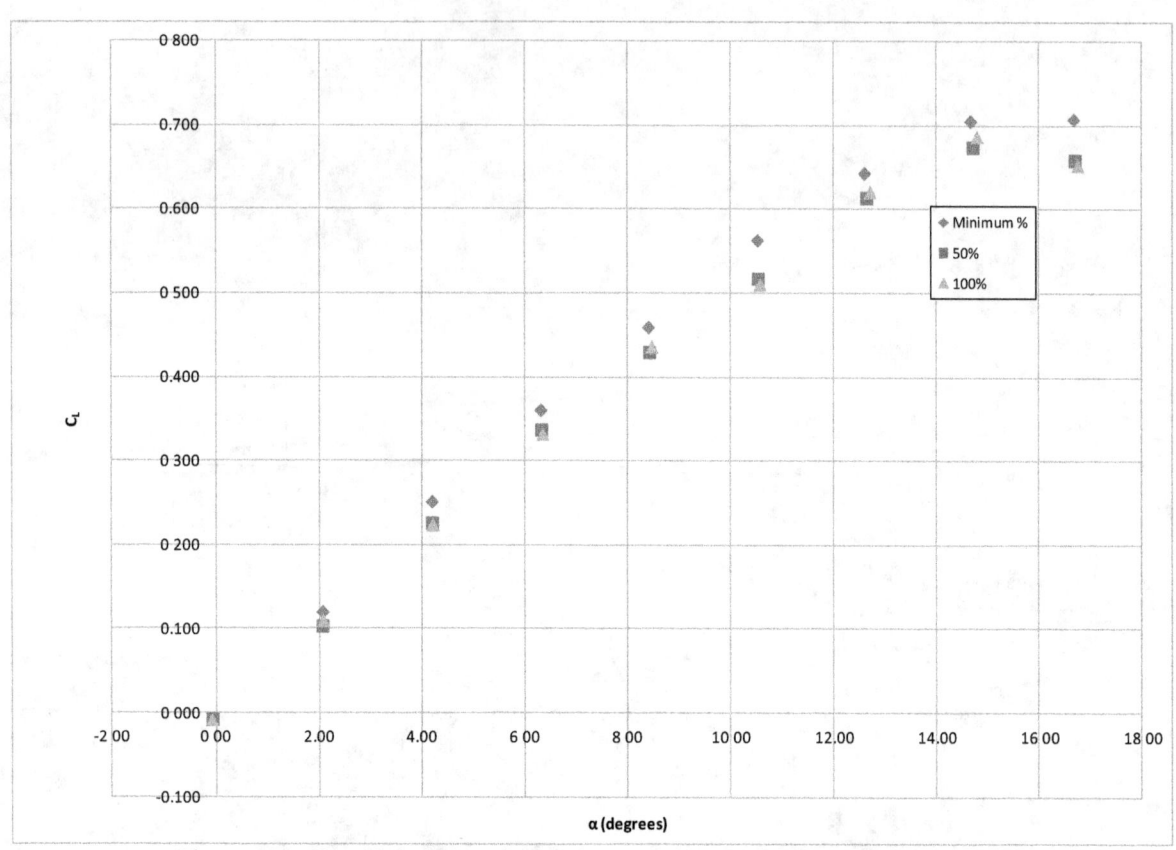

Figure 18. Effect of chord length variation on ogival lift coefficient at four knots.

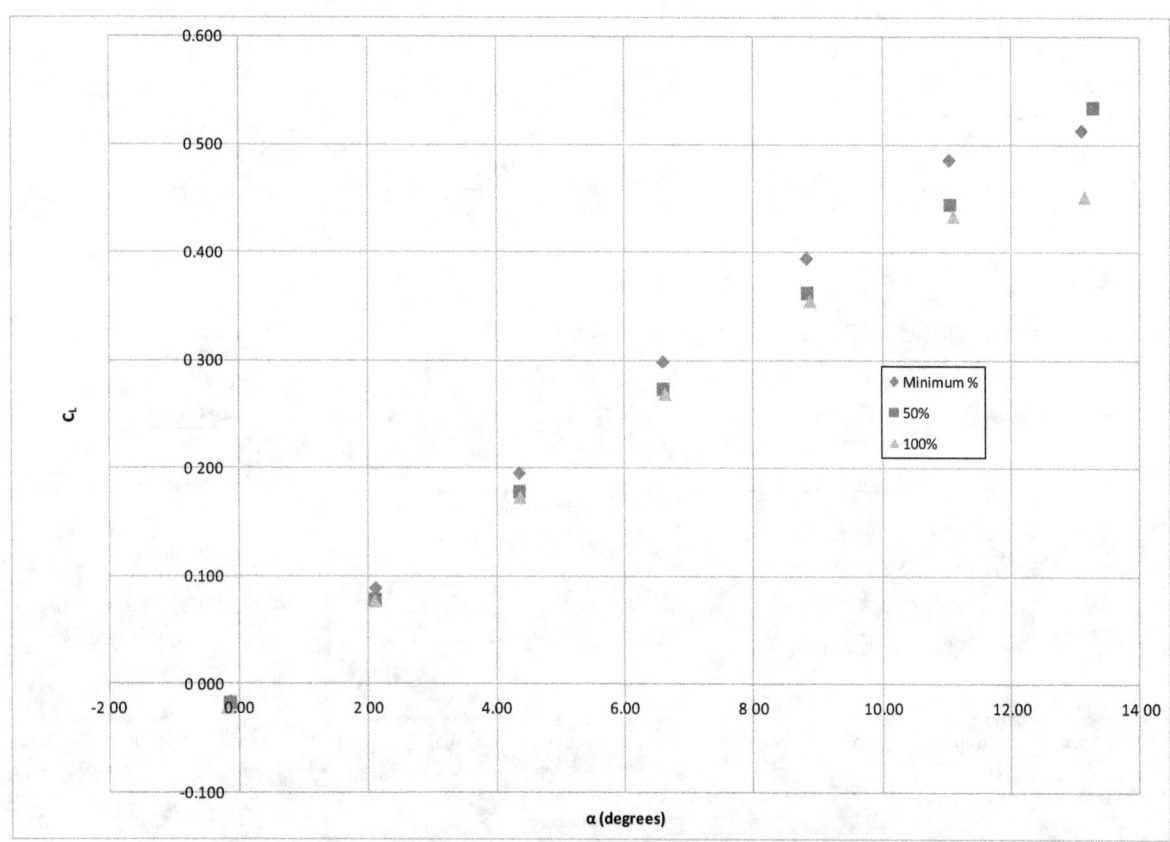

Figure 19. Effect of chord length variation on ogival lift coefficient at six knots.

Figures 20-25 show the impact of chord length on the C_L/C_D ratio versus C_L. Although not as strong a benefit as when comparing C_L versus the angle of attack, there is still the trend that reducing the chord length improves the foil efficiency as speed increases. At the lowest speeds however there is quite a bit of scatter and at high lift coefficients the lift-to-drag ratio is quite low for both sections. This is likely due to laminar transition effects at the lower speed and separation effects at the higher speed. The SD8020 section showed slightly higher lift-to-drag ratio than the ogival section at four and six knots.

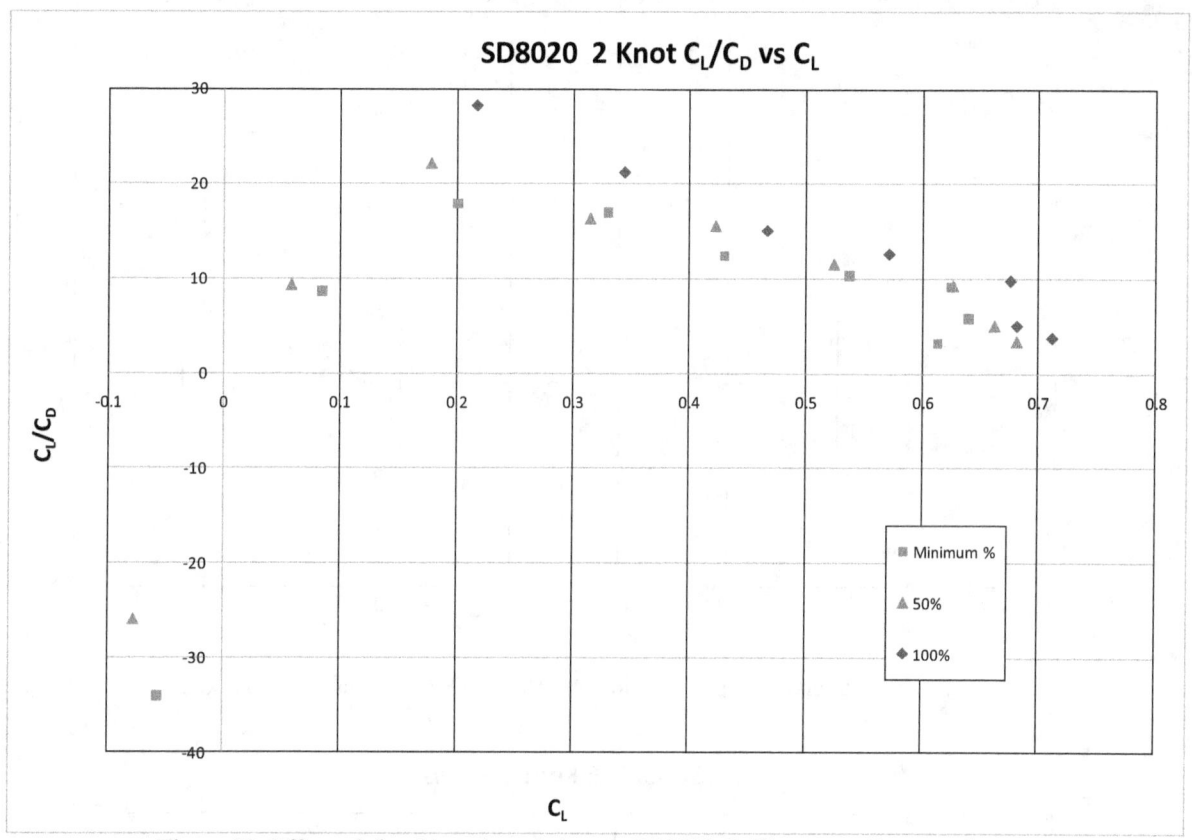

Figure 20. Effect of chord length variation on SD8020 C_L/C_D versus C_L at two knots.

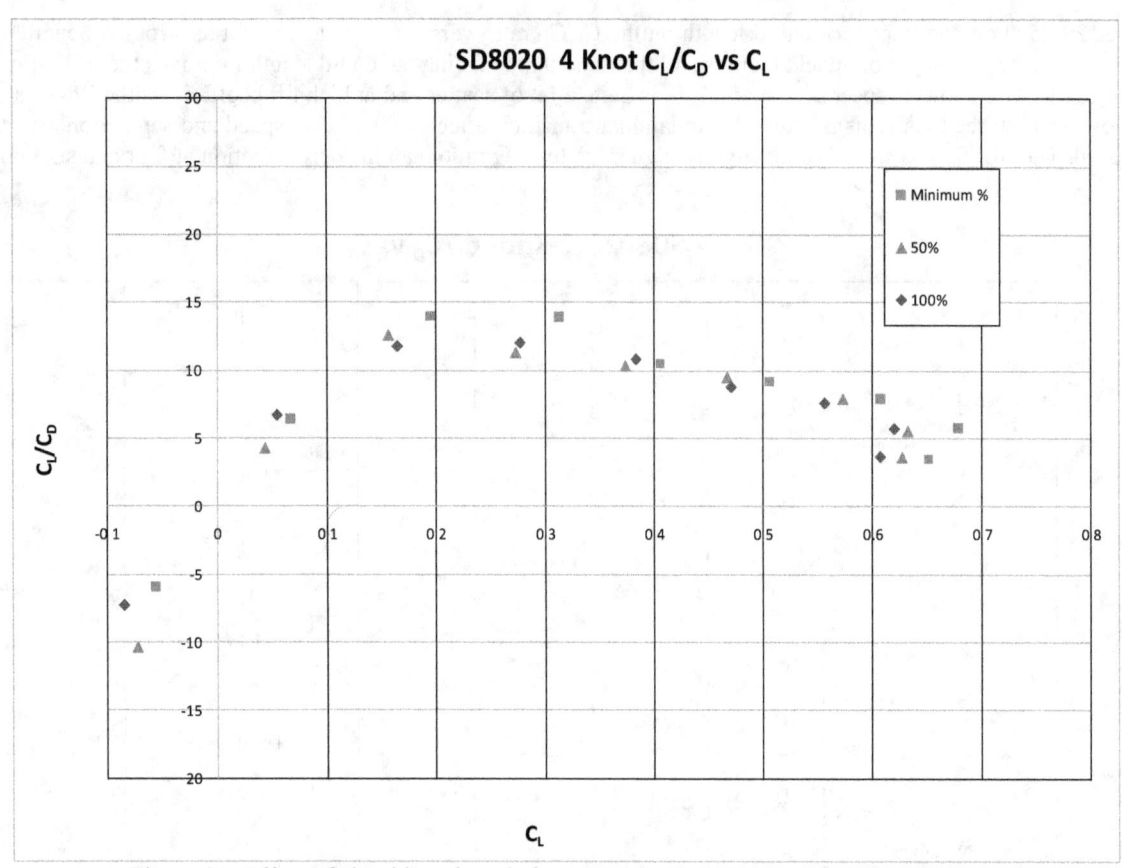

Figure 21. Effect of chord length variation on SD8020 C_L/C_D versus C_L at four knots.

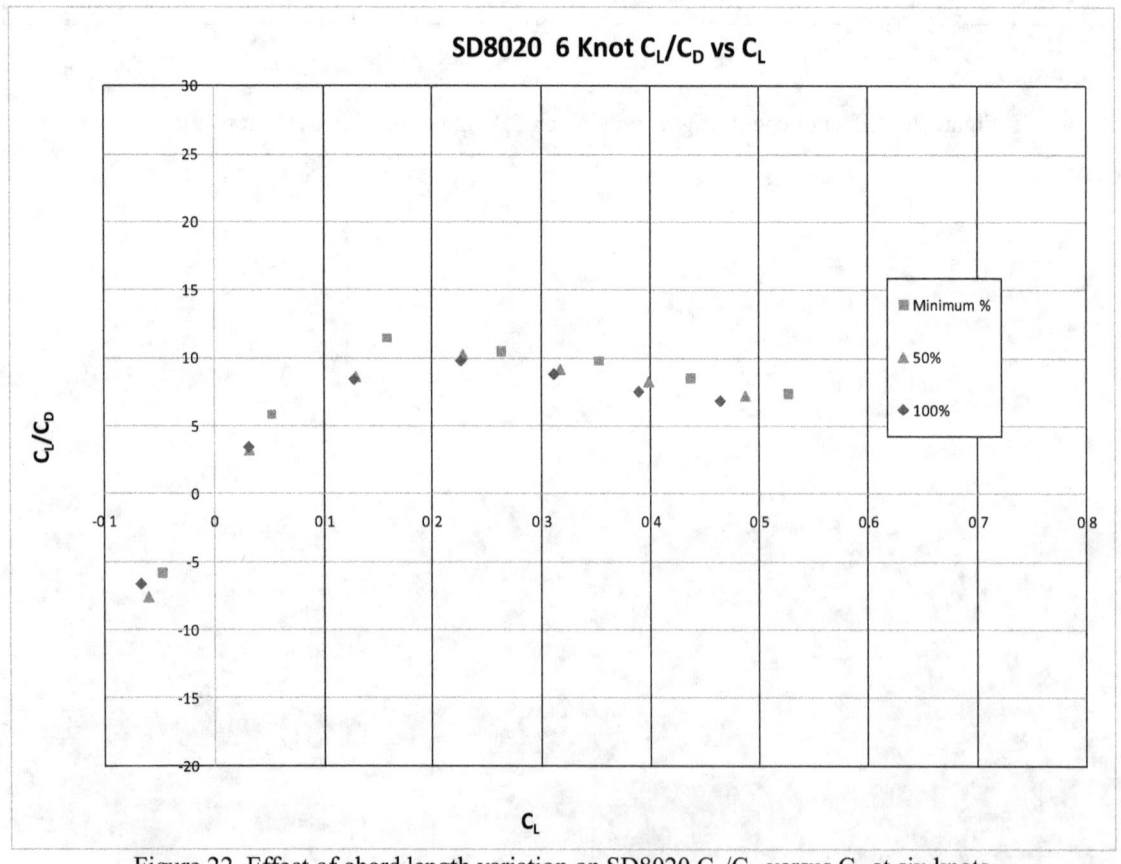

Figure 22. Effect of chord length variation on SD8020 C_L/C_D versus C_L at six knots.

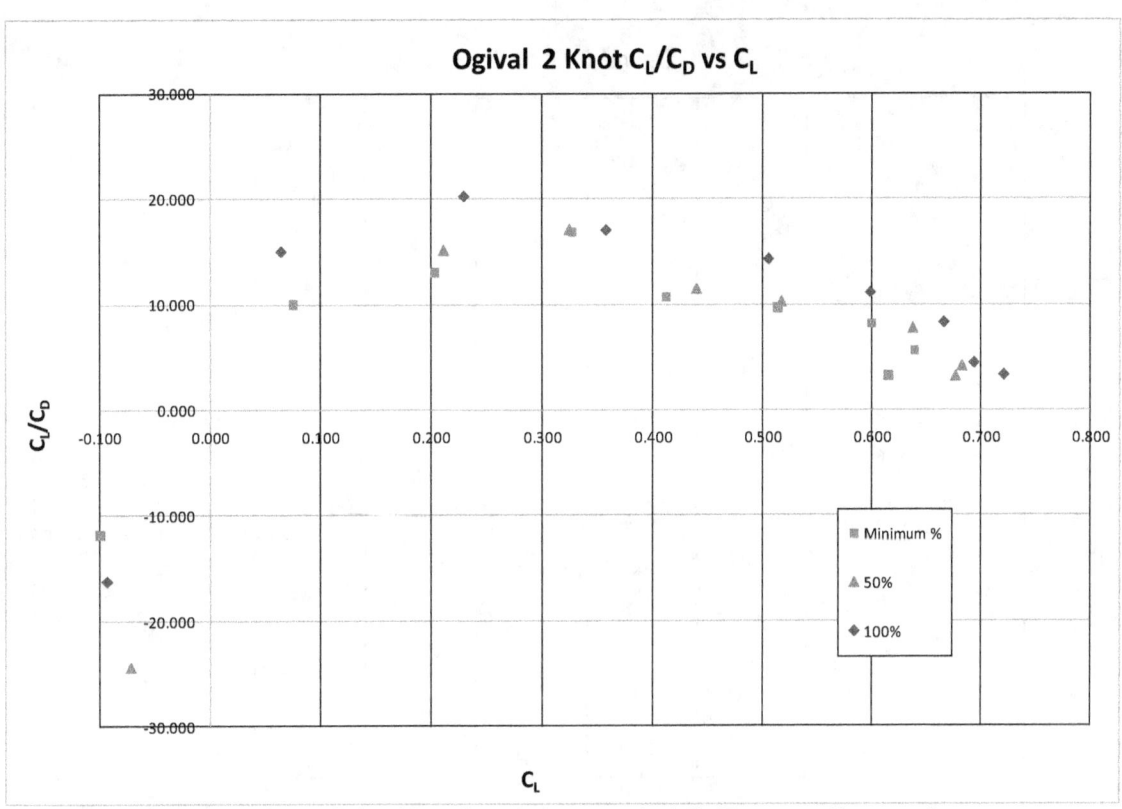

Figure 23. Effect of chord length variation on ogival C_L/C_D versus C_L at two knots.

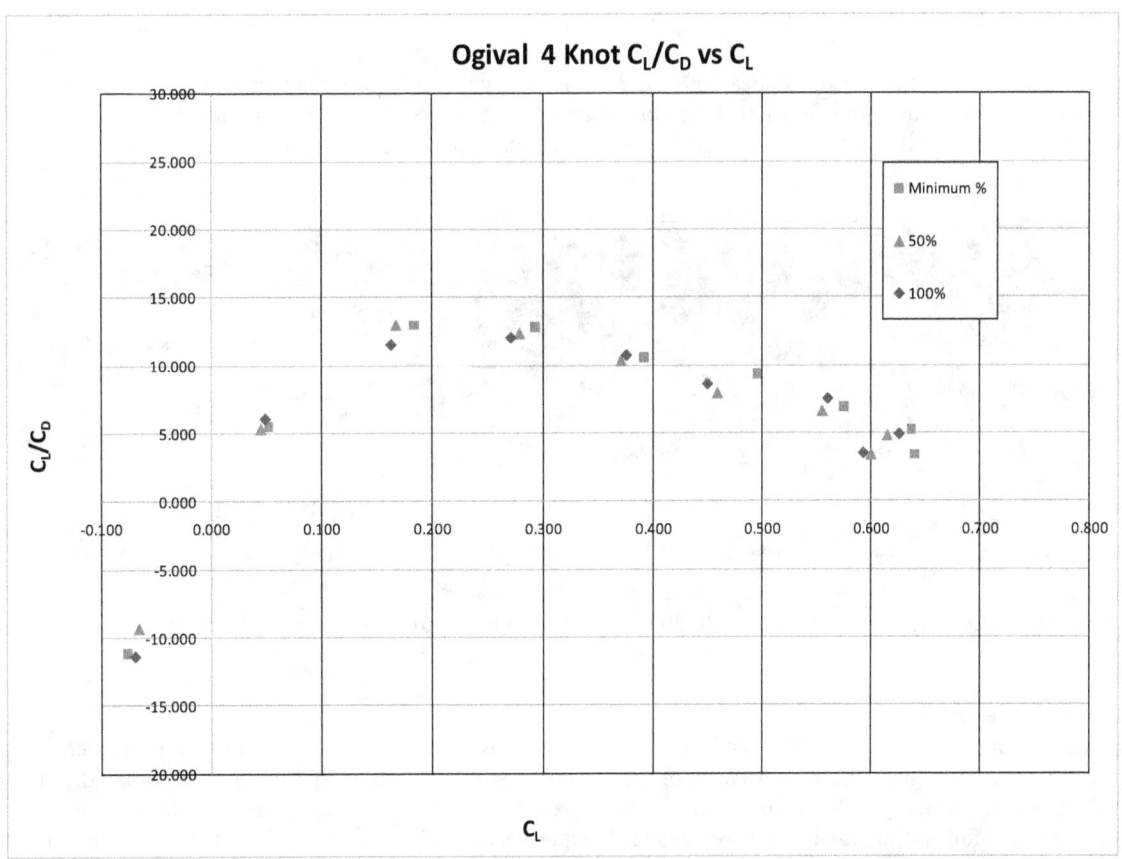

Figure 24. Effect of chord length variation on ogival C_L/C_D versus C_L at four knots.

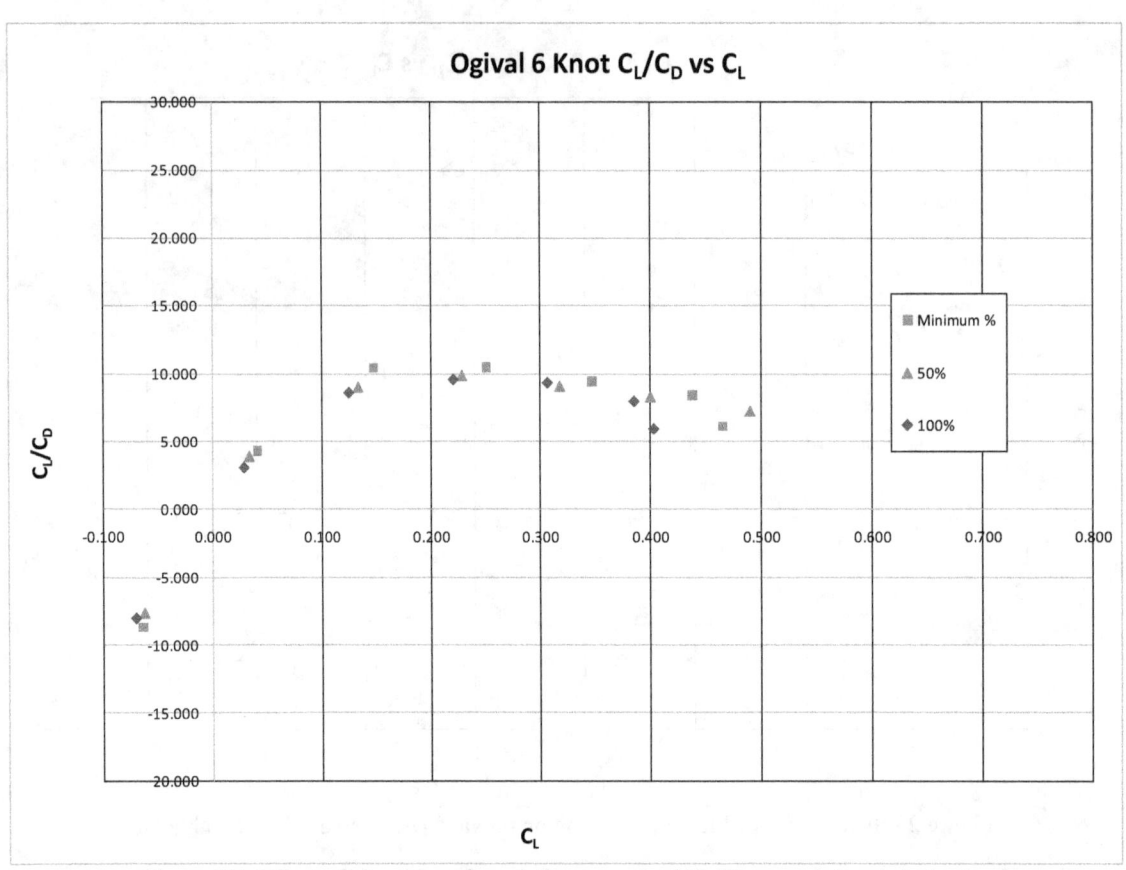

Figure 25. Effect of chord length variation on ogival C_L/C_D versus C_L at six knots.

The chord length experiment results appeared to substantiate the general hypothesis that minimizing the chord length at the free surface improved the foil efficiency at the speeds tested. Figure 26 shows photographs of the shortest and longest chord SD8020 sections at six knots and six degrees angle of attack. The larger waves and spray are evident in the longer chord rudder.

Figure 26. Shortest and longest chords of SD8020 section at six knots and six degrees angle of attack. Note the greater spray and waves on the longer chord.

Section Shape Experiments

As seen in Figure 3, the two sections tested in addition to the baseline section have significantly different shapes. The SD8020 has its maximum thickness well forward of the midpoint and has a relatively blunt edge compared to the ogival's sharp leading edge. While the SD8020's shape provides more gradual pressure recovery and shows lower drag characteristics when fully submerged, its shape leads to greater wave generation at the free surface. In contrast, the fore-and-aft symmetric ogival section has a sharp leading edge, reducing wave making but making the pressure recovery steeper and leading to separation problems at higher angles of attack. As the rudder is a control surface, having separation at a low angle of attack will lead to control problems. While perhaps the ideal would be to have the SD8020 fully submerged and the ogival at and above the waterline, in reality, the waterline changes and a transition is needed. The question is whether transitioning to a surface-piercing shape such as the ogival would be worth the fabrication complexity and cost.

One way to compare the potential section effect is to look at the general shape and peaks of the two sections' C_L/C_D versus C_L curves at a given speed. Using this measure, for the nine pairings from Figures 20-25, the SD8020 had a higher peak and generally higher values in six cases and the two shapes were effectively equal in the three other cases. In no case did the ogival section have higher values.

The differences between the two sections however are not large, likely because the majority of the rudders were similar as only the small section at the free surface was different between any of the rudders. The trend did show a slight advantage to the SD8020, although it is possible that a shorter taper region where the ogival extends only slightly below the free surface may be beneficial.

Design Implications

The results indicate some improvement may be possible in surface piercing sailboat rudder design. Specifically, a rudder with a forward sweep of 5 degrees and the shortest possible chord at the free surface might be an improvement over the baseline design. To gain an idea of how much improvement might be possible in practical units, a velocity prediction program (VPP) was used to compare the baseline and the potentially improved rudder designs.

PCSAIL, an Excel-based VPP written by David Martin (2001) was used for the analysis. Characteristics of the *Fireball* were taken from the Fireball Class Rules (2011) and by measuring existing boats. To stay within the limits of the code (displacement speeds) and the experiments, the speeds considered were at six knots and below, which roughly corresponds to sailing in winds less than 10 knots. While this stayed within the test results, the trends seen in the tank indicate the advantage of the smaller chord may be even more pronounced at higher speeds due to the increased importance of spray at higher speeds. As written, the code calculates rudder drag using the ITTC friction method with the inputs limited to the rudder's physical dimensions. This implies a zero angle of attack and no induced drag. This simplified method leads to a reduction in drag coefficient with increasing speed, while the data showed an increase in the rudder drag coefficient due to wavemaking and spray. In the case of rudders located under the hull the error is probably small unless the rudder becomes exposed due to heel. The ITTC approach also is somewhat unrealistic as a dinghy rudder is generally in constant motion to make small heading changes.

To account for the changes seen in the experiments the code was changed to use the as-measured total rudder drag coefficients using a curve fit through the results from the SD8020 chord and sweep experiments added together (which is probably a bit optimistic but indicates the best-case situation). Although the VPP did not include yaw-balance, the drag coefficients used were those at two degrees angle of attack to better represent the typical rudder condition of slight weather helm. This also agreed with input from experienced *Fireball* sailors that the boat was generally set up with small rudder loading. As a high-performance dinghy the *Fireball* was predicted to reach six knots rather easily, particularly when sailing off-wind. Figure 27 is the Polar Diagram predicted by the VPP using the Winder rudder with no sweep for wind speeds of four, six and eight knots, which generally kept the boatspeed below six knots.

To compare the baseline and proposed rudder the VPP was used to predict the time required to sail around a one-mile windward-leeward course in four, six and eight knots of wind. Table 1 shows the upwind and downwind velocity made good (VMG) in knots, seconds to sail the course and difference between the two rudder designs. On average the proposed rudder with the shortest chord and forward sweep was predicted to be 19 seconds per mile faster. This is a significant speed improvement for a competitive boat class and surprised the researchers. Part of the reason can be traced to the relatively flat resistance curve of a lightweight planing dinghy where small improvements can create large changes, but the result must be taken with the understanding of the proposed design's negative features that were not considered by the VPP. This includes an increase in rudder weight, which will increase the pitch gyradius, and therefore the added resistance in waves. Additionally, the forward sweep may lead to unfavorable steering characteristics due to the unstable yaw moment caused by having the center of pressure forward of the pintle axis and the typically unfavorable separation pattern seen in forward swept airfoils. This will certainly require greater skill by the helmsman. Finally, the forward sweep will increase the likelihood of catching weeds.

Table 1. VPP estimates for the velocity made good for the baseline and proposed rudder and the change in time required to sail a one-mile windward-leeward course.

Windspeed	Baseline Rudder			Proposed Rudder			
	Upwind VMG	Downwind VMG	Secs/mi	Upwind VMG	Downwind VMG	Secs/mi	Delta
4	1.64	1.94	2011	1.66	1.95	1993	-19
6	2.37	2.87	1376	2.42	2.90	1355	-21
8	2.94	3.75	1076	3.00	3.80	1059	-17

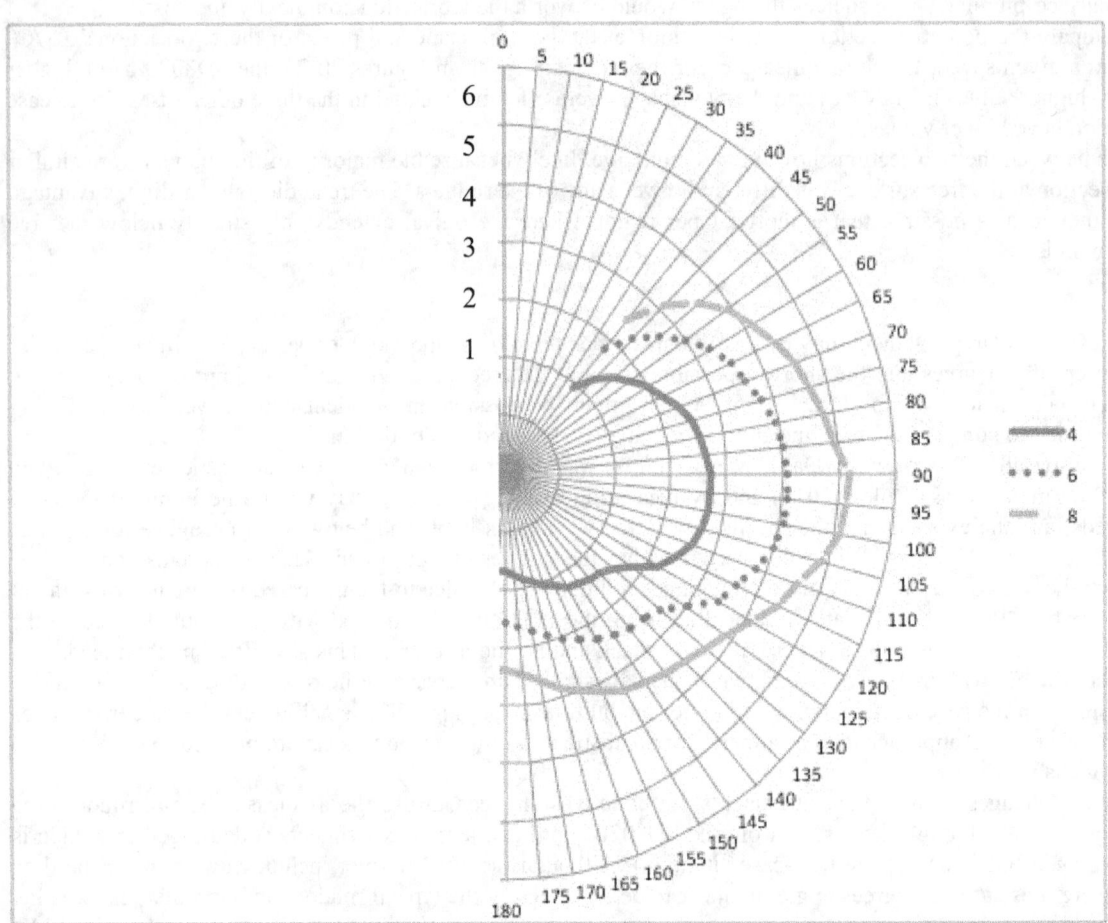

Figure 27. PCSail Polar Plot for the *Fireball* for wind speeds of 4,6 and 8 knots of breeze. The boat speed is in knots. An overlay of the improved performance is small enough that it would not be discernable on the plot.

CONCLUSIONS

This study presented experimental investigation of potential modifications to transom-hung small sailboat rudders.

It was shown that reducing the chord length near the free surface may result in an increase in lift-to-drag ratio for a given rudder. The key limits to reducing the chord length as the rudder passes through the free surface are the structural capacity of the rudder and the impact of the additional weight in the stern brought on by the increased weight of the strengthened rudder.

Forward sweep also provided an increase in lift-to-drag ratio. Forward sweep may be limited by practicality of weed shedding and rudder balance.

Modification of the section shape over the portion of rudder that passes through the free surface had little measureable effect in the range of test conditions. The ogival section traditionally used for surface piercing struts and the airfoil section gave similar results.

In most cases the measured differences were slight, although trends indicated that small amounts of forward sweep and shorter chord lengths increased the lift-to-drag ratio in the speed range most commonly seen. In real terms, a hypothetical *Fireball* rudder with a 13% SD8020 section, 84 mm chord at the free surface and five degrees of forward sweep showed an average of 19 seconds per mile improvement over the baseline rudder in calm water. While there is uncertainty based on the precision of the experiments, this predicted speed increase is substantial and warrants further investigation.

The findings of this study indicate that there may be value in investigating the effects of reducing the chord length at the free surface and adding forward sweep to a transom-hung rudder.

ACKNOWLEDGMENTS

While only one name appears on this paper it truly was a team effort. The author would like to thank the generous assistance provided by America's Cup hydrodynamicists Paul Bogataj and Joe Laiosa in developing the original hypothesis and developing the test parameters. Bill Beaver and John Zcelesky of the USNA Hydromechanics Laboratory contributed significantly in developing and producing the experimental setup and data acquisition system and Tom Price of the USNA TSD Model Shop was instrumental in fabricating the test sections. Dr. Micahel Morabito of the USNA NAOE Department provided invaluable advice on the data analysis, editing and motivation to write the paper. Funds to purchase the Fireball rudder and other materials were provided by the United States Naval Academy. Most importantly however was the motivated and tireless hard work by the student researcher, Clark Hayes, who truly made this project happen.

REFERENCES

Beaver, W.; Zseleczky, J. "Full Scale Measurements on a Hydrofoil International Moth", proceedings of The 19th Chesapeake Sailing Yacht Symposium, Annapolis, Maryland, March 2009

"INTERNATIONAL FIREBALL CLASS RULES 2011", Fireball International, 2011. Accessed from http://www.fireball-international.com/media/1171/Fireball%20Class%20Rules.pdf, 17 January 2013

Hoerner, S. F. (1965). *Fluid-dynamic drag*. Brick Town, NJ: Hoerner Fluid Dynamics.

Huebner, T. (2001). "Performance evaluation of sailboat rudders". Unpublished report for EN495 at USNA. A summary of the report can be found in, Miller, Paul H., "Student Research Projects for the New Navy 44 Sail Training Craft", 16th Chesapeake Sailing Yacht Symposium, Annapolis, MD, March 21-22, 2003

Kroo, Ilan, Wing Geometry and Lift Distribution. *Applied Aerodynamics: A Digital Textbook*. from www.desktop.aero/appliedaero/wingdesign/geomnldistn.html (accessed May 07, 2011)

Keuning, J. A. and Verwerft, B. "A New Method for the Prediction of the Side Force on Keel and Rudder of a Sailing Yacht Based on the Results of the Delft Systematic Yacht Hull Series" - 19th Chesapeake Sailing Yacht Symposium, Annapolis, MD, March 20-21, 2009

Keuning, J. A. Katgert, M. and Vermeulen, K. J. "Further Analysis of the Forces on Keel and Rudder of a Sailing Yacht", 18th Chesapeake Sailing Yacht Symposium, Annapolis, MD, March 2-3, 2007

Larsson, L, Eliasson, R. & Orych, M.(2014). *Principles of Yacht Design*. Camden, ME: International Marine/McGraw-Hill. Pg. 110-111

Marchaj, C. A. (1979) *Aero-Hydrodynamics of Sailing*. Dodd, Mead & Company, New York

Martin, D., Beck, R., (2001) "PCSail, A Velocity Prediction Program for a Home Computer," in the Proceedings of the 15th Chesapeake Sailing Yacht Symposium, Annapolis, MD, USA, January 2001

Miller, P. H. ,"Dynamic Lift Coefficients for Spade Rudders on Yachts", 18th Chesapeake Sailing Yacht Symposium, Annapolis, MD, March 2-3, 2007

Millward, A. "The induced drag of a vertical hydrofoil." PhD Thesis, University of Southampton, UK, 1967. *This reference was cited by Molland and Turnock (below) although a copy was not available for confirmation.*

Molland, A, & Turnock, S. (2007). *Marine rudders and control surfaces*. Burlington, MA: Butterworth-Heinemann.
Introduction to SAS. UCLA: Academic Technology Services, Statistical Consulting Group. From http://www.ats.ucla.edu/stat/mult_pkg/faq/general/coefficient_of_variation.htm (accessed May 07, 2011)

AP Statistics Tutorial. Stattrek.com. from http://stattrek.com/Lesson3/TDistribution.aspx?Tutorial=AP (accessed May 07, 2011)

Science of the 470 Sailing Performance

Yutaka Masuyama, Kanazawa Institute of Technology, Kanazawa, Japan
Munehiko Ogihara, SANYODENKI AMERICA, Torrance CA, USA

ABSTRACT

To sail a 470 faster, authors consider the sailing performance of the 470 from the various measured data and the simulated results of Velocity Prediction Program (VPP). This is a summary of TOBE-470 [1] presented by the author.

NOTATION

A	Sail area
B	Breadth at design waterline
C_L	Lift force coefficient
C_D	Drag force coefficient
C_N	Yaw moment coefficient
D	Design draft (including center board 1.08m)
Fn	Froude number
L	Length on design waterline (4.4m)
$K_{Trapeze}$	Righting moment of Trapeze
K, N	Moments about x- and z-axis in horizontal body axis system
S	Wetted area (4.52m^2)
T	Resistance
U, V	Velocity components of boat along x- and y-axis in horizontal body axis system
U_A	Apparent wind speed (AWS)
U_T	True wind speed (TWS)

V_B	Boat velocity
VMG	Velocity Made Good
X, Y	Force components along x- and y-axis in horizontal body axis system
β	Leeway angle
γ_A	Apparent wind angle (AWA)
γ_T	True apparent wind angle (TWA)
δ	Rudder angle
φ	Heel angle or roll angle
ψ	Heading angle
ρ_a	Density of air
ρ_w	Density of water
$\Delta\,\overline{GZ}$	Righting moment (without trapeze)

INTRODUCTION

The 470 (Four-Seventy) was designed in 1963 by the Frenchman Andre Cornu as a double-handed mono-hull planing dinghy. The name comes from the overall length of the boat in centimeters (470cm). The 470 is a World Sailing International Class has been an Olympic class since the 1976 games. In Japan, the 470 is used in university championships and in National Athletic meets. So, it is in the most popular dinghy race in Japan. An ideal crew weight of skipper and crew is 130kg, it is suitable for Japanese who are smaller than Europeans and Americans. This paper progresses the science of the sailing performance of the 470 with respect to aspects such as hull performance, sail performance, steady sailing performance and maneuvering performance.

Specification of the 470

Table 1 shows the technical details [2]. Figure 1 shows the Sail plan [2]. Figure 2 shows the hull shape [3]. The 470 has a very flat hull form. This implies that the hull form in the wetted area and submerged area variations are greatly dependent on the trim angle (pitch angle) and heel angle, which is considered to affect the performance as well.

Table 1 Technical detail of 470.

Length:	4.7m
Length of waterline:	4.4m
Weight:	120kg
Mast:	6.76m
Main:	$9.12m^2$
Jib:	$3.58m^2$
Spinnaker:	$13.0m^2$

Figure 1 Sail plan of 470.

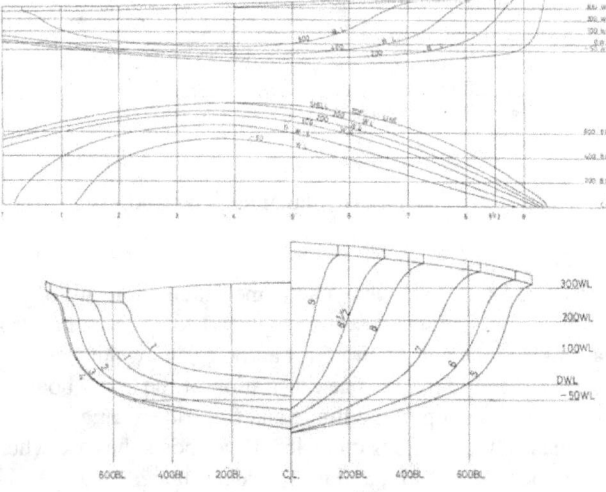

Figure 2 Hull shape of 470[2].

Shape of Hull

Figure3 shows result of the calculation of displacement and wetted area variations for draft depth from hull shape of Figure 2. The horizontal axis is the draft depth. The vertical axis shows displacement $\triangle$ and wetted area ◆. When the gross weight including the crew and skipper is 250kgf, the draft depth is about 0.15m, and the wetted area is about 3.8m^2. Also, when the gross weight variation is 35kgf, the draft depth variation is 1cm and the wetted area variation is 0.25m^2.

Figure 4 shows calculated stability. The horizontal axis is the heel angle. The vertical axis shows stability. The gross weight including the crew (70kgf) and the skipper(60kgf) is 250kgf. However, skipper can hike out. So, CG of crew is 2m away from the hull centerline as of 90kgf (70kgf x 1.3). Dashed line shows without trapeze. Solid line shows full trapeze.

O: TWS 6m/s, Sail Full Power, TWA: 60°, AWA: 40°, AWS:8.5m/s

△: TWS 8m/s, Sail 70% Power, TWA: 60°, AWA: 40°, AWS:10.7m/s (heel moment was calculated as 70%)

X : TWS 8m/s, Sail Full Power, TWA: 60°, AWA: 40°, AWS:10.7m/s

From the Figure 4, without trapeze, the maximum righting moment is achieved at a heel angle of about 30°, but it is small as 60kgf-m. On the other hand, when trapeze is performed, it becomes maximum righting moment of about 220kgf-m at a heel angle of 25°. However, since the righting moment decreases at a further heel angle, the range of the heel angle that can be restored is surprisingly narrow. In the case of cruiser, since the ballast keel is around 40% of the hull weight on the bottom of the ship, the righting moment will continue to increase to about 50° in heel angle. This is the reason why dinghy can capsize more easily than cruiser. Heeling moment decreases with heel angle because heel decreases effective angle of attack of sail. The cross point of the heel moment curve is obtained under this condition when the righting moment curve of the hull becomes a balance point, that is, a steady sailing state. In the case of TWS 6m/s, heel angle is balanced at about 10°. The symbol of X indicates TWS 8m/s and the sail is full power. X intersects the righting moment of the blue line at heel angle 45°, righting moment is already decreasing, the boat will capsize at once with kinetic energy that makes it heel to 45°. In the case of power down sail to 70% (△), the heel angle is balanced at about 20°. Based on the above, the following can be concluded, a 470 dinghy with righting moment by trapeze has little increase in righting moment by heel, the maximum righting moment is at about a heel angle of 25° (side deck touches the water). As soon as the side deck begins to touch the water, it is necessary to immediately reduce the power of sail or luffing up to avoid capsize. In actual sailing, heeling moment which acts to increase heel is generated by the force of water acting on the hull (especially center board). For this reason, it is necessary to consider the moment of restoration of the hull by 10 to 20% lower than the value shown in Figure 4.

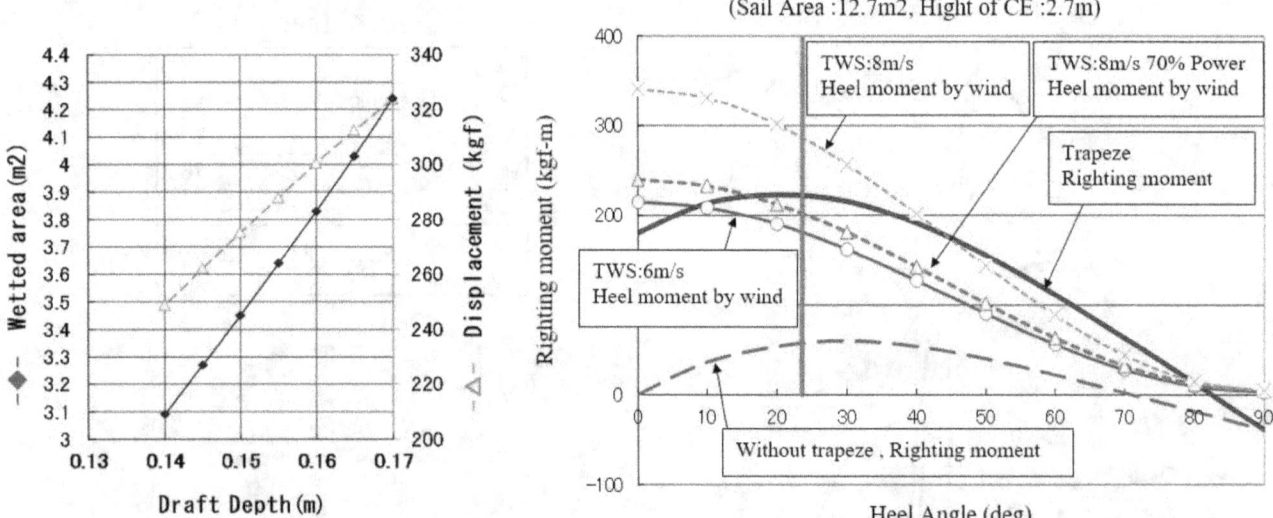

Figure 3 Draft depth/Displacement/Wetted area.　　　　**Figure 4 Righting moment/Heel moment.**

Tacking Maneuver

Figure 5 and 6 shows dynamic measurement result of tacking maneuver. Figure 5 (a) and 6 (a) shows the variation of the rudder angle (δ) and heading angle (ψ) for 25 seconds from 5 seconds before tacking. Figure 5 (b) and 6 (b) shows the variation of the V_B. Figure 5 (c) and 6 (c) shows the variation of the boat trajectories. Circles indicate the position of center of boat at each second. The illustration of the small boat symbol indicates the heading angle every two seconds. The wind blows from the top of the figure and the grid spacing is taken as 10 meters. Red shows the case where the rudder used gently, and max rudder angle is up to 45°. Blue shows the case where the rudder used quickly up to 45°. Green shows the case where the rudder used gently, and max rudder angle is up to 30°. U_T in the figure are average value for 25 seconds, but it is slightly different for each tacking. Since the situation before and after tacking is somewhat different, superiority or inferiority is hard to judge, however, the following is a summary.

(1)In case of blue, speed reduction during turning is large.

(2)In case of green, the time towards the wind is long, so the speed reduction may be large.

(3)In case of red, this seems to be the most reasonable operation.

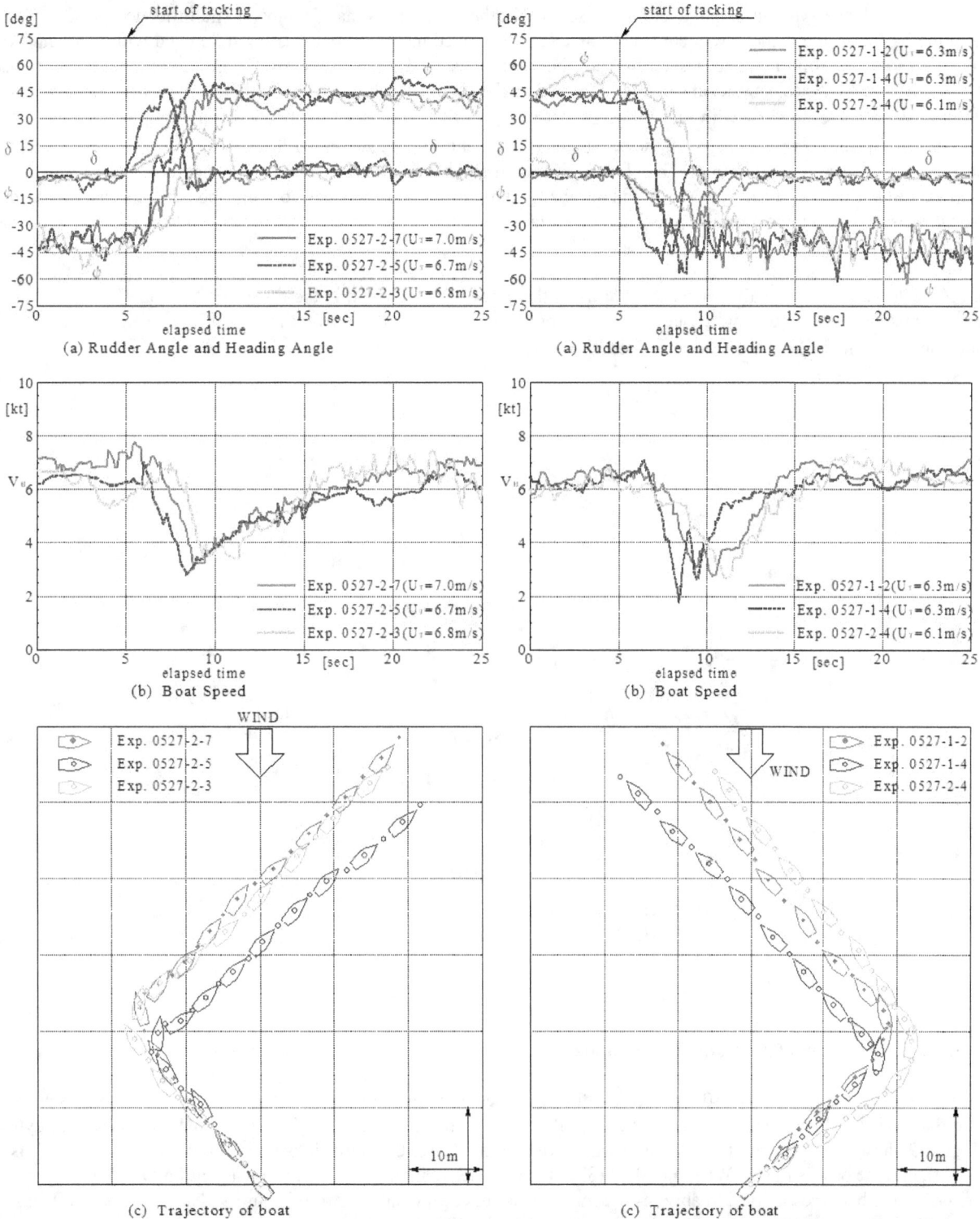

Figure 5 Tacking from starboard to port tack.　　　Figure 6 Tacking from port to starboard tack.

The Full-scale 470 Towing Test

Figure 7 shows the result of towing by motor boat with skipper and crew aboard Yamaha's 470 onboard and equipped with all the necessary equipment for sailing (total weight 260kgf). Although the tow point should originally be subtracted from the wind pressure center height of the sail, in consideration of the danger at the time of high speed, the towing rope was tied to the height of 1.5 m from the cockpit floor surface of mast. For the same reason, the rudder is set, and the skipper is steering, but the center board is raised. Towing speed is measured with a handy GPS, resistance is measured by motor boat using a spring balance. Measurements were made in the Northern Bay of Nanao in Japan. The influence of tide and tidal wave in the bay was small, and it was confirmed that accurate speed measurement was possible even with the handy GPS. The curves shown by solid line and dashed line are those obtained by Tatano Hisayoshi [3] in a towing tank test at Osaka University using a 1/2.5 scale model and converted into actual ship. The dashed red line is the result of without center board, which is in good correspondent with the full-scale test results (without center board) within the range of velocity 7kt($\fallingdotseq$ 3.6m/s) or less. At the V_B=6kt($\fallingdotseq$3.1m/s), the resistance of the center board is about 1.9kgf, which is about 10% of the total resistance 19kgf. Even at V_B=6kt, more than half is frictional resistance. Since residual resistance is almost determined by the hull shape and displacement, the sailor cannot change anything other than the boarding position to change the pitch. Frictional resistance can be reduced by polishing the surface.

Planing

In Figure 7, the resistance value increases sharply in the range of 5.5kt to 8kt in the full-scale 470, but the rate of increase seems to be dull in the higher speed range beyond that. To clarify the above, the resistance coefficients and the Froude number in the form of them are indicated by ● and ○ points in Figure 8.

The resistance coefficient (C_T) and Froude number (Fn) are defined as:

$$C_T = \frac{T}{\frac{1}{2}\rho_w V_B^2 S} \quad (1) \qquad\qquad F_n = \frac{V}{\sqrt{gL}} \quad (2)$$

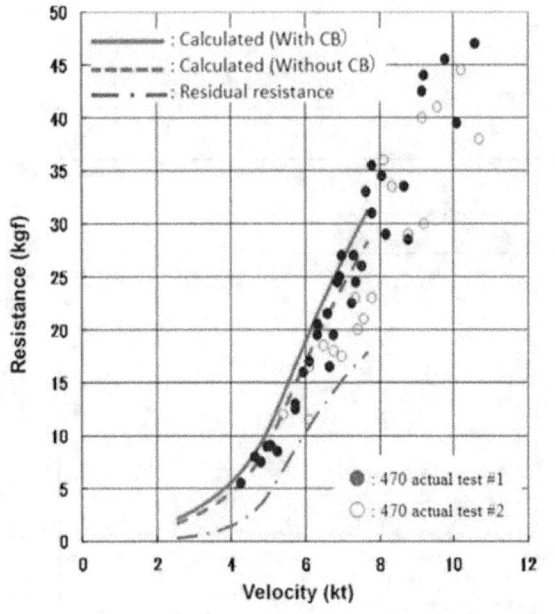

Figure 7 Variation of resistance (without heel).

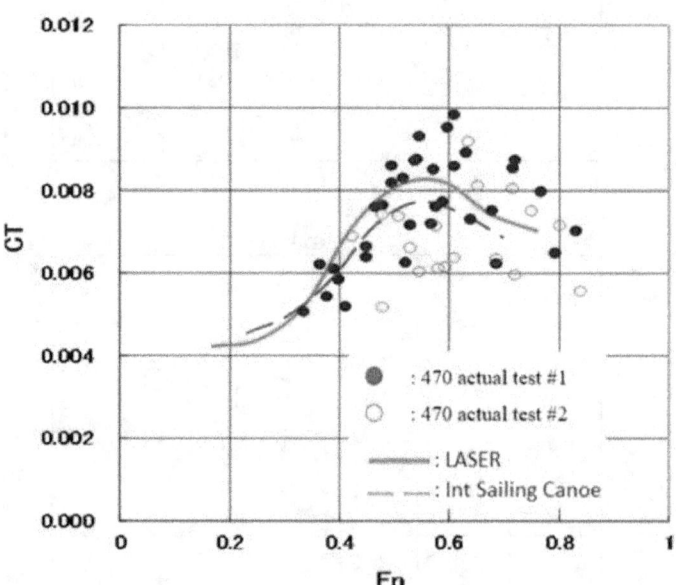

Figure 8 Variation of C$_T$ and F$_n$.

There are peaks in the range of Fn=0.5 to 0.6 for all three types of boats, and the resistance coefficient decreases as it goes beyond that. When it becomes faster than the above, it becomes semi-planing state [8]. The whole hull rises with dynamic lift. Figure 9 shows the attitude of the 470 at towing. Also, in this case, the stern sinks considerably from Fn=0.5 (6.4kt) to 0.6 (7.7kt), and the bow is raised. When Fn=0.63 (8kt), the total heave increases. And then reached Fn=0.8 (10.2kt), the stern also rises, and it becomes a considerably level trim. At this time, you can see from Figure 9 that the spray has come out considerably from the rear. As described above, in the case of a 470, exceeding 8kt is a measure for shifting to high-speed sailing. From the above, we call semi-planing boat velocity at 8kt (Fn=0.63) to 12kt (Fn=0.94), and further boat velocity the planing state.

Figure 9 Attitude of the 470 at towing.

Tank Test

We measured the hydrodynamic force acting on the hull by tank test using 1/5 model of the 470. Figure 10 shows the tank test. Figure 11 shows definition of coordinate system and forces and moment. Figure 12 shows the variation of hydrodynamic coefficients (X ', Y' and N ') acting on the hull by cross flow (rudder angle 0°).

The hydrodynamic coefficients are defined as:

$$X' = \frac{X}{\frac{1}{2}\rho_w V_B^2 LD} \qquad Y' = \frac{Y}{\frac{1}{2}\rho_w V_B^2 LD} \qquad N' = \frac{N}{\frac{1}{2}\rho_w V_B^2 L^2 D}$$

(3)

Figure 10 Tank test 1/5 model.

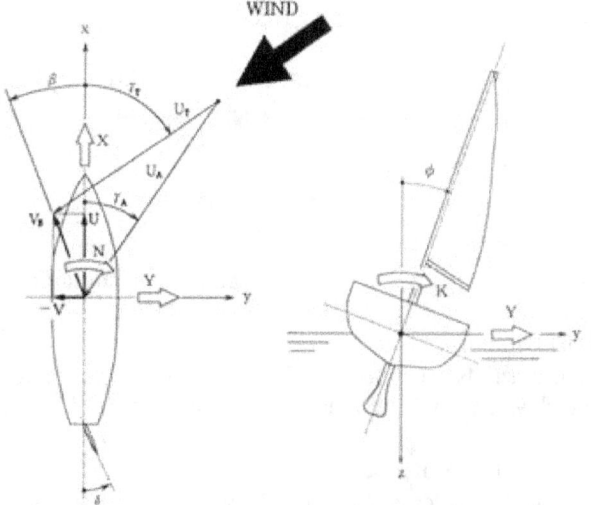

Figure 11 Definition of coordinate system and forces and moment, (+)-ve is indicated direction.

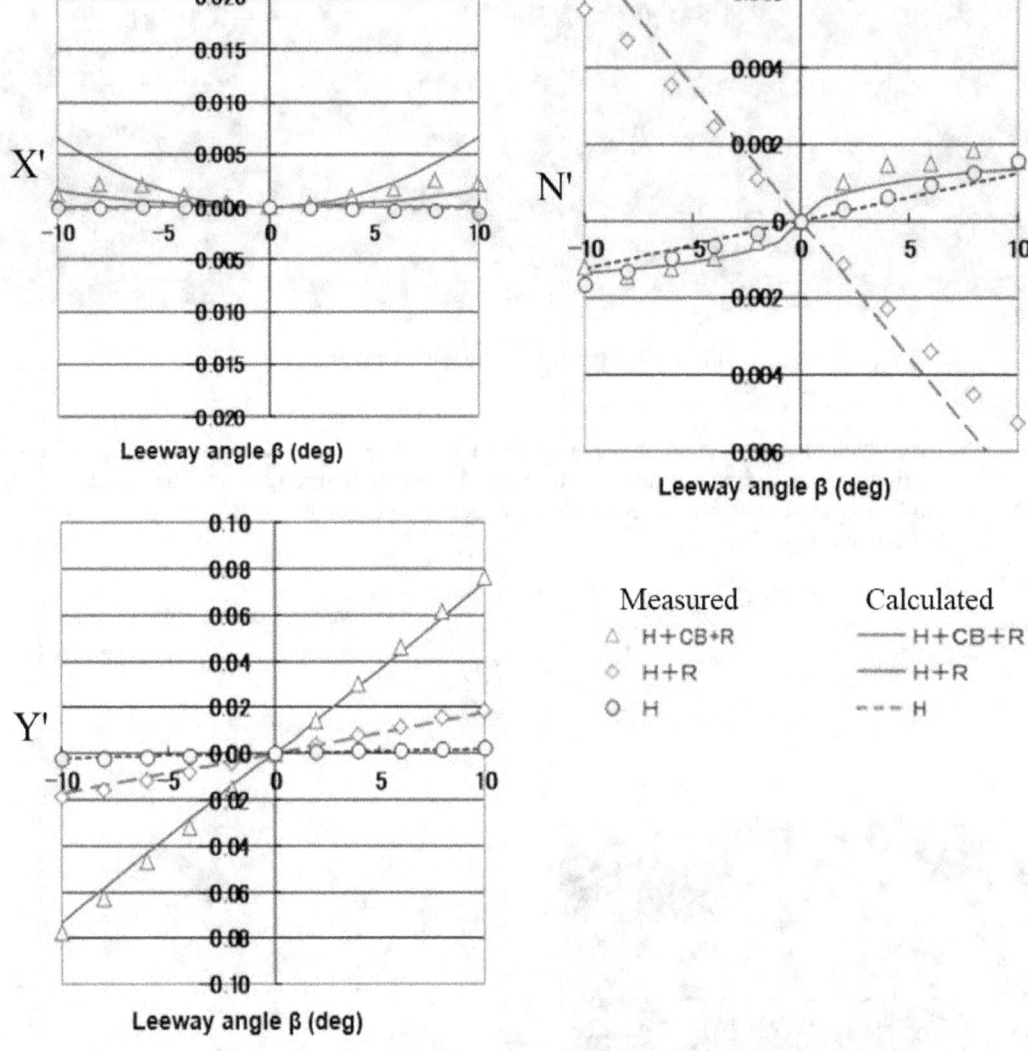

Figure 12 Variation of aerodynamic coefficients of hull.

(1)H (Hull: O)
Moment is generated even by hull alone. The fact that the graph rises to the right indicates that the yaw moment is positive (clockwise), that is, weather helm when the leeway angle is plus (receiving flow from the left front of the bow).

(2)H+R (Hull + Rudder: ◇)
When a rudder is attached, it becomes a graph going down to the right. This is because lateral forces acting on the rudder cause a large counterclockwise moment at right turn. In other words, when turning to the right, it acts so that it turns back to the left by rudder. This is the same effect that weathercock always shows wind direction, which is called "weathercock stability".

(3)H+CB+R (Hull + Rudder + Center Board : △)
When the center board is attached, it turns out that it returns to the right-rising graph like the case of only hull (1). This is because the center board is attached slightly ahead of the hull center, the lateral force acting on this is to counter the moment by the rudder. When rudder and center board are installed, yaw moment acts in a slight direction, but it further strengthens the turn (weather helm).

Center Board Angle
Figure 13 shows the variation of the position of the force application point relative to the angle of the center board when the leeway angle is 4°. When the center board angle is 90° (full down), the force point is about 0.2m ahead from the hull

center (2.25m ahead from transom in the case of the actual 470). Figure 14 shows the variation of the center board angle and lateral force coefficient Y '. The slope at 90° is the largest (good performance), and at 60 ° it decreases by about 17%. When close hauled (AWA about 25°), the CE of sail is approximately 0.2m ahead of the hull center at heel angle 0 ° [1]. Therefore, in close hauled when the center board angle is 90° (full down) and the heel angle is 0°, the force application point of hydrodynamic force acting on sail is almost the same position. It is almost just helm (rudder angle≒0°).

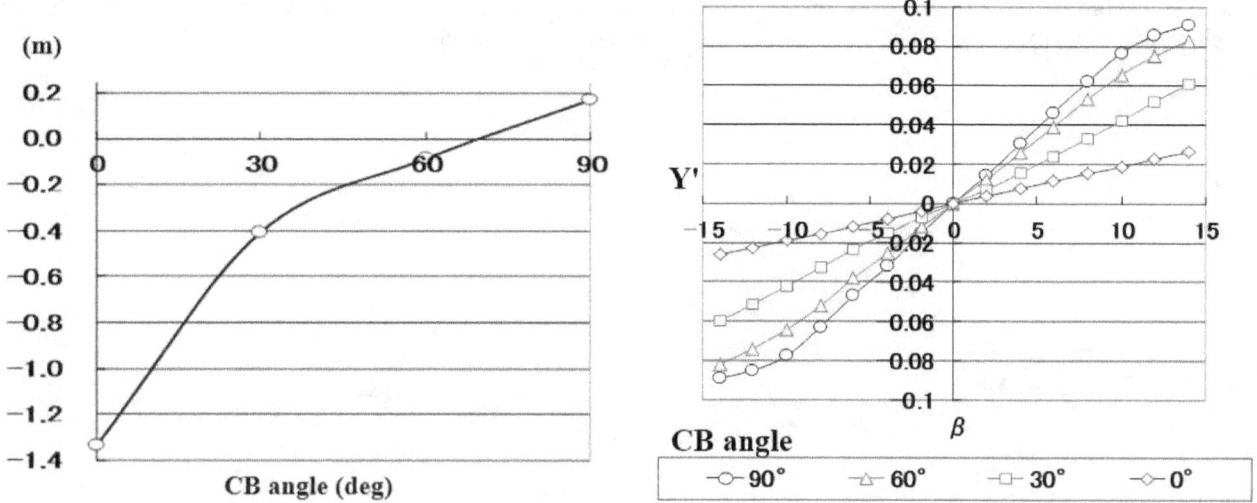

Figure 13 CB angle and force application point.   Figure 14 CB angle and Y'.

Rudder Angle (δ)

Figure 15 shows the variation of δ and hydrodynamic coefficients of hull (X ', Y', N '). Y 'and N' are nearly the same as the measured values when the δ is within the range of ± 15°. The minus of X ' means that it is a drag. When the δ becomes ± 15° or more, the measured values of Y 'and N' do not become larger anymore and stay at a constant value, whereas the minus value of X ' continues to increase. This is because the rudder stalls. In other words, even if δ is larger than ± 15°, the effect as a rudder does not change, it simply means the increase of drag. This drag can calculate from equation (3). EX:V_B=6kt:5kgf (δ=±10°), 12kgf (δ =±15°).

From Figure 7, the total resistance of straight (heel 0°) is about 19kgf at V_B=6kt. Comparing the resistance of the above with the drag of the rudder, it can be understood that rudder has a large drag. People who are not balanced between sailing attitude and sail and who are always sailing with a large rudder angle, should adjust mast rake and sail trim, and it is better for them to make rudder angle smaller.

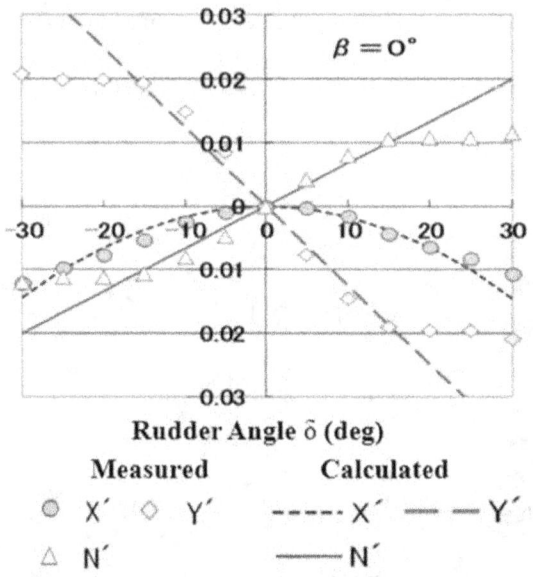

Figure 15 δ and hydrodynamic coefficients.

Heel Angle

Figure 16 shows the wetted area in the case of heel angle 0° and 25° at displacement of 250kgf [3]. The heel angle of 25° is the angle at which the leeward deck starts contacting with water. With heel the wetted area becomes quite asymmetric. Due to this asymmetry, a moment to turn upwind (weather helm) will occur even with only hull. Figure 17 shows the wetted area variation when only hull heels. At a heel angle of 25°, the wetted area decreases by about 20%. From figure 7, since most of the hull resistance at low speed is frictional resistance, you can see the magnitude of the effect of decreasing the wetted area by heeling it at breeze. On the other hand, since the wave resistance is considered not to change so much by heel, at the time of high speed, the resistance reduction effect by heel cannot be expected so much.

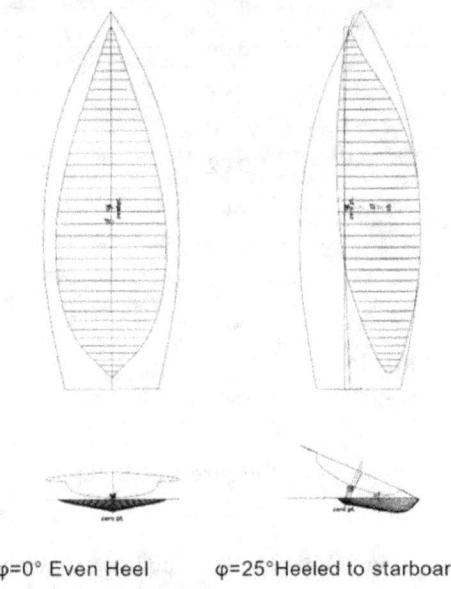

φ=0° Even Heel   φ=25°Heeled to starboard

Figure 16   Wetted area.   **Figure 17 Heel angle and Wetted area.**

Wind Tunnel Test

To clarify the sail performance of the 470, a model test was conducted using the wind tunnel equipment of Kanazawa Institute of Technology [7]. The wind tunnel device used for the experiment is a blow-off type, usually a shrinkage nozzle is attached, and the blowout port is a measuring section of 0.5m square. In this experiment, to make the sail model as large as possible, the shrinkage nozzle was removed, and it was used as a measuring section of 1.5m square. In this case, the maximum wind speed is about 6m/s, and the velocity distribution of the cross section of the measurement part is about 5% at the maximum. The model of sail is 1/6 scale, and the test was carried out with mainsail and jib, and spinnaker combined. Figure 18 shows the combination of "mainsail + jib" and "mainsail + jib + spinnaker". Figure 19 shows the definition of aerodynamic force acting on sail.

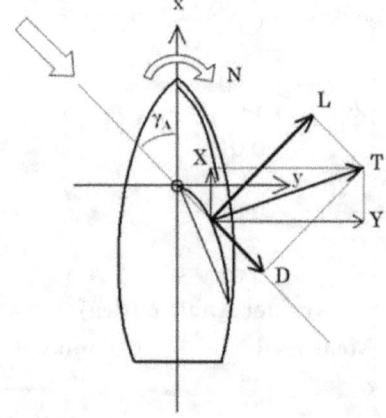

Figure 18 Wind tunnel (Main+Jib)/(Main+Jib+Spin)   **Figure 19 Definition of coordinate system and forces and moment acting on sail.**

These forces and moment are made dimensionless and shown as follows.

$$C_L = \frac{L}{\frac{1}{2}\rho_a U_A^2 A} \quad , \quad C_D = \frac{D}{\frac{1}{2}\rho_a U_A^2 A} \quad , \quad C_X = \frac{X}{\frac{1}{2}\rho_a U_A^2 A} \quad , \quad C_Y = \frac{Y}{\frac{1}{2}\rho_a U_A^2 A} \quad , \quad C_N = \frac{N}{\frac{1}{2}\rho_a U_A^2 A^{\frac{3}{2}}}$$

(4)

Performance of Mainsail + Jib

Figure 20 shows the wind tunnel test results of mainsail + jib. Since the apparent wind angle at close hauled is 25° to 30°, C_X is about 0.3 to 0.4 when heel angle is 0°, C_Y is about 1.2 to 1.4, and there is a difference of about 4 times. The maximum value of C_Y is around $\gamma_A=35°$, which is 50° to 60° with true wind angle, which is the same as the sense of actual sailing that heel becomes tightest at an angle slightly bear away from close hauled. The yaw moment coefficient C_N has a minus value means that it is going to turn counterclockwise (weather helm) with port tack. This weather helm becomes strong as the heel angle increases. This is because the force application point of Sail moves to the leeward side by heel and thrust also contributes to yaw moment.

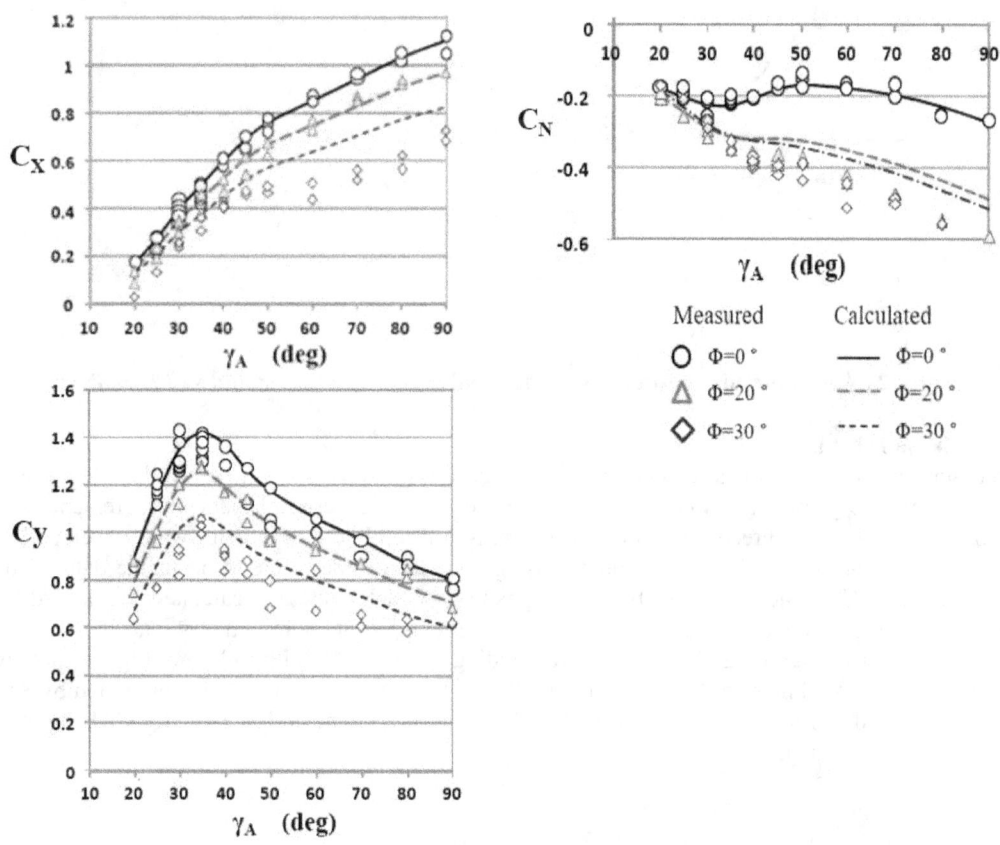

Figure 20 Aerodynamic coefficients of Mainsail+Jib and γ_A (Port Tack).

Performance of Mainsail + Jib + Spinnaker

Figure 21 shows the wind tunnel results of mainsail + jib + spinnaker and without jib. Thrust with jib is larger than without jib in the case of $\gamma_A \leq 100°$. Thrust without jib is larger than with jib in the case of $\gamma_A \geq 120°$. That is, while using spinnaker jib is effective up to about 100° in apparent wind angle, but it is only an obstacle to become more down wind. In this area the jib sheet is slack, and the jib is playing. As a result, because it is supposed to disturb the flow to the spinnaker, it seems to be better to take some measures such as narrowing jib.

220

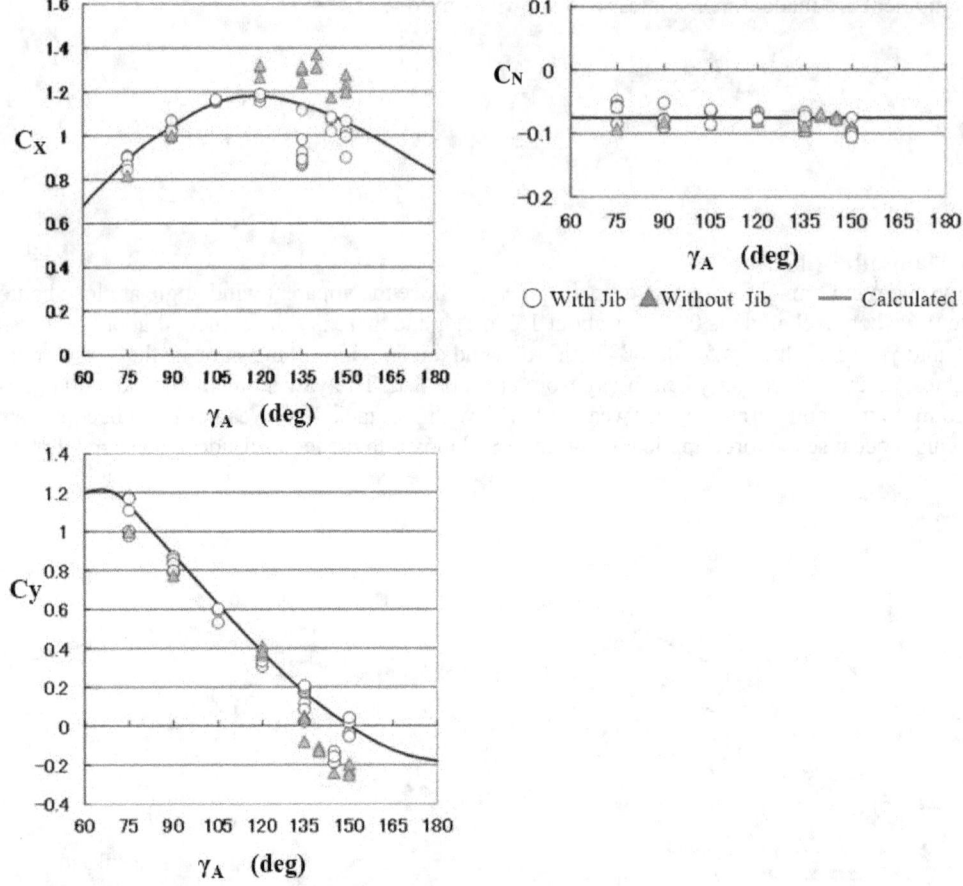

Figure 21 Aerodynamic coefficients of Mainsail＋Jib+Spinnaker and γ_A (Port Tack).

Measurement of the sail shape

For shape measurement of sail with large curvature like spinnaker, it is impossible to grasp the whole shape by the two-dimensional method by photography from mast top or boom. For this reason, these sail shape measurements were made using SolidFromPhoto which is a photo three-dimensional software. SolidFromPhoto is free software developed by Mr. Shizuo Hara [4], it can calculate three-dimensional coordinates using epipolar geometry based on image data of multiple digital cameras and output it as a CSV file. In this software, the position of each camera is calculated backward by clicking the corresponding point (mark) attached on the target object on plural (three or more) images, and based on this, the mark coordinates can be calculated. Figure 22 shows the corresponding points by SolidFromPhoto. Figure 23 shows the output result of three-dimensional coordinates of SolidFromPhoto. Since the three-dimensional data outputted by SolidFromPhoto is a CSV file, it is converted to a smooth curved surface (NURBS surfaces) to be an IGES file for CAD input, and it is taken as input data for numerical calculation.

Figure 22   Input of SolidFromPhoto.

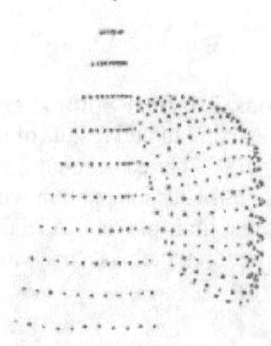

Figure 23   Output of SolidFromPhoto.

Computation Fluid Dynamics (CFD)

As a method to analyze computation fluid dynamics, we used a solver called FLOWPACK developed by Mr. Yusuke Tahara [5] of the National Maritime Research Institute. FLOWPACK is a comprehensive numerical fluid calculation tool built using multiblock NS/RaNS method for mainly applying to the field of ship design. For pre-post processing, Advanced Aero Flow (AAF) created by Mr. Masanobu Katori [6] of North Sails Japan was used. AAF is a fluid analysis tool developed mainly with the calculation of sail in mind. By inputting the IGES file representing the sail shape, a sail mesh is automatically generated, and after performing the calculation using FLOWPACK, it is possible to display the results such as streamlines. Figure 24 shows the mesh model of sail. Figure 25 shows the result of numerical calculation.

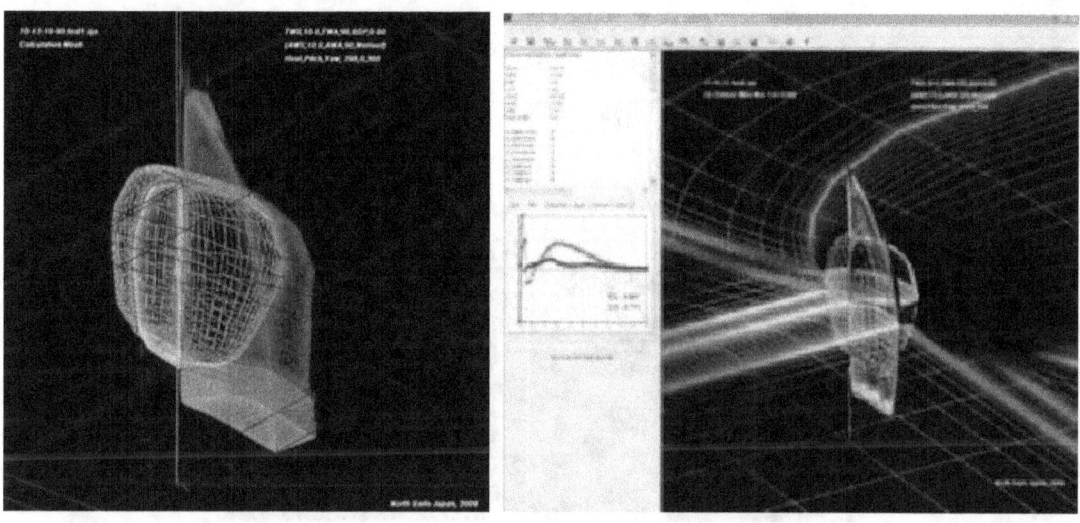

Figure 24 Mesh model by AAF.   Figure 25 Calculation result of FLOWPACK.

Comparison of Flow on Sail Surface

Figure 26 shows the comparison of the results of numerical calculation with the observation of the flow by attaching tuft to the surface of sail. This is an example of wind angle 90° (beam reach of port tack), it shows the leeward side of spinnaker. The wind is blowing from the right side of Figure 26, in the experiment the right tuft which is the luff of sail flows along the surface, but the left side and the upper side are peeling off. In the calculation result, the whitened part represents the region not desquamated, and almost corresponds with the region where the tuft of the experiment flows.

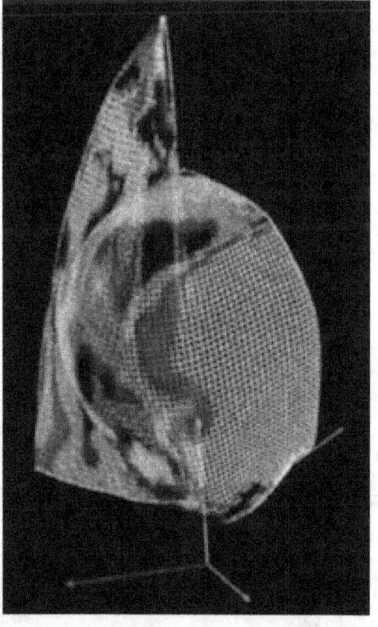

Figure 26 Comparison of surface flow.

Comparison of Streamlines

Figure 27 shows the results of visualizing flow using smoke-wire technique using smoke and comparison of streamlines by calculation. This is an example of the wind direction 120° (broad reach of port tack), and Figure 27(a) shows the top view. The wind is blowing from the left side, and the flow is greatly bent by the spinnaker. The calculation result also shows the appearance of peeling on the back side of spinnaker and the appearance that flow on the ventral side crosses the flow of mainsail. Figure 27(b) shows the comparison of the same situation when viewed from the side. Experimental results and calculation results show that the three streams from the back side and the ventral side of the spinnaker and from the ventral side of the mainsail are twisted. Please also note that the influence of Sail is considerably widespread.

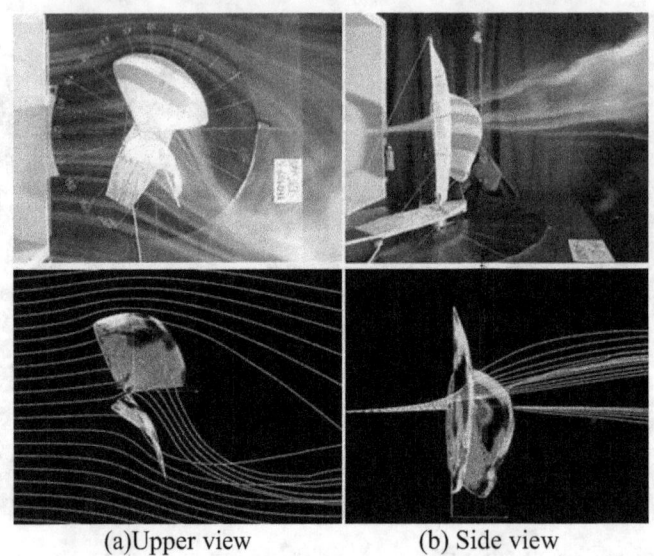

(a)Upper view   (b) Side view

Figure 27 Comparison of streamlines.

Comparison of Aerodynamic Coefficients

Figure 28 shows the result of aerodynamic forces acting on the sail converted to dimensionless number from equation (4). The horizontal axis is the angle between hull centerline and wind angle. Solid points indicate experimental data and hollow points indicate calculated values. The experimental data of C_L decreases with increasing wind angle, and it drops significantly at 165°. The calculated value corresponds well with experimental data at 75° to 120°, but it is lower than experimental values at 135 ° or more. This is thought to be because the influence of separation in this area is larger in the calculated value. Compared to the C_D value, the C_L value is larger than the C_D value up to nearly 120°. This means that the force (lift) generated in the unseparated part of the sail surface is greater than the force (drag) produced by separation, and even though it is a spinnaker, it shows the importance of flowing the wind smoothly. On the other hand, as for the C_D value, the calculated value is below the experimental data. The model of sail is made by laminating a thin sail cloth, and in fact there are delicate irregularities on the bonding surface. It is thought that the roughness of such a sail cross surface affects drag force coefficients, but its size is not clear yet. These are task of future [9]. The C_N value is the yaw moment coefficient around mast. The experiment is port tack, the negative value means weather helm, but the value is not so large. Further considering around the hull center, the degree of weather helm is further reduced. Although the absolute value of experimental data and calculated values are small, it showed good agreement. This means that both force application points are good agreement, and it can be said that the calculation result correctly expresses the state of flow over the entire surface of sail.

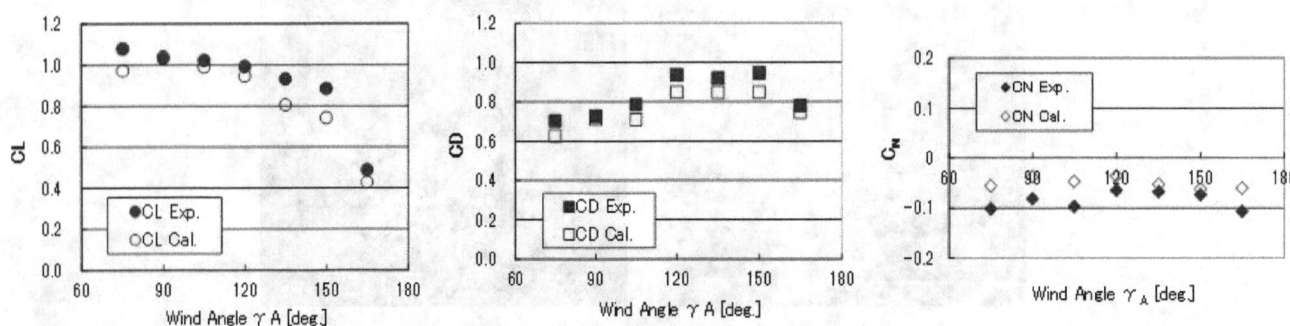

Figure 28 Comparison of aerodynamic coefficients.

Velocity Prediction Program (VPP)

For equilibrium calculation of an aircraft and so on, it is necessary to think about these three axial forces and all moment around three axes. However, in the case of a ship traveling in a horizontal plane, it can be considered that the vertical direction (z axis direction) is already balanced by gravity and buoyancy. Also, pitching moment (y-axis) can also be handled as constant since the longitudinal restoring force of the hull is very large. For this reason, sailing yacht's equilibrium can be considered with respect to the following four degrees of freedom.

(a)X: Thrust of sail and hull resistance
(b)Y: Side force of sail and leeway drag of hull
(c)K: Heel moment of sail and restoring moment of hull
(d)N: Yaw moments of sail and hull

A method for finding a point where two such forces and two moments are balanced is called VPP. VPP is a program for solving a four-element simultaneous equation. Aero/hydro-dynamic forces acting on hull and sail are defined as:

$$\frac{1}{2}X'\rho_w V_B^2 LD + \frac{1}{2}C_X \rho_a U_A^2 A - T = 0$$

$$\frac{1}{2}Y'\rho_w V_B^2 LD + \frac{1}{2}C_Y \rho_a U_A^2 A = 0$$

$$\frac{1}{2}K'\rho_w V_B^2 LD^2 + \frac{1}{2}C_K \rho_a U_A^2 A^{3/2} - \Delta\overline{GZ} + K_{Trapeze} = 0$$

$$\frac{1}{2}N'\rho_w V_B^2 L^2 D + \frac{1}{2}C_N \rho_a U_A^2 A^{3/2} = 0$$

$$(5)$$

In case of a small dinghy, the heel angle can be controlled considerably by trapeze and power-down of sail. In the equation (5), the heel angle is one of the unknowns, but in this case the heel angle is set to a constant value of 10° (starboard tack: -10°). Also, the position of the center of gravity of trapeze was set so that K moment and righting moment are in equilibrium. The maximum of the center of gravity position of trapeze is 2m from the hull centerline, but when K moment does not balance at high winds, aerodynamic coefficients of sail are reduced (power-down) so that heel angle becomes 10°. When powering down, X, Y, K, and N are uniformly reduced. Therefore, the unknown value of K moment becomes a trapeze distance or a power down factor instead of a heel angle. Aerodynamic force acting on sail in equation (5) must be calculated using apparent wind speed (U_A) and apparent wind angle (γ_A) observed on the ship. But what we want is the performance on true wind speed (U_T) and true wind angle (γ_T). Therefore, from the relation of wind speed triangle shown in Figure 11, it is necessary to convert to U_A and γ_A by using equation (6). Figure 29 shows flowchart of VPP to solve equation (5).

$$U_A = \sqrt{U_T^2 + V_B^2 + 2U_T V_B \cos(\gamma_T + \beta)}$$

$$\gamma_A = \sin^{-1}\left\{\frac{U_T}{U_A}\sin(\gamma_T + \beta)\right\} - \beta$$

$$(6)$$

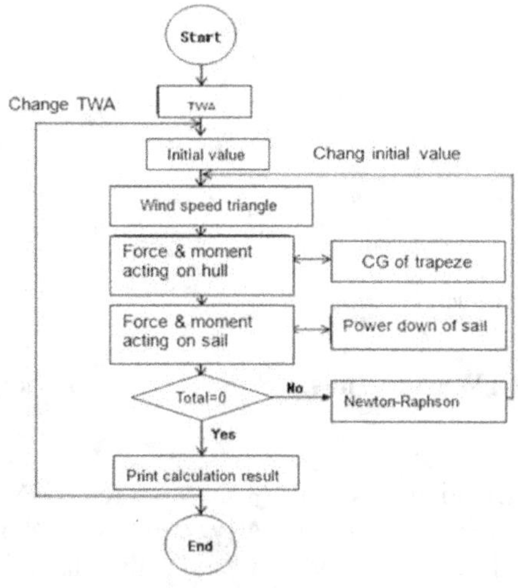

Figure 29 Flowchart of VPP.

RESULTS OF VPP AND MEASURED DATA

Sailing Performance of TWS 5～7m/s

Figure 30 shows a polar diagram showing the performance of TWS 5 to 7m/s from the result of VPP. In the case of Main + Jib, the maximum V_B is obtained from TWA 75° to 85° (between close reach and beam reach). The maximum V_B in the case of Main + Jib + Spin is obtained from TWA 100° to 120° (between beam reach and broad reach). The intersection of the solid line and chain thin line becomes the standard of TWA which starts using spinnaker, and in this wind speed range TWA is about 90° to 100°. ○ symbols are the measured data of the top three boats obtained from the No10 race result of the All Japan 470 Championship 2007. Symbols other than ○ are measured data of Osaka University. The measured data of close hauled showed good agreement with the result of VPP. The data of the All Japan Championship, V_B has reached a maximum of 9.6kt at TWA about 110°. In the case of Mainsail + Jib + Spin, it is not impossible speed with a blow of about 7m/s, and the top three boats continue semi-planing of 8 to 9kt or more for more than 2 minutes. However, if the TWA is slightly shifted, the speed suddenly decreases, indicating that it is difficult to continue the semi-planing state at this wind speed. In this wind speed range, it was found that the maximum value of VMG in the leeward direction can be obtained with TWA = 150°.

Sailing Performance of TWS 8～10m/s

Figure 31 shows a polar diagram showing the performance of TWS 8 to 10m/s. Measured data ○ symbols are 2004 Olympic send-off race data. On this July 10, winds of 8 to 11m/s were blowing through all the races, and valuable strong wind data was obtained. In the VPP calculation result, TWA which obtains the maximum V_B is TWA=95° to 110° in the case of Main + Jib and TWA=120° to 130° in the case of Main + Jib + Spin (TWA are leeward side than TWS 5 to 7m/s). In addition, it can be thought that TWA=105° ～ 120° as a guide for starts using spinnaker. TWA of course from windward mark to gybe mark is about 120° and there are many data. It will be difficult to decide at which timing to use spinnaker in this course during this strong wind. In the running, the VMG to the leeward has a large wind angle range of TWA=140° to 150° and the angle is wider than in weak wind. On the other hand, most of measured data at race are leeward than 150°. Although it may be uneasy to sail away from the group going direct to the leeward mark, it is better to luffing up to broad reach at least TWA 150°.

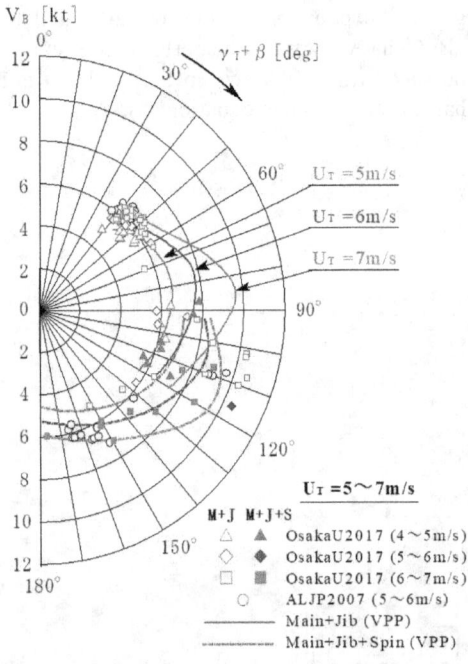

Figure 30 Polar diagram at TWS 5 to 7m/s.   Figure 31 Polar diagram at TWS 8 to 10m/s.

Variation of V_B by TWA and AWA

Figure 32 shows a summary of VPP calculation results of V_B variation for TWA (γ_T +β) shown in Figure 30 and 31. Figure 33 shows the variation of V_B represented in Figure 32 by AWA (γ_A). Figure 32 shows the course running on the sea surface, and Figure 33 shows the wind angle sensed by skipper and crew, indicating that it is turning to the windward side. TWA and AWA are represented by the wind speed triangle shown in Figure 11, which can be calculated by equation (6), varies greatly depending on V_B and TWA. Points A to H in Figure 32 correspond to points A' to H' in Figure 33. In the example of U_T=8m/s shown in Figure 32, there is a width of about 45° from close hauled ($\gamma_T+\beta$=45°) to the point A of the beam reach that is the maximum speed, but in the AWA of Figure 33 there is only a change of about 25°from γ_A=30° to A ' point (γ_A=55°). The point B (γ_T +β=120°), which is the maximum speed when spinnaker is used, becomes B ' point of γ_A=75° and sensationally feels that it is closer to beam reach than broad reach. On the other hand, the points C and D where the VMG toward the leeward side is the maximum are the C ' and D' points in the AWA, and the AWA changes drastically with a slight difference in the TWA. The same is true for points E to H of U_T=10m/s, please refer to your sail trim.

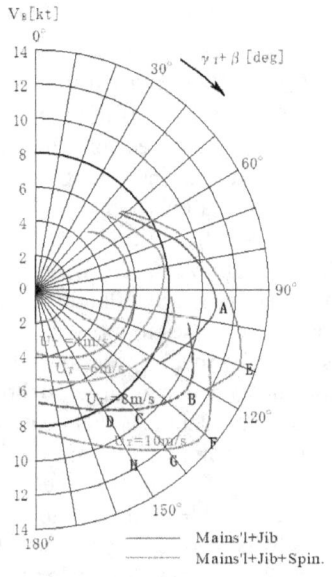

Figure 32 TWA and V_B.

Figure 33 AWA and V_B.

Variation of AWS(U_A) by TWA and AWA

U_A changes greatly depending on the wind angle. Figure 34 shows the variation of U_A and TWA. Figure 35 shows U_A and AWA. Close hauled in Figure 34, U_A is about 1.2 to 1.4 times larger than U_T. Since the force acting on sail is proportional to the square of wind speed, the force is about 1.5 to 2 times as large. It turns out that the U_A becomes the same as U_T when TWA is about 110°, it becomes surprisingly leeward side. U_A between broad reach and running decreases rapidly. U_A of TWA 180 ° is about 0.6 times of U_T. Since V_B reaches the maximum speed in Main + Jib is A ' and E', U_A is still larger than U_T. In Main + Jib + Spin, V_B reaches maximum speed at B ' and F', where U_A drops considerably to about 0.8 times of U_T. This is one reason why the maximum speed is not so different from that of A ', E', despite adding a big sail called spinnaker. Figure 35 shows the wind angle and wind speed that the skipper and crew feel on boat, as well as Figure 33, AWA is quite windward side.

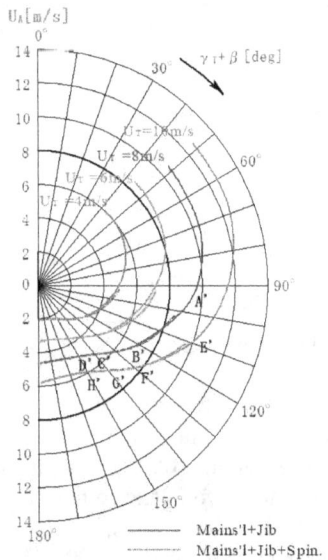

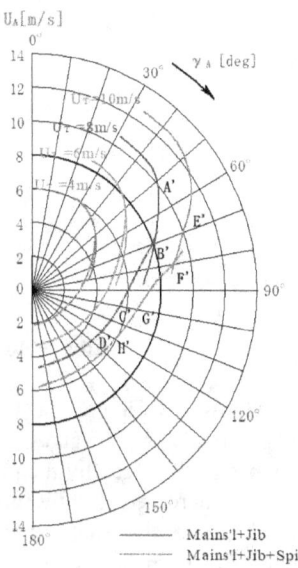

Figure 34 TWA and U_A.   **Figure 35 AWA and U_A.**

Mechanism of High-Speed Sailing

Figure 36 and 37 shows measured data of high-speed sailing and the results of VPP. Figure 36 and 37 (a) shows the variation of true wind angle (γ_T) and AWA(γ_A) for 30 seconds. (b) shows the variation of true wind speed (U_T), AWS(U_A) and boat velocity (V_B). (c) shows the result of VPP. (d) shows the boat trajectories. Circles indicate the position of center of boat at each 3-seconds. The wind blows from the top of the figure and the grid spacing is taken as 25 meters. Variation are divided into A to F to explain.

High-Speed Sailing of Mainsail + Jib (Figure 36)

In the range of A to B, γ_T changes from -90° to -105° and U_T rises from 6m/s($\fallingdotseq$12kt) to 7.5m/s($\fallingdotseq$14.5kt). V_B by VPP is 7.5kt at γ_T=-90°, U_T =6m/s (point A in Figure (c)), 9.0kt at γ_T=-105° and U_T =7.5m/s (point B in Figure (c)). The measured data of V_B shown in Figure (b) has a time lag of the GPS output and the response is delayed for 2 to 3 seconds due to the inertia of the ship (deviated to the right). V_B is accelerating from about 8kt to 10kt with increasing wind speed. In B to C, TWA is bear away to 130°, so U_A decreases to 5m/s and V_B=6.7kt in the balance calculation (point C in Figure (c)). Measured data of V_B also shows deceleration a little late. It is luffing up from C to D (γ_T=-75°). From D to E, you can see that it is receiving blow of U_T =8 m/s($\fallingdotseq$15.5kt). The balance condition values at this time are V_B=9.4kt, U_A=10.3m/s, γ_A= -48° and thrust of sail is 44kgf (point D in Figure (c)). Measured data also shows that it reaches nearly 10kt during this time. At E, the blow ends and U_T returns to 7m/s, and γ_T is about -100°. From this, thrust of sail decreases to thrust 38kgf, and V_B=8.5kt by balanced calculation (point E in Figure (c)). However, measured data does not show much speed down. Here, if V_B of D is maintained at 9.4kt, γ_A becomes smaller than the value of balanced calculation but U_A becomes large, so thrust of sail becomes 39kgf, which is slightly larger than point E. Since this is not enough, it is considered that the actual boat maintains the V_B of 9.5kt by increasing the force of sail in some way (point F in Figure (c)). From the point F, it seems that it has been able to maintain about 10kt since U_T also increases with luffing up again to nearly γ_T=-90°. Although the above is temporarily accelerated by blow of 8m/s, it is a mechanism that V_B 10kt of data is obtained with U_T of about 6 to 7m/s. By the way, in the graph of Figure 36 (c), if U_T is 8m/s or more, it is shown that V_B exceeds 8kt in the range of γ_T=60°~65°. When V_B exceeds 8kt, it becomes semi-planing state. Even if it's sailing to upwind, it can be expected to accelerate as shown between D and E in Figure (b). When it is close hauled with strong wind, if V_B does not glow due to wave, it is worth to try about 10° to 15° bear away and go into semi-planing.

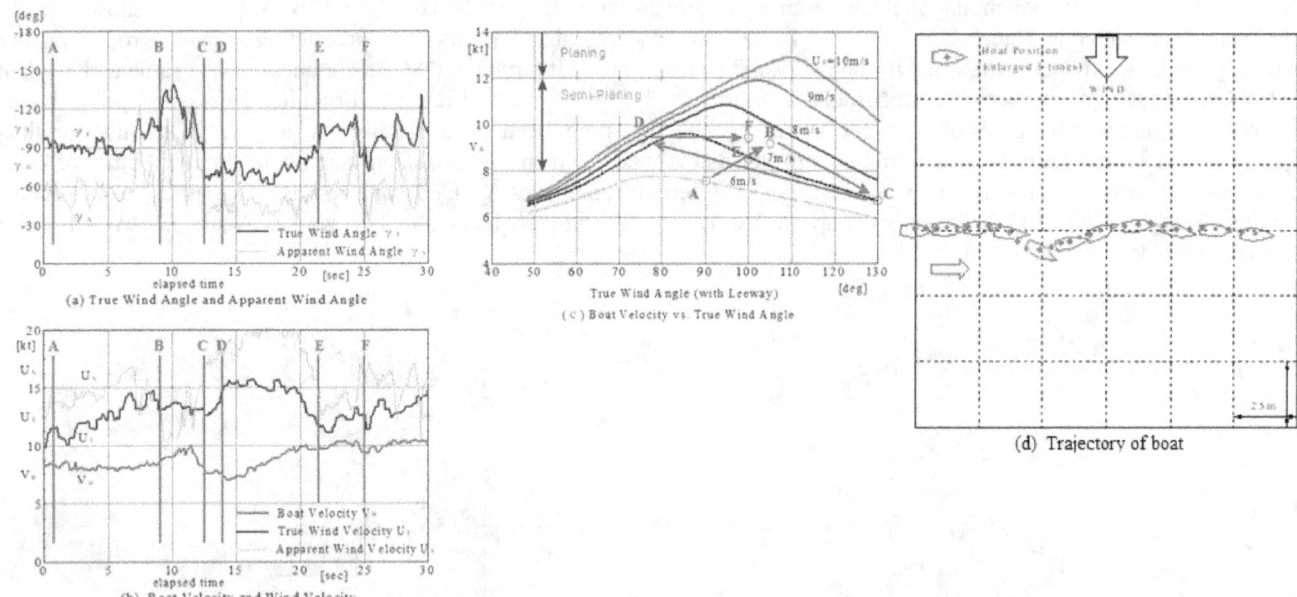

(a) True Wind Angle and Apparent Wind Angle

(c) Boat Velocity vs. True Wind Angle

(d) Trajectory of boat

(b) Boat Velocity and Wind Velocity

Figure 36 High-speed sailing (Main + Jib).

High-Speed Sailing of Mainsail + Jib + Spin (Figure 37)

In A, sailing is performed at γ_T =130° and U_T=7m/s($\fallingdotseq$14kt), and the balanced calculation V_B is 8.6kt(point A in Figure (c)). Measured data has been influenced by the previous speed, already reached about 10kt. B to C, luffing up to γ_T =115° and the wind speed has also increased slightly (point B in Figure (c)). From C to E, you can see that it is receiving blow of U_T =8.5m/s($\fallingdotseq$17kt). At this time, the balanced calculation values are V_B=10.8kt, U_A =7.8m/s, γ_A=75° and thrust of sail is 53kgf (point C in Figure (c)). D to E, slightly bear away to γ_T =125°, the values of balanced calculation at this time are

V_B=11.8kt, U_A=7.0m/s, γ_A=80° and thrust of sail is 60kgf (point D in Figure (c)). Measured data also shows 12kt during this time. At E, blow finishes and goes back to U_T =7.5m/s and γ_T =120°. Values of balanced calculation are V_B=10.5kt and thrust of sail is 51kgf (point E in Figure (c)). However, also in this case measured speed does not show such a speed drop, it seems that V_B of 12kt or more is maintained while picking up a little blow (F point in Figure (c)).

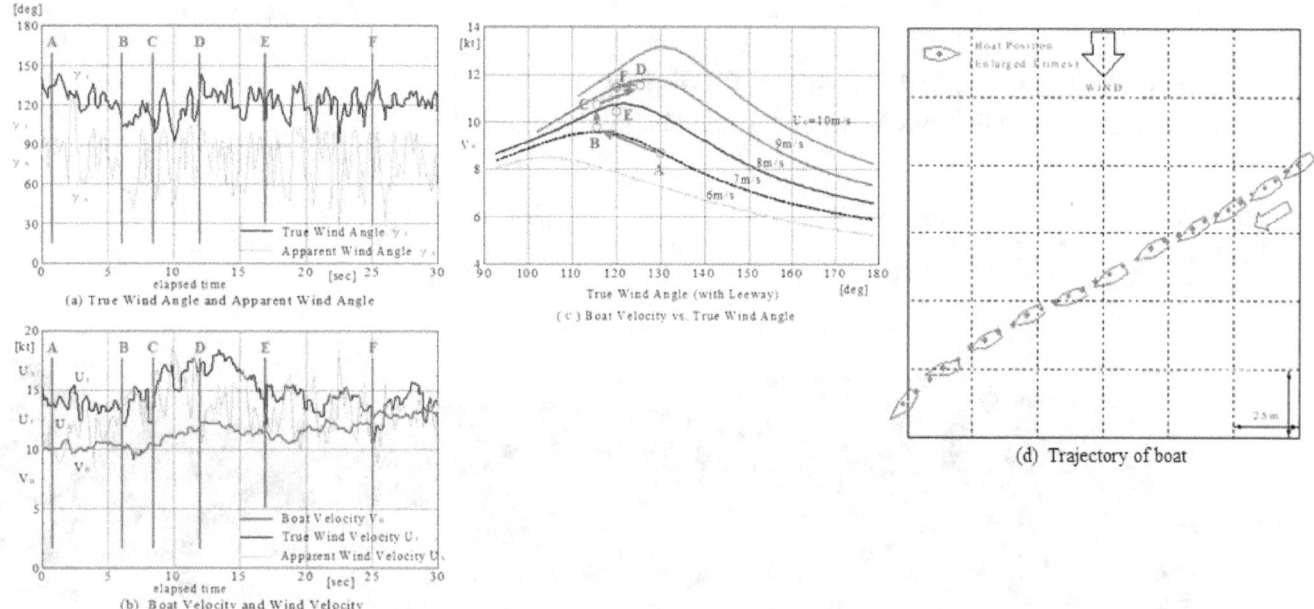

(a) True Wind Angle and Apparent Wind Angle

(b) Boat Velocity and Wind Velocity

(c) Boat Velocity vs. True Wind Angle

(d) Trajectory of boat

Figure 37 High-speed sailing (Main + Jib + Spin).

CONCLUSIONS

From the results of the above analysis, it was found that even a (static) balance calculation by VPP can nearly explain the outline of the mechanism of high-speed sailing exceeding 10kt. However, in VPP, there are still many unknown parts, such as sometimes the thrust of sail temporarily drops, while measured data sometimes maintains high speed. For these phenomena, it will be necessary to advance the analysis through simulation of dynamic motion in the future. Utilizing this report, 470 sailors can run the 470 physically faster. This report will be useful for 470 sailors aiming for the Tokyo Olympic Games in Japan in 2020.

ACKNOWLEDGEMENTS

The authors wish to thank Kanazawa Institute of Technology, Osaka University, QPS laboratories, Japan 470 class association, Digital Data Supply, North Sail Japan, KIT Actual Seas Ship and Marine Research Laboratory and Photo Wave for their cooperation in carrying out this research.

REFERENCES

[1] Y.Masuyama., "TOBE 470,", Syutei Kyokai Japan, 2017.

[2] Japan 470 class association: http://www.470jpn.org/

[3] H.Tatano., "Hydrodynamic Analysis on Sailing:6st Report, Journal of KSNA No193,", Japan, 1984.

[4] S. Hara.,"SolidFromPhoto,", Japan

[5] Y. Tahara., "A Reynolds-Averaged Navier-Stokes Equation Solver for Prediction of Ship Viscous Flow with Free Surface Effects, Proceedings of NAPA,", Japan, 2008.

[6] M. Katori ., "Advanced Aero Flow Software Manual", North Sails Japan, 2009.

[7] H. Suito, Y.Masuyama, Y.Tahara, M.Katori, "Sail Performance Analysis for Downwind Condition by Wind Tunnel Test and CFD Calculation,", Nov 2011.

[8] C.A. Marchaj: "Sailing Theory and Practice", Dodd, Mead & Co. New York, 1964.

[9] VIOLA, I. M. and FLAY, R. G., Sail Aerodynamics: On-Water Pressure Measurements on a Downwind Sail, Journal of Ship Research, SNAME, Vol.56, No.4, pp.197-206, 2012.

Sailing Catamarans: Design for Cruising

Albert Nazarov, Albatross Marine Design, Thailand

Figure 1 - IS41 catamaran(design by AMD) pictured at Charming islands of Thailand
Main design parameters: length of hull/WL 11.98/11.90m; beam 7.02m; max draft 1.43m; displacement light craft/full load 8100/10000kg; sail area main/jib 53.6/29.5m^2, 4 cabins, 2 bathrooms, saloon.

ABSTRACT

Statistics are provided for number of sailing catamarans and approaches to craft dimensioning are reviewed. Styling trends and typical catamaran arrangements are featured. Weight components are studied for number of catamarans of different sizes and levels of comfort on board. The effect of catamaran architecture on performance is studied by combining VPP predictions with CFD modeling of various deck/cabin configurations. A summary of safety requirements specific to catamarans is given. Case studies are presented of a number of cruising catamarans designed by AMD and a new prospective concept of 44' catamaran featured.

INTRODUCTION

Today catamarans comprise a large share of pleasure craft and yacht market in range of lengths L_H>10m. Their presence is remarkable in marinas located in tropical and subtropical areas, where they count up to 25% of fleet of yachts. During last 12 years Albatross Marine Design (AMD) has developed a large number of sailing and power catamaran concepts and designs in the 18 to 120' range, of which about 50 are launched and some of them are in series production featured by Nazarov 2015.

The main advantage of catamarans as yachts is the large usable areas and comfort resulting from excellent accommodations, especially in public spaces. Unfortunately, this obvious advantage would often cause overloading of available spaces with structures, outfit and equipment, making the craft relatively heavy and sacrificing hydrodynamic efficiency. Meanwhile, for properly designed catamaran, performance due to reduced hydrodynamic resistance, fuel efficiency and the ability to maintain high speed on seaway are very attractive features. Other attractive points are motion characteristics and increased general safety due to higher freeboard, shallow draft, stability/buoyancy and duplicate systems.

The main disadvantages of catamarans are higher construction costs and docking fees due to their surfaces and large beam, if compared with monohull yacht of the same length. There is also a mentality factor with a certain prejudice about catamarans, mostly regarding their seaworthiness, windward performance and appearance. Though penetration of catamarans into certain yachting areas can convince naysayers and cause a 'chain reaction' with a subsequent increasing number of catamarans. Today, major on-water boat shows have dedicated 'catamaran berths', as well as such berths in marinas located in popular catamaran cruising areas.

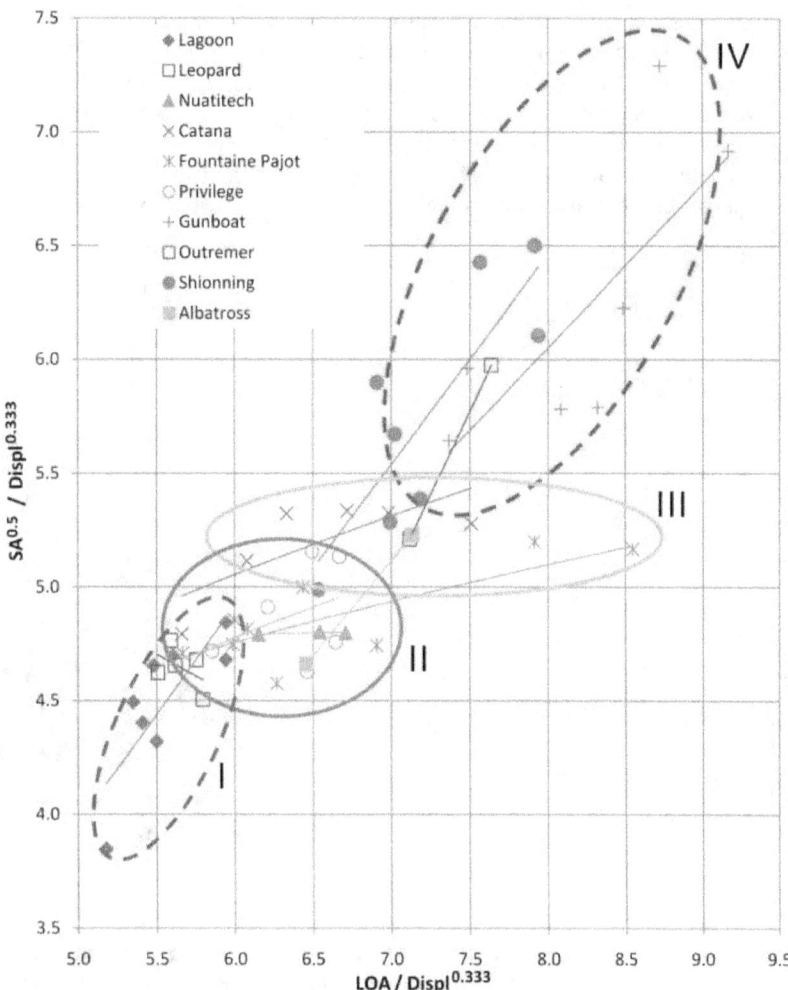

Figure 2 - Some statistical data on sailing catamarans showing non-dimensional characteristics
Group (I) covers relatively heavy catamarans with modest sail area mostly intended for charter or motorsailing. Groups (II) and (III) are catamarans attractive for performance-oriented cruising sailors. Group (IV) are catamarans with high sailing properties and Spartan accommodations; those are used mostly for racing and weekend cruising

CATAMARAN PARAMETERS AND STATISTICS

Most practicing designers start work on new designs with a study of prototype craft. Talking in numbers, for a sailing catamaran there are a few key design factors where 'lightness' is the priority; in simple words, catamarans should be light to realize their advantage in performance. From the point of view of hydrodynamic efficiency, catamaran platforms are justified when slenderness ratio of one demihull $L/V_d^{1/3}>(5...6)$, where L and V_d are the craft's length and a demihull's volumetric displacement. This gives the slenderness ratio of the whole catamaran $L/Displ^{1/3}>(4.0...4.8)$, where $Displ$ is displacement of craft in tons. If the craft is heavier than a monohull concept may be a better solution.

The beam of sailing cruising catamarans usually comprises B_H=0.43...0.56L_H. Excessive beam increases the stability and ability to carry sails, but incurs an increase in the structural loads and weights. Larger catamarans have beam restrictions from practical considerations such as transportation and docking. Beam is related to the horizontal clearance c - the distance between inner hull sides, and also to the distance between demihull centerlines B_{CB}.

One more upmost parameter of a sailing catamaran is the vertical clearance t, i.e. the distance from underside of bridge deck to the water plane; it defines the seakeeping being the critical for wet deck slamming. For a displacement catamaran, minimal t would be (0.04...0.06)L_H; this ratio is generally used for power catamarans. Higher t is needed for sailing catamarans which are sailed heeled; a minimal reasonable value would be B_H/10. In this respect, wider catamarans would need higher t and less pronounced bridge deck-hull side junction/chamfer, to cater for heeled condition. Sailing heel angles can be estimated by the stability assessment formulas from ISO12217:2-2015 or from other considerations. Detailed discussion of these and other hull shape parameters is presented by Nazarov, 2015.

On figure 2, statistical data on sailing catamarans with hull length L_H=10...24m is presented showing groups for few different manufacturers and designers presented by Nazarov 2015. Many sailboat designers would specify light craft displacement (or measurement displacement) rather than full load displacement to ISO8666 thus making the boat look more 'attractive'; so on figure 2 *Displ* corresponds to light craft displacement. The graphs show dimensionless parameters such as slenderness ratio (length overall to displacement) $L_{OA}/Displ^{1/3}$ and sail area to displacement $SA^{1/2}/Displ^{1/3}$. These ratios are performance indicators: for a craft's hull resistance and its ability to reach speed, and for the power of sails available to deliver the required thrust. In general, there are a variety of approaches and each sailing catamaran design should be evaluated with respect to its target group of users before being labeled as 'too heavy' or 'lacking comfort'.

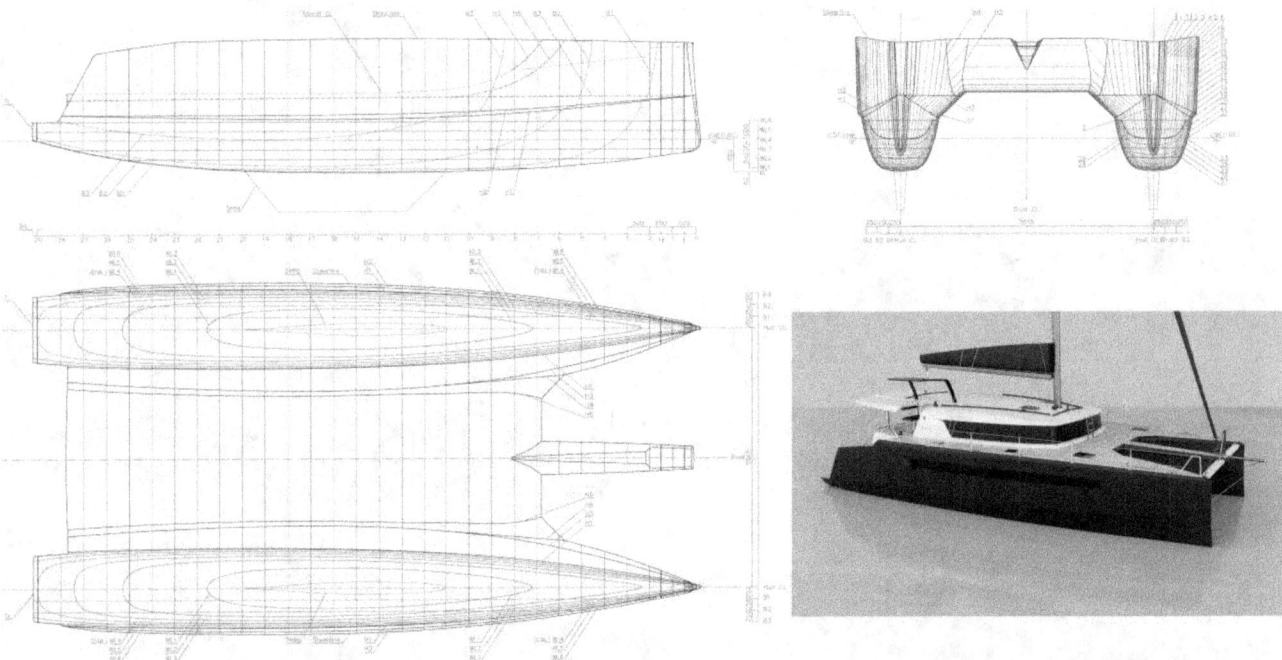

Figure 3 - Hull shape and exterior of14.7m sailing catamaran of PS153 design (concept by AMD)
Note the careful expansion of the above water part - for better habitability, but without extensive drag in waves.

In AMD's practice, a parametric design approach is used to dimension the sailing catamaran in the early design stages. This type of analysis is usually made in parallel with preliminary sketching.

An example of application of this parametric method is presented in table 1. During the preliminary design stage a few approximations are made to achieve the convergence of the weight and the volumetric displacement, and the satisfaction of other design parameters. Weight factors are shown in the table; reliable estimate of weights is critical at early stages of design, especially on catamaran craft which are sensitive to load and where exceeding design weight can badly impact vertical clearance required for seakeeping.

ARCHITECTURE OF CATAMARANS

HULL GEOMETRY:

Length of hull, m	L_H	14,7	
Length of WL, m	L_{WL}	14,553	
Beam of hull, m	B_H	7,6	
Beam of demihull @WL, m	B_{WL}	1,6	
Beam between centers of bouyancy, m	B_{CB}	5,52	
Canoe body draft, m	T_C	0,7	
Volume displacement, m³	V	15,26	From TW:
Weight displacement, kg	M_{LDC}	15638	15727 <Match?
Immersion, kg/cm	q	326	

Additional data for mass calculation:

Beam of demihull @bridge deck, m	B_{HWD}	2,08	
Depth of demihull @middle, m	D	2,7	
Bridge deck clearance, m	h	1	
Bridge deck length, m	L_{WD}	9	
Bridge deck beam, m	B_{WD}	3,44	
Bridge deck height, m	H_{WD}	1,00	
Cabin length, m	L_{SUP}	4	
Cabin beam, m	B_{SUP}	4	
Cabin height, m	H_{SUP}	1,3	
Volume of demihull (box), m³	V_{DH}	82,56	
Volume of bridge deck structure, m³	V_{WD}	30,96	
Volume of cabin, m³	V_{SUP}	20,80	

DESIGN FACTORS:

Length to beam ratio (overall)	L_H/B_H	1,934	
Length to beam ratio	L_{WL}/B_{WL}	9,096	
Beam to draft ratio	B_{WL}/T_C	2,286	one hull
Slenderness ratio	$L_{WL}/0.5V^{1/3}$	7,398	one hull
Block coefficient	C_B	0,468	one hull
Midship area coefficient	C_M	0,78	one hull
Waterplane area coefficient	C_{WP}	0,7	one hull
Prismatic coefficient	C_P	0,6	one hull

POWERING:

Motoring speed, kts	v_S	9,3	
Recommended engine power, h.p.	P_S	94	
Engine power (installed), h.p.	P_S	70	
Range under power, nM	R	300	
Fuel requried (20% margin, 80% MCR), L	v_F	338	

SAILING RIG:

Mainsail hoist, m	P	20,0	
Mainsail foot, m	E	7,0	
Foretriangle height, m	I	17,0	
Foretriangle base, m	J	5,0	
Roach factor		1,3	
Gennaker area factor		1,65	
Mainsail area, m²	SAM	91	
Foretriangle area, m²	SAJ	43	
Gennaker area, m²	SAS	140	
Sail area/MOC displacement ratio	$SA/DSPL^{2/3}$	23	>20

STABILITY:

Size factor (old ISO12217-2)	SF	220 516	A >40,000; B >15,000
Sail area/footprint factor	$SA/L_{WL}B_{CB}$	1,66	
Max angle of stability curve, deg	ϕ_{GZmax}	12,09	
VCG (taken at sheerline), m	VCG	2,70	
Max transverse righting momentum, N*m	LM_T	280 494	MOC load
Max transverse righting arm, m	GZ_M	2,04	A >1,85; B >1,30

Structural mass factors, kg/m³

FRP sandwich demihull	17...34
FRP sandwich bridge deck	22...44
FRP sandwich cabin	12...17

Deck equipment, systems, electrical mass factors:

	Deck	Systems	Electrical
Min	0,3	0,3	0,3
Average	0,37	0,39	0,35
Max	0,5	0,5	0,5

WEIGHTS:

Structure:

Structural mass factor for demihull, kg/m³	m_{DH}	25
Structural mass of demihull, kg	M_{DH}	2064
Structural mass factor bridge deck, kg/m³	m_{WD}	35
Structural mass of bridge deck, kg	M_{WD}	1084
Structural mass factor for cabin, kg/m³	m_{SUP}	20
Structural mass of cabin, kg	M_{SUP}	416
Structural mass margin, kg	$M_{STRSpare}$	250
Mass of structure, kg	M_{STR}	5877

Interior & outfit:

Interior mass factor, kg/m³	m_I	18
Mass of interior & outfit, kg	M_I	3904

Machinery:

Number of engines	n	2
Installed engine power (one engine), h.p.	P_E	35
Mass of engines	M_E	378
Machinery mass factor	k_M	1,4
Mass of machinery	M_M	530

Deck equipment:

Deck equipment mass factor	m_D	0,35
Mass of deck equipment, kg	M_D	496

Systems:

Systems mass factor	m_{SYS}	0,35
Mass of systems, kg	M_{SYS}	496

Electrical:

Electrical system mass factor	m_{EL}	0,35
Mass of electrical system, kg	M_{EL}	496

Sailing rig:

Sailing rig mass factor	m_R	0,35
Mass of sailing rig & hardware, kg	M_R	1112

Other:

Additional equipment	M_A	0
Weight margin	M_{MAR}	400
Light craft mass, kg	M_{LCC}	13690
Safety equipment	M_{SE}	149
Non-consumable stores, kg	M_{NCS}	0
Crew at MOC	n_{MOC}	2
Mass of crew at MOC	M_P	150
Minimum operation mass, kg	M_{MOC}	13989

Crew/passengers/personnel:

Number of crew/passengers/personell, kg	n	12
Mass of one crew/passenger	m_P	75
Mass of crew/pasengers, kg	M_P	900
Mass of personal items, kg	M_{PI}	240

Fuel:

Fuel tanks capacity (used), L		500
Mass of fuel, kg	M_F	438

Other luquids:

Fresh water capacity (used), L	M_F	300
Waste tank capacity (used), L	M_W	100
Other liquids	M_L	0
Stores, luggage, etc.	M_{CRG}	300
Payload	M_{MTL}	2038
Loaded displacement mass, kg	M_{LDC}	15727

Machinery mass factors:

All types	1.3...1.7

Interior & outfit mass factors, kg/m³:

Luxury yachts	18...25
Sport cruisers, cabin cruisers	7...20

Crew at MOC:

L_H<8m	1
L_H=8...16m	2
L_H>16m	3

With the increasing number of catamaran yachts in the recreational craft market, the manufacturers are trying to be distinct and pay significant attention to the exterior shapes (see figure 1) of their products. Habitability and luxury requirements for catamarans are also growing, and to accommodate these trends the designers are often forced to increase the volumes of catamaran structures (see figure 4). Recent developments include large vertical windows (see figures 3,4), walk-through saloons with forward cockpits and huge overhangs of cabin roofs. Unfortunately, not all these styling experiments result practical and efficient shapes, and many sailing catamarans are overloaded with superstructures of unfavorable configurations.

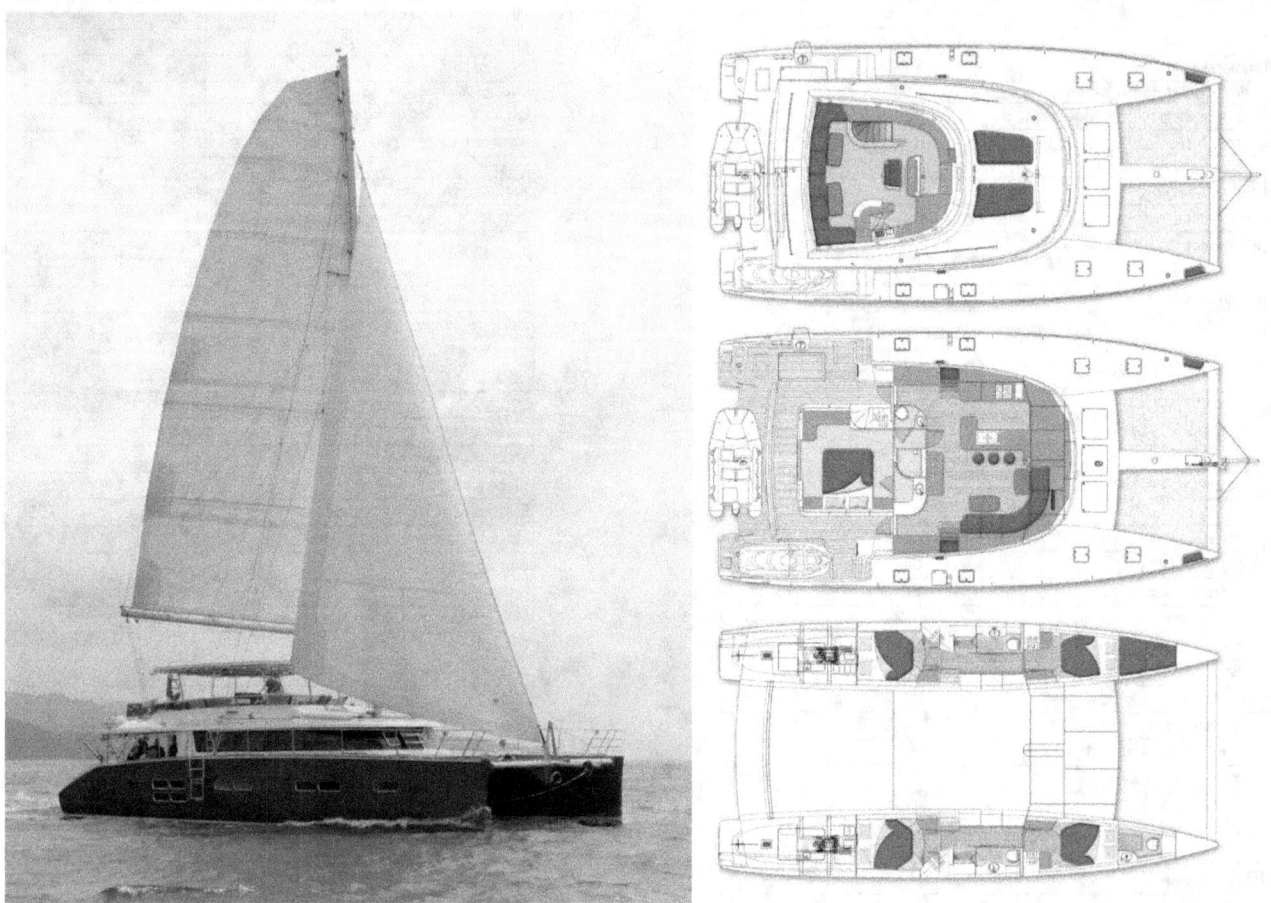

Figure 4 - Sailing catamaran of J1800 design (design by AMD)
Custom catamaran with luxury liveaboard accommodations, though with good performance indicators. A unique feature of this boat is the master bedroom location at main deck, giving natural airflow and a panoramic view. Main design parameters: length of hull/WL 18.05/17.61m; beam 9.02m; max draft 1.51m; displacement light craft/full load 17300/23200kg; sail area main/jib 108.3/74.2m², master cabin,4 guest cabins, crew cabin, 4 bathrooms, saloon, flybridge.

This approach is justified in terms of saloon interior requirements, which are much larger than the volumes available on monohulls. On other hand, performance expectations are still high as for many sailors especially with dinghy racing backgrounds the word 'catamaran' is used as a synonym of speed. These performance expectations are not true for the average charter catamaran which is designed for a combination of comfort, safety and cost saving.

Cruising catamarans are 'tall' by their nature due to requirements for vertical clearance – distance from wet deck to water surface. This is the reason of expected pronounced effect of sailing catamaran aerodynamics on its performance, especially on closed-hauled courses. In response to this some catamarans focus on improved performance withlow profile 'aerodynamic' catamaran cabins which can result in insufficient headroom.

Velocity prediction programs (VPP) used for sailboat design usually require the input of windage elements, but detailed data on aerodynamic coefficients of above water structures of sailing catamarans is lacking. As a result, windage drag is often a guess when doing performance predictions for a sailing catamaran and can cause serious discrepancies between estimated and actual performance, especially at apparent wind angles $\psi=0...60°$. In Claughton 1998, Fossati 2007, Larsson 2014, Slooff, 2015 methods of air drag prediction are presented, but they mostly cover monohull craft. In this paper we focus on catamaran specifics and review some of the superstructure design considerations.

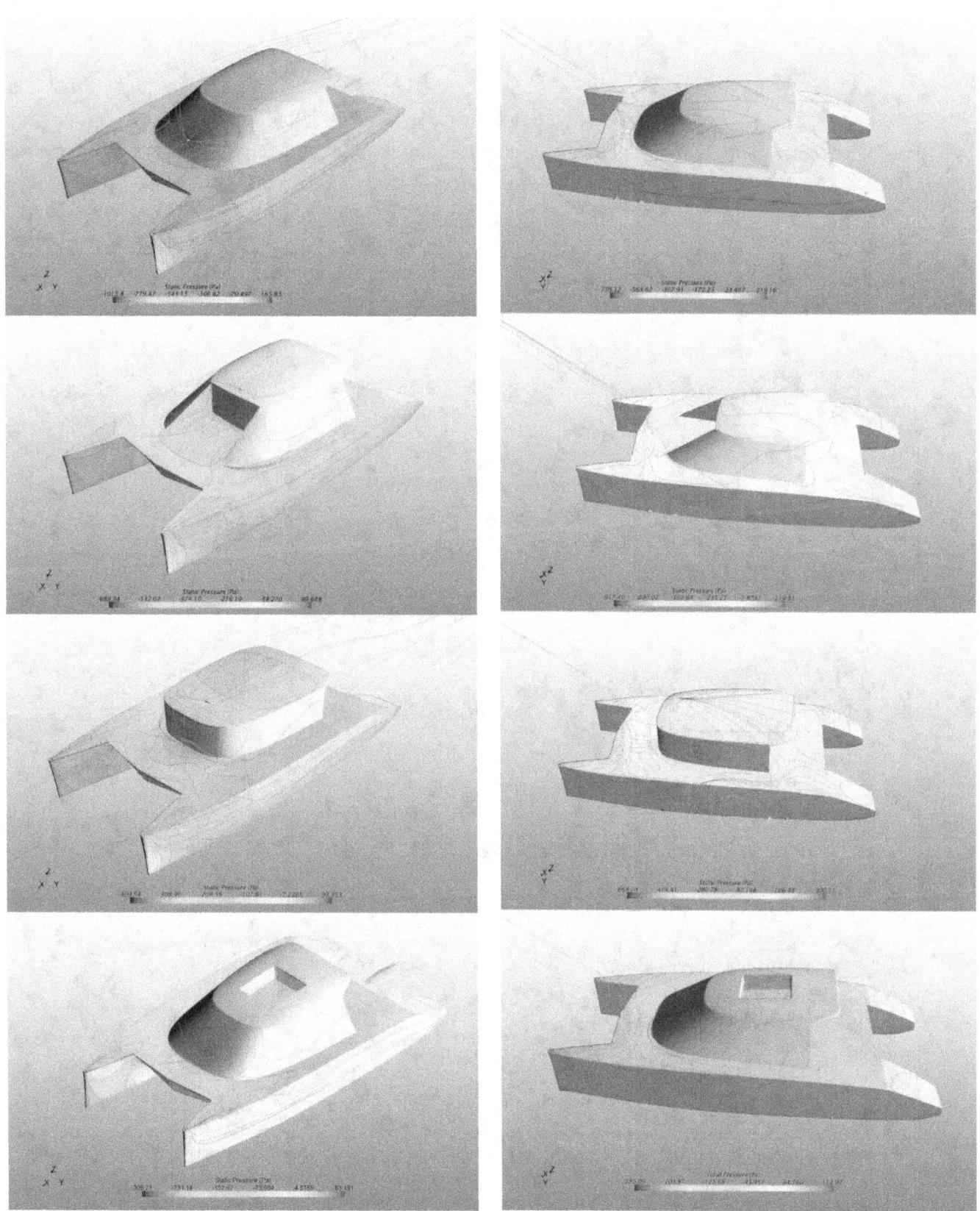

Figure 5 - Illustrations of airflow and pressures at ψ=0° and ψ=40° for four superstructure configurations

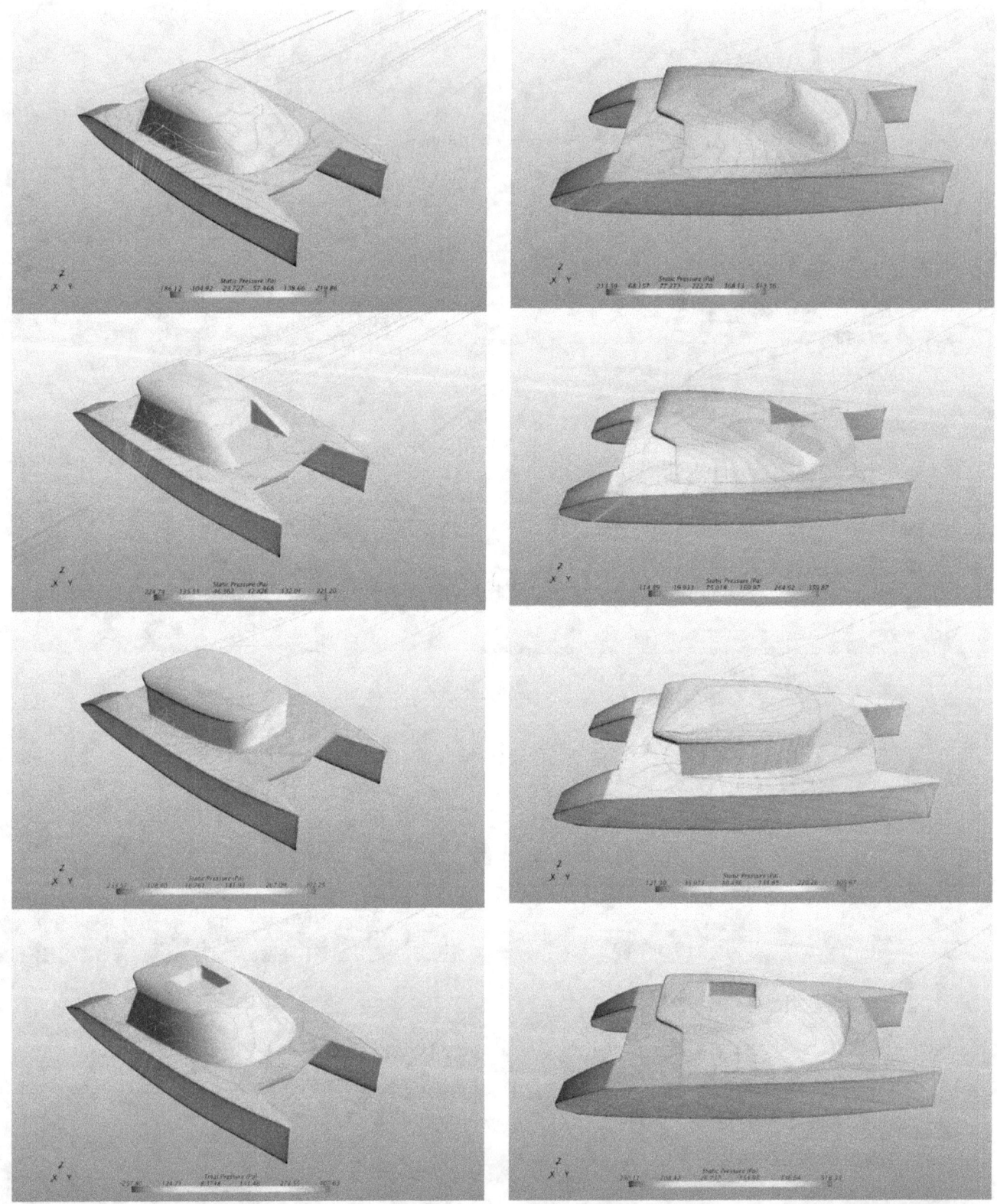

Figure 6 - Illustrations of airflow and pressures at ψ=90° and ψ=135° for four superstructure configurations

CATAMARAN AERODYNAMICS

A numerical study of the aerodynamics of catamaran superstructures was made using the OpenFOAM computational fluid dynamics (CFD) code.Four different superstructures presenting typical trends in catamaran design were modelled on the same hull of a 38-foot sailing catamaran. The studied configurations are presented on figures 5-6 including a streamlined shape, streamlined shapes with front and top recesses and a straight cabin side shape.

To estimate the effect of Reynolds number, the calculations were made for range of wind speeds 15, 20 and 25m/s. Variation of flow incidence angle corresponds to apparent wind angles of $\psi=0,20,40,60,90,135,180°$ without effect of gradient and sailing rig.

To evaluate the contribution of the cabin and hull, the hull with a bridge deck structure and bare demihull have been calculated separately as well. The results of study are presented on figures.5-7 in form of aerodynamic coefficients in ship's system C_X, C_Y related to front and side profile areas of the craft. The moments are defined around axis with origin at centre of buoyancy of the craft; the moments are of no practical interest for present research. In general, the influence of cabin shape within studied configuration on C_X, C_Y of above water part is rather small. Notable difference exists in range of $\psi=40...60°$ where model #3 with straight sides showing some disadvantage in C_X. As expected, not the shape itself but freeboard height and cabin profile areas would have predominant influence on catamaran aerodynamic drag.

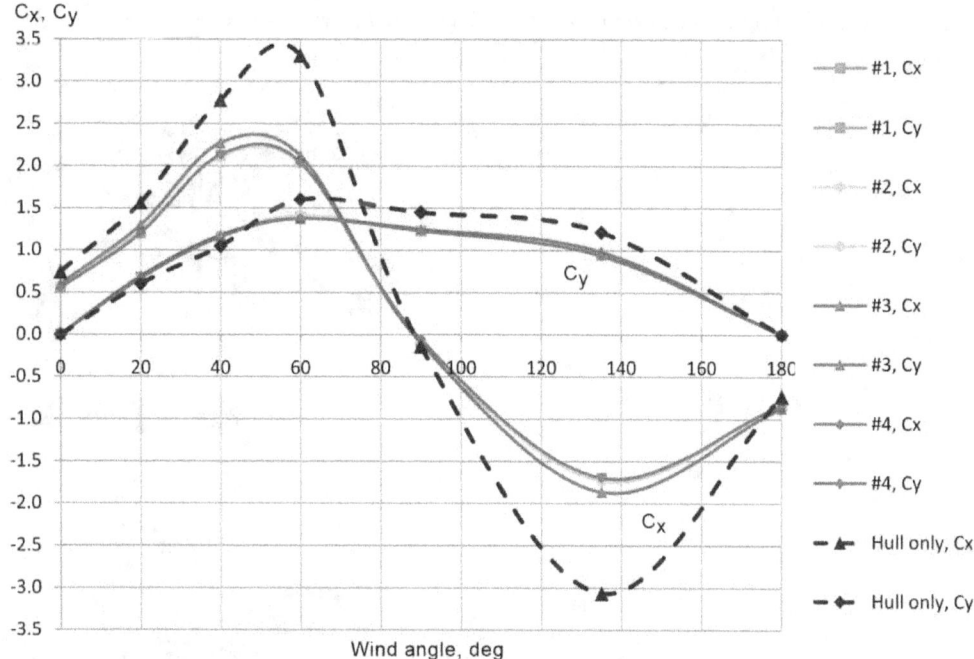

Figure 7 - Aerodynamic coefficients C_X, C_Y for 25m/s flow in range of $\psi=0...180°$ for four superstructure configurations

Form of approximation for C_X, C_Y can be recommended:

$$C_X = (k_m - 0.5k_0)\sin 2\Psi + k_0 \cos\Psi$$
$$C_Y = m_m \sin\Psi$$

where k_0 and k_m are equal to C_X at $\psi=0$ and $60°$ respectively;m_m is equal to C_Y at its maximum $\psi\approx90°$. This form correctly represents the behavior of C_Y, C_Y in range of $\psi=0...180°$ and provides accurate estimate of these coefficients first of all at close-hauled courses. For the sample above $k_0=0.55$; $k_m=2.2...2.4$; $m_m=1.4$.

IMPACT OF CATAMARAN EXTERIOR ON SAILING PERFORMANCE

Next question to consider would be: what is the contribution of aerodynamic properties of above water part of a catamaran to total sailing performance? For monohull keelboats in close-hauled sailing conditions, for the share of above water part on drag is as large as 25% of the total drag, while on lift is only 3% (Sloff, 2015).

Most of mathematical models of sailing craft aerodynamics treat above water part of the hull and cabin as source of 'parasitic drag' in rather simplified manner (Larsson 2014). In general, these models were created for monohull sailing craft and drag coefficients of the above water part C_D are often taken as constants; this does not allow enough flexibility for catamarans where aerodynamics is of more importance.

The impact of design configurations on the sailing performance of 39-foot catamaran was studied with use of the WinDesign VPP (figure 8) developed by Wolfson Unit, University of Southampton (Claughton, 1998).

In particular, practical interest was the performance 'cost' of increasing cabin height to some reasonable level. The studied options included:

- Two options of cabin – low (C2) and high (C1); ergonomics study for these cabins featured on figure 10. Difference in cabin projection areas: frontal - 29%; side – 39%; difference in height of cabins is 250mm.
- Two options of appendages – daggerboards (D) and low aspect ratio (LAR) keels of some typical configurations.

The results of calculations are presented on figures7,8; some useful conclusions can be made:

- The difference in performance between low and high cabins (see figure 10) results only 0.9...1.4% of 'velocity made good' to upwind (VMG) while thedifference in absolute speed does not exceed 1 % close-hauled;
- The difference in upwind VMG between catamarans with daggerboards and LAR keels options exceeds 20%;

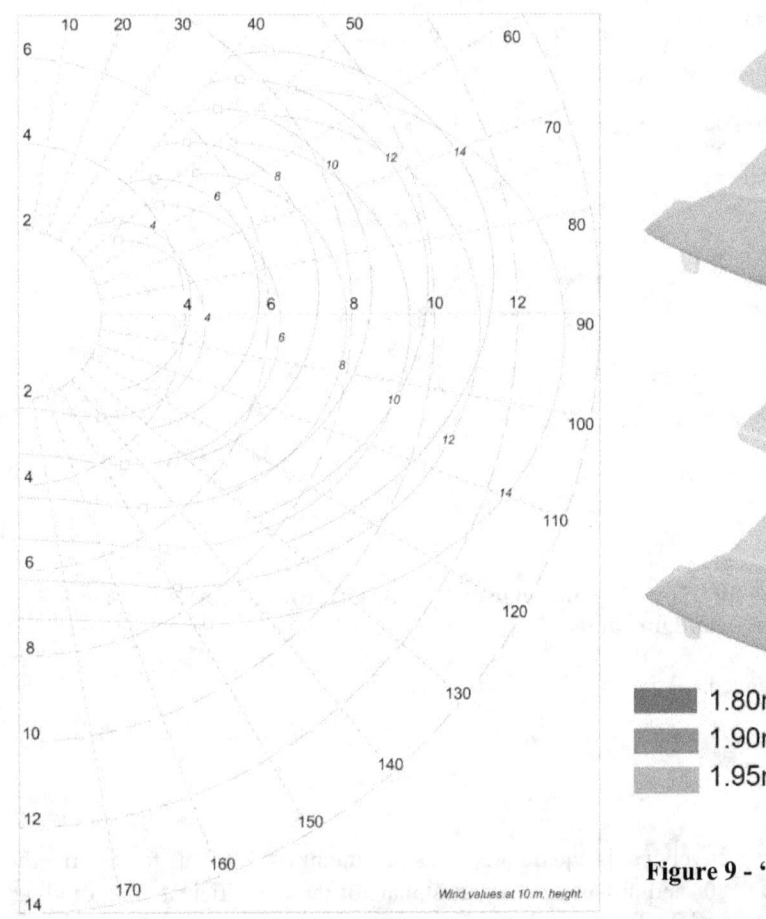

Figure 8 - Polar diagrams of speed for 39-foot catamaran, C2-D option

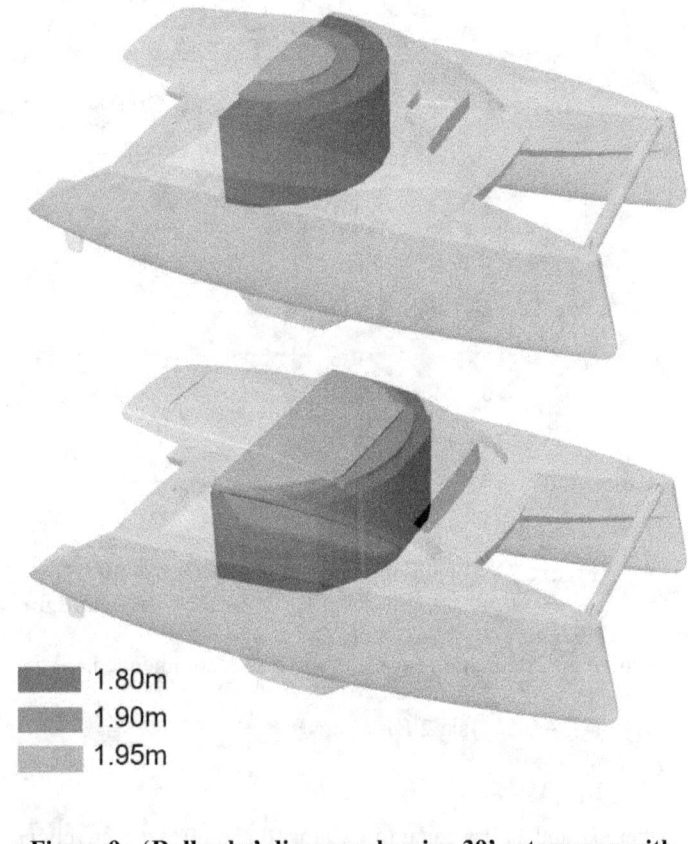

1.80m
1.90m
1.95m

Figure 9 - 'Roll cake' diagram showing 39' catamaran with two cabins of different heights
Above – low (C2), initial version with insufficient overall headroom, limited headroom above stairway to demihull below – high (C1), final version with cabin raised total by 250mm. Note ridges 50mm high on final version allowing recess for solar panels, additional headroom and increased roof stiffness.

Selection of LAR keels for most of cruising catamarans is justified due to simplicity, ease of manufacturing, less obstruction in the interior, shallow draught. For an average cruising catamaran with LAR keels, the improvement of performance due to low profile or streamlined cabin shape is negligible, though degradation of comfort is obvious (see figure 9).

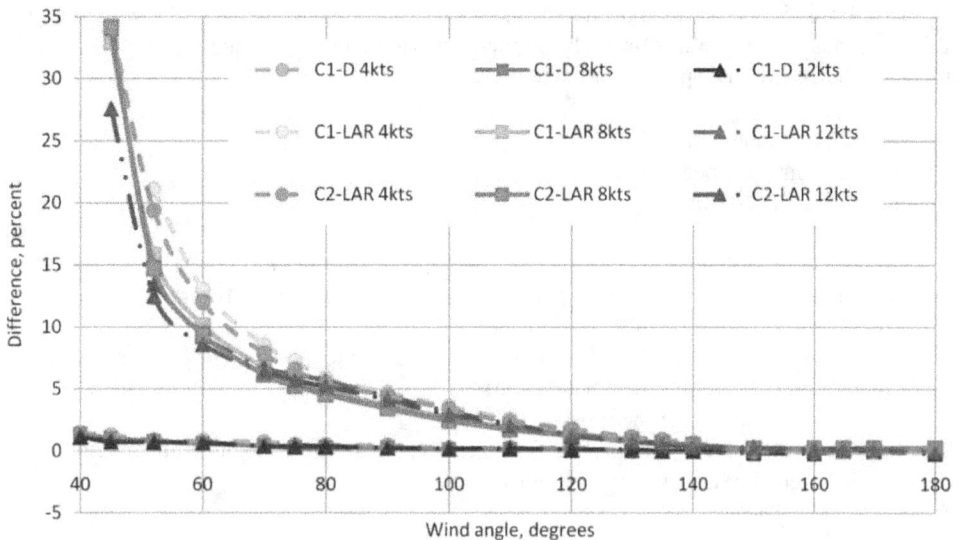

Figure 10 - Comparison of speed for different configurations of 39-foot catamaran, in percent with C2-D option

SAFETY REQUIREMENTS

Most of recreational catamarans are now designed to comply with European recreational craft directive 2013/53EU. The Directive covers safety requirements in relation to ISO Small Craft group of standards, though alternative standards can be also considered. All ISO requirements can be subdivided into few groups, some of them specific for catamaran craft are reviewed below:

- Downflooding/freeboard, draining, watertightness
- Stability
- Flotation when inverted
- Escapes
- Structures
- Requirements to systems and equipment

The stability requirements for sailing multihulls in the current version of ISO12217-2 are quite complex and are not easy to analyze at early design stages without performing detailed calculations and a full assessment to the standard. Nevertheless, stability 'automatically' complies for most cruising catamarans of normal proportions, with 'footprint ratio' $SA/L_{WL}B_{CB}<1.7$, where SA is sail area, L_{WL} is waterline length. Stability requirements to catamarans can be grouped into:

- Bare poles factor, which indicates ability of catamaran to survive in extreme wind conditions without sails, in transverse and longitudinal directions.
- Transverse rolling and capsize on breaking wave, in the standard related to maximum righting lever GZ_{max}. De-facto, this factor limits minimum overall beam of a catamaran for each category.
- Longitudinal stability, i.e. pitchpoling over bow in breaking waves is assessed by the area under the longitudinal righting moment curve. This factor is strongly related to the length of a catamaran.
- Diagonal stability, which is an assessment of the transverse righting moment for a catamaran excessively trimmed by the bow and stern.
- Wind speed limits for different sail combination, in transverse and longitudinal directions.

Requirements for inverted floatation are aimed at providing catamaran safety following capsize. If floatation is provided by compartments, only non-habitable compartments would count as buoyancy volumes. In our designs, such volumes are forepeaks and engine rooms, both accessible from deck. One might additionally consider permeability factors for flooded spaces - in composite sandwich structures they comprise 0.75...0.85 and can significantly affect the results of the calculations.

The structures of recreational catamarans will soon be covered by the standard ISO12215-7 which is not yet enforced and still in development. Preliminary versions of the standard are contradictory and it is likely to be significantly updated. Say, it does include rig loads which are not properly defined for catamarans. In general, the standard covers local loads which are similar to monohull craft, and specific catamaran loads:

- Local loads on wet deck and inner sides
- Global strength such as transverse bending, shear and twisting
- Loads from sailing rig and equipment

However, global strength is by default satisfied for most of cruising catamarans L_H<24m of normal proportions. Local strength in our practice is defined using current ISO12215-5 standard for monohulls. If required, wet deck loads and global loads can be evaluated using rules of classification societies.

The ISO standards assume that multihull craft can be capsized and escape hatches are required for multihulls for evacuation in case of inversion. However, escape hatches are usually located low to waterline and can themselves present weakpoints as they are subject to wave and other possible impacts.

In general, the certification process for multihull craft often depends on the designer's understanding of the background of the standards, and the experience of a surveyor with this type of boat.

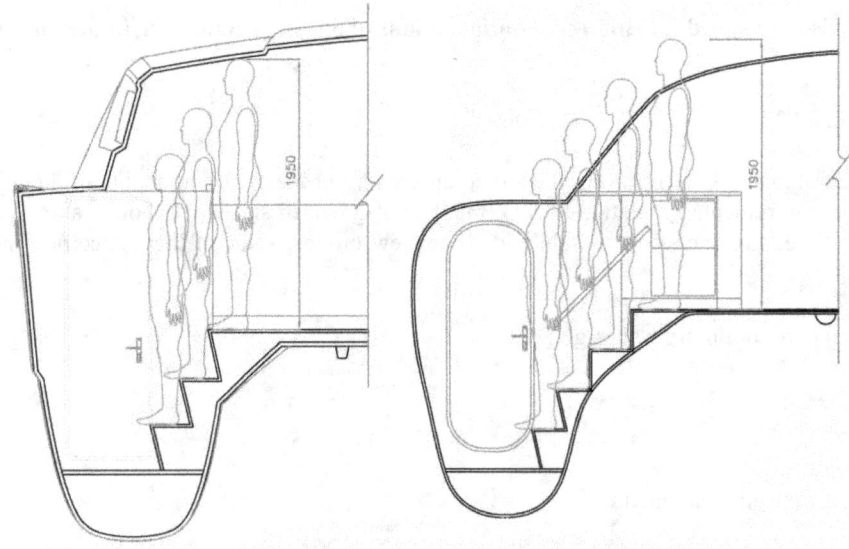

Figure 11 - Headroom issues on two catamaran designs
Left – excessive headroom though giving more spacious saloon feeling from inside; right – lack of headroom tough looking 'sleek' from outside.

CONSIDERATIONS FOR EXTERIOR AND INTERIOR DESIGN

Working on catamaran exterior and superstructure in particular, the designer should have in mind broad range of considerations other than aerodynamics (White, 1997).

One of them is habitability achieved by headroom in saloon and above the stairway from the bridge deck down to the demihull. Interestingly, we found that on some 'sleek-looking' 40-50-foot catamaran models the standing headroom was provided only at the saloon entrance (figure 9,11) and the headroom at the periphery and above the stairways is lacking. Usually, 95-percentile male data is used to define acceptable headroom, but one should have in mind that available anthropometric statistics might be dated and would also depend on regional factors. For instance, a 95-percentile male from the Netherlands would be 1.96m tall. At the 3D modelling stage, it is helpful to draw the headroom volume – the box representing the area where a person should be able to stand; this volume should cover main passageways and activity areas, and also the galley (figure 9).

The feeling of space and views are important emotional design factors. On a catamaran, a seated person should preferably be able to look outside through the windows.

Deck safety and ergonomics are critical, it is common knowledge that sloped and rounded surfaces can be extremely uncomfortable to walk on. For formal safety assessment ISO15085 standard should be followed. For complicated cabin shapes it is helpful to make a full size mock up prior to final manufacturing. Such a mock up is mandatory to get comfortable access to the mast, winches and steering.

Equipment and hardware positioning on the cabin roof requires its proper shaping and support lugs. As common practice, headsail tracks should be placed about 12...14° to catamaran's centreline. Another consideration is solar panels – they should not block access to the boom, and often require a gap from underside for cooling – this means either recess or installation brackets should be provided.

Visibility from the helm station on many sailing catamaran models is rather poor and restricted by the massive cabin. Ideally, a helmsperson should enjoy all-round visibility (excluding sectors naturally closed by the headsails), and view of the sails for efficient steering. The ability to see alongside is desirable for mooring operations and maintaining sight of anchor equipment – for anchoring. Standard ISO11591 currently does not cover visibility on sailing boats but can be a useful guideline. Note that according to USCG statistics about 60% of all small craft accidents are collisions with other boats and objects mostly due to poor lookout, and so visibility is a priority factor for safe navigation.

Tents, arches, windshields and other add-on structures - experience shows that these items should be drawn by the designer from beginning, otherwise they will be added by the owner detracting the craft's appearance. These items provide protection of crew from the sun (essential for tropics), placement of mainsheet traveler, etc. but can also add some considerable windage.

Construction limitations for shaping are related to certain materials and construction methods. Say, cabin assembled in plates or pre-fabricated panels would need developable surfaces.

Ventilation is essential for the comfort and natural airflow - there is a trend to place large opening windows and hatches to provide cross flow through the saloon. Such opening windows usually require flat surfaces to be installed.

Natural lighting of cabin space is important, but it is desirable to avoid direct sunlight into saloon from large sloped windows, as such sunlight is a source of heat transfer and can make stay on board a catamaran in tropics extremely uncomfortable.

Structural design considerations, as the cabin participates in the global transverse bending/twisting of the catamaran. It is also experiencing considerable concentrated loads (for instance, from headsail tracks). For headsail sheet loads acting in direction of sheet F_H in N, the following empirical formula originating from sailing hardware suppliers can be used $F_H=0.2A_HW^2$, where A_H is area of headsail in m^2; W is wind speed in knots. Structural issues are also related to thermal deformations of the superstructure and glazing.

Last but not least is the aesthetics of a cabin – the area of expertise of a stylist. Unfortunately, cabin shape often becomes client's fetish and very often his subjective demand to make catamaran's cabin 'look like a racing car' might significantly impair other important qualities. As it has been shown in the studies above, such 'racing look' has negligible effect on overall sailing performance.

It is surprising how many yacht stylists/designers (and their clients) do not think about all the spectrum of above mentioned considerations concentrating only on 'edgy look'; as a result – these designs can inherit problems with headroom, practicality, comfort and safety. In general, despite the mentioned restrictions there is a lot of space for a designer's creativity in developing nice shapes and new types of functionality.

NEW CONCEPTS

The design shown in figure 12 features a rather unusual layout with the de-facto open saloon under a hardtop roof, seating and pantry – designed for a hot climate to allow the airflow through the public space. The P44 catamaran has minimalistic accommodation and is intended mostly for racing and weekend cruises in the tropics and is inspired by sailing experiences in the Andaman Sea. For the European RCD certification, the 'saloon' is formally treated as an open space/cockpit and openings down to the demihulls are fitted with doors with the required degree of watertightness. Swing steering wheels allow changing of the steering position for racing and cruising. Split into mouldings is shown allowing shipment in 40' containers with subsequent assembling on structural bonding compound.

CONCLUSIONS

A parametric approach to cruising catamaran designs has proven to be useful and allows a catamaran designer to deal with real numbers in terms of displacement, performance indicators and safety factors.

Catamaran architecture is always a compromise of styling and comfort, performance and safety, as well as practical considerations.

The shape of the cabin itself has a minor effect on the aerodynamics of above water part and the performance of a cruising catamaran. The size of the windage area does matter and if reasonable proportions are exceeded, such a cabin can cause a significant loss of windward performance, especially on catamarans with low aspect ratio keels.

On other side, calculations indicate that the cost of the compromise between 'sleek look' and increasing headroom to acceptable/comfortable limits is very small and more than affordable. In the case of a catamaran with LAR keels where upwind performance is already compromised, aerodynamics of deck and cabin are of secondary importance.

Safety standards for recreational catamarans are still in development and are developing in direction of more complicated approaches. A designer is required to understand the physics behind each criteria, and often required to negotiate/guide certifying authorities.

Figure 12 - P44 catamaran design (concept by AMD)
This design was awarded silver prize at European Product Design 2018 contest.

REFERENCES

Claughton A.R., Wellicome J.F., Shenoi R.A. Sailing Yacht Design: Theory. Longman, 1998.
Fossati F. Aero-Hydro Dynamics and Performance of Sailing Yachts. International Marine, 2007.
Larsson L., Eliasson R.E., Orych M. Principles of Yacht Design. 4th Edition, International Marine, 2014.
Nazarov A. Small Catamarans: Design Approaches and Case Studies//International Journal of Small Craft Technology, RINA, Vol.157, Part B1 2015.
Nazarov A., Jabtanom A., Leeprasert A., Phormtan T., Suebyiw P. Catamaran Yachts: Styling Trends and Design Practices// Marine Design 2015, 2-3 September 2015, London, UK
Slooff J.W. The Aero and Hydromechanics of Keel Yachts. Springer, 2015.
White C. The Cruising Multihull. International Marine, 1997.

Bibliography of Prior Chesapeake Sailing Yacht Symposia Papers

The First CSYS- January, 1974
• On the Handicapping of Distance Racing Yachts – A Proposal for IOR IV. Fisher, Bennett
• Handicapping Rules and Performance of Sailing Yachts. Letcher, John S., Jr.
• Analysis of Chesapeake Bay Racing Results 1972 and 1973 Seasons. Peach, Robert W.
• Yacht Rating. Strohmeier, Daniel D.
• Measurement Parameters of the I.O.R. Rule. Stephens, Olin J., II
• Booms Are Obsolete. MacLear, Frank R.
• A Breakthrough - Slotted Headsail Luff Support Systems. MacKenzie, Alan
• Some Observed Effects of Foil Control on Hydrofoil Sailing Vehicle Performance. Bradfield, W. S.
• Scale Experiments with the 5.5 Metre Yacht Antiope. Kirkman, Karl L.
• The Performance of Sailing Yachts in Oblique Seas. Pedrick, David R.
• Directional Stability and Control of Sailing Yachts. Scott, Walter H., Jr.
• Aerodynamics of High Performance Wing Sails. Scherer, Otto

The Second CSYS - January, 1975
• Theory Of Sailing Applied to Ocean Racing Yachts. Myers, Hugo A.
• America's Cup 1974 - An Overview- Organizing to Win. du Moulin, Richard T.
• America's Cup 1974- An Overview- Design and Construction- The First Race. Pedrick, David R.
• A Cruising Boat. Curtze, Charles A., RADM.
• Vane Self-Steerers for Cruising Yachts. Ratcliffe, Gererd
• Extended Cruising - An Overview. Court, Kenneth E.
Seakeeping and the Sailing Yachtsman. Compton, Roger, Johnson, Bruce, and Van Duyne, Carl
• Yacht Keels - An Experimental Study. DeSaix, Pierre
• Kevlar 49 Aramid, A New Material for Boat Hull Construction. Miner, Louis H., Wolffe, Robert A., and Woodrick, James V.
• Flotation for Ballasted Sailing Yachts?. Wilson, Vance O. and Reuter, Wolfgang
• America's Cup 1974 - An Overview- Racing for the Cup. Herreshoff, Halsey Chase

The Third CSYS - January, 1977
• Aluminum Construction. Wyland, Gilbert
• Sailing Yacht Construction in Fiberglass. Goman, William J.
• Surfing: Motions of a Vessel Running in Large Waves. Letcher, John S., Jr.
• The Preservation of Chesapeake Bay Watercraft. Baker, William A.
• Ocean Racing. Strohmeier, Daniel D.
• Principles of Sail Design. Haarstick, Stephen
• Wing Sail Versus Soft Rig: An Analysis of the Successful Little America's Cup Challenge of 1976. Bradfield, W. S. and Madhavan, Suresh
• Analyzing a Yacht for Hydrodynamic Characteristics that Affect What Type of Sails and Rigs Will Work Best. Doyle, Robert E.

The Fourth CSYS - January, 1979
• Evolution of Offshore Ratings - To the Limit. Pedrick, David R.
• A Summary of the H. Irving Pratt Ocean Race Handicapping Project. Kerwin, Justin E., Newman, J. N.
• The Measurement Handicapping System of USRYU. Strohmeier, Daniel D.
• Selecting a Keel Appendage for a Cruising Yacht. Berman, Deborah W.
• Yacht Structural Design for Light Scantlings. Herreshoff, Halsey C.
• Theoretical Estimation of the Influence of Some Main Design Factors on the Performance of International Twelve Meter Class Yachts. van Oossanen, Peter
• Photographic Essay: Ship Training on the Gazella Primeiro. Roewe, George J., Jr.
• A Microcomputer Beats to Windward. Clauser, Milton U.
• A Computer-Based Method for Analyzing the Flow Over Sails. Thrasher, D.F., Mook, D.T., and Nayfch, A.II.
• The Evolving Role of the Towing Tank. Kirkman, Karl L.

The Fifth CSYS - January, 1981
• Design Development of a 40m Sailing Yacht. Benford,Jay R.
• Yacht Performance Analysis with Computers. Pedrick, David R. and McCurdy, Richard C.
• Geometry of Sailmaking. Andresen, Ted
• Sailing Yacht Capsizing. Stephens, Olin J., II, Kirkman, Karl L., and Peterson, Robert S.
• Kinetics in Small Boat Racing. Smith, Peter G.
• RED HERRING, High Performance Cruising Ketch. Hubbard, David W.

Bibliography of Prior Chesapeake Sailing Yacht Symposia Papers

• Mathematical Hull Design for Sailing Yachts. Letcher,John S., Jr.
• A Design Guide for Estimating Speed Made Good. Berman, Deborah W.

The Sixth CSYS - January, 1983
• Analysis of Steady Flow Over Interacting Sails. Register, David S. and Irey, Richard K.
• Yacht Design With Computers: New Methods for New Tools. Hazen, George S. and Killing, Steve
• Marine Electronic Navigation- A General Overview. Closs, Thomas H., Jr.
• Sailing Yacht Capsizing. Kirkman, Karl L., Nagle, Toby J., and Salsich, Joseph O.
• Rowing and Sailing Craft of the Chesapeake Bay. Tilp, Frederick
• Design and Engineering Aspects of Free-Standing Masts. Sponberg, Eric W.
• FRP Bottom Blistering. Fraser-Harris, A.B.F., COM and Kyle, James H.

The Seventh CSYS - January, 1985
• Experimental Analysis of Five Keel-Hull Combinations. Gerritsma, J. and Keuning, J. A.
• Sailboat Bow Impact Stresses. Ward, Lawrence W.
• Selection Criteria for Plastics Used in Through-Hull Fitting. Fraser-Harris, A.B.F.,COM and Leyden, Jerry J.
• Sailboards, Inventions, Yachts, and Exotic Craft. Russell, Diana
• Extended Cruising the Second Time Around. Court, Kenneth E.
• The Calculation of Sail Panels Using Developed Surfaces. Clemmer, George
• Stress Analysis for Light Alloy 12M Yacht Structures Comparison Between a Transverse and a Longitudinal Structure. Boote, Dario, Ruggiero, Vincenzo, Sironi, Nicola, Vallicelli, Andrea, and Finzi, Bruno
• The Development of the 12 Meter Class Yacht Australia II. van Oossanen, Peter

The Eighth CSYS - March, 1987
• The Application of VPPs to Practical Sailing Problems. Kirkman, Karl L.
• An Assessment of the Progress in Yacht Design Through an Examination of Model Yacht Characteristics. Claughton, Andrew, Howlett, Ian, Stollery, Roger
• Dinghy Design and the International Fourteen. Ames, Robert M. and Weiss, Paul F.

• The Comparison of Potential Driving Force of Various Rig Types Used for Fishing Vessels. Marchaj, C.A.
• Brushfire - An Experience in Building a Masthead Cutter. Williams, John J.
• Keel Design for Low Viscous Drag. Obara, Clifford J. and Van Dam, C.P.
• The Interpretation of Results from Tank Tests on 12m Yachts. Campbell, Ian and Claughton, Andrew
• The Analysis of Wave Resistance in the Design of 12 Meter. Scragg, Carl A., Chance, Britton, Jr., Talcott, John C., and Wyatt, Donald C.
• Stars & Stripes '87; Computational Flow Simulations for Hydrodynamic Design. Boppe, Charles W., Rosen, Bruce S., Laiosa, Joseph P., and Chance, Britton, Jr.
• Data Collection and Analysis for the 1987 Stars & Stripes Campaign. Letcher, John S. and McCurdy, Richard C.

The Ninth CSYS - March, 1989
• Fiber Reinforced Plastic Sailing Yachts - Some Aspects of Structural Design. Curry; Robert
• Guides to the Approximation of Sailing Yacht Performance. Stephens, Olin J.
• Scientific Sail Shape Design. Greeley, David S.,
• Kirkman, Karl L., Drew, Alan L., and Cross-Whiter, John
• The Design and Construction of the Second Pride of Baltimore. Gillmer, Thomas C.
• The Planning, Design and Construction of the 44-foot Offshore Training Craft for the U.S. Naval Academy. McCurdy, Ian and Bonds, John
• The Effect of Counter Length on Hull Resistance. Claughton, Andrew R.
• Performance Prediction Method for Multihull Yachts. Oliver, Clay

The Tenth CSYS - February, 1991
• Gyradius Measurements of Olympic Class Dinghies and Keel Boats. Hinrichsen, Prof. Peter F.
• Structural Design and Construction of America's Cup Class Yachts. Reichard, Prof. Ronnal P.
• The Delft Systematic Yacht Hull (Series II) Experiments. Gerritsma, Prof. ir. J., Keuning, Ir. J., and Onnink, A. R.
• A New Technique for Testing a Sailing Yacht in Waves. Kapsenberg, G. K.
• Magic III - An Old Man's Day Sailer. Miller, Capt. Richards T. and Whitacre, Harold M., III
• A Numerical Approach to the Design of Sailing Yacht Masts. Boote, Dario and Caponetto, Mario

Bibliography of Prior Chesapeake Sailing Yacht Symposia Papers

• Model Test Techniques Developed to Investigate the Wind Heeling Characteristics of Sailing Vessels and Their Response to Gusts. Deakin, Barry
• Sailboat Performance in a Current. Nolan, James P.
• Sailboat Hydrodynamic Drag Source Prediction and Performance Assessment. Boppe, Charles W.
• The Effect of Pitch Gyradius on Added Resistance of Yacht Hulls. Moran, James F.

The Eleventh CSYS - January, 1993
• Applications of Relational Geometric Synthesis in Sailing Yacht Design. Letcher, John S., Jr. and Shook, D. Michael
• Refinements in the Techniques of Tank Testing Sailing Yachts and the Processing of Test Data. Teeters, James R.
• SPLASH Free-Surface Flow Code Methodology for Hydrodynamic Design and Analysis of IACC Yachts. Rosen, Bruce S., Laiosa, Joseph P., Davis, Warren H., and Stavetski, David
• IACC Appendage Studies. Tinoco, E. N., Gentry, A. E.,Bogataj, P., Sevigny, E. G., and Chance, B.
• Modeling IACC Sail Forces by Combining Measurements with CFD. Milgram, Jerome H., Peters, Donald B., and Eckhouse, D. Noah
• Towards a Rational Upwind Sail Force Model for VPPs. Euerle, Steven E. and Greeley, David S.
• Numerical Approach to Aeroelastic Responses of Three-Dimensional Flexible Sails. Fukasawa, Toichi and Katori, Masanobu
• A Review on I1 Moro di Venezia Design. Caponnetto, Mario
• The Nippon Challenge America's Cup 1992-Progress in Hull Development. Nagami, Yoshihiro.
• How to Go Cruising. Hays, James O. and Hays, Anne M.
• Notes on Sailing Ship History: Academy Versus Shipyard. Stephens, Olin J., II
• The A-Class Catamaran: Development of Serious Fun. Beadling, Robert G. and Beadling, Walter H.
• Dynamic Performance of Sailing Cruiser by Full-Scale Sea Tests. Masuyama, Yutaka, Nakamura, Ichiro, Tatano, Hisayoshi, and Takagi, Ken
• Hazards and Challenges of Cruising the Northeast Coast of North America. Jordan, Edwin C. and Jordan, Mary K.
• The Partnership for America's Cup Technology: An Overview. Gretzky, James A. and Marshall, John K.
• Stars and Stripes Design Program for the 1992 America's Cup. Todter, Chris, Pedrick, David, Calderon, Alberto, Nelson, Bruce, Debord, Frank, and Dillon, Dave
• Elements of Resistance of IACC Yachts. Milgram, Jerome H. and Frimm, Fernando C.
• Sailing Yacht Performance in Calm Water and in Waves. Gerritsma, Prof. ir J., Keuning, Ir. J. A., and Versluis, A.
• Seakeeping and Added Resistance of IACC Yachts by a Three-Dimensional Panel Method. Sclavounos, P. D. and Nakos, D. E.
• The Effects of Flare and Overhangs on the Motions of a Yacht in Head Seas. Kuhn, John C. and Schlagcter, Eric C.
• Analysis of Lift and Drag on a Surface-Piercing Foil. Kuhn, John C. and Scragg, Carl A.
• Performance Prediction Software for IACC Yachts. Schlageter, Eric C. and Teeters, James R.

The Twelfth CSYS - January, 1995
• Scoring IMS Regattas - An Empirical Study of Alternative Methods. Cane, John W.
• Drawing with Performance Prediction. Schwenn, Peter and Hazen, George
• Design Criteria for Composite Masts. Miller, Paul
• The Development of the B&R Rig, Structural Space Frame and Tripod Support System with Integrated Boom. Bergstrom, Lars and Ridder, Sven O.
• The Alexandria Class Dinghy - A Design For Change. Hunley, William H.
• Design, Construction, and Performance of a 27' MORC Boat. Jones, Brian A.
• Imagine- an Open Class 60 BOC Racer- Design and Program Management- Lessons Learned. Court, Kenneth E. and Kaufman, F. Michael III and Whitacre, Harold M III
• The Design of Yacht Sailplans for Maximal Upwind Speed. Day, Dr. Sandy
• Tacking Simulation of Sailing Yachts - Numerical Integration of Equations of Motion and Application of Neural Network Technique. Masayuma, Yutaka and Fukasawa, Toichi and Sasagawa, Hiroshi
• Wing - Body Interaction on a Sailing Yacht. Keuning, Prof. ir. J. A. and Kapsenberg, Ir. G. K.
• Improvement of Sailing Yacht Performance Prediction by Including Force-Moment Equilibrium for the Calculation of Helm Angle in a Velocity Prediction Program. van Oossanen, Dr. Peter
• YACHT97: A Fully Viscous Nonlinear Free-Surface

THE 23ʳᵈ CHESAPEAKE SAILING YACHT SYMPOSIUM

ANNAPOLIS, MARYLAND, MARCH 15-16ᵗʰ, 2019

Bibliography of Prior Chesapeake Sailing Yacht Symposia Papers

Analysis Tool for IACC Yacht Design. Farmer, J. and Martinelli, L. and Jameson A.

The Thirteenth CSYS - January, 1997
• On Test Measurements in Full Scale Sailing Test Programs. Howard P. Grant and Olin J. Stephens
• Full Scale Measurement of Sail Force and the Validation of Numerical Calculation Method. Yutaka Masuyama, and Toichi Fukasawa
• An Investigation of Full Scale Forces Produced by a Sail. Nathan Bossett and Ian Mutnick
• Optimisation of a Sailing Rig using Wind Tunnel Data. IMC Campbell
• Model Tests in Support of the Design of a 50 Meter Barque. Barry Deakin
• The Restoration of AVEL. Clark Poston
• BATOPERF, A Performance Prediction Software and Its Influence on Modern Yacht Design. Sylvain Fargeas and Juan Kouyoumdjian
• Development of Proposed ISO 12217 Single Stability Index for Mono-Hull Sailing Craft. Dr. Peter van Oossanen
• The Cogito Project: Design and Development of an International C-Class Catamaran and Her Successful Challenge to Regain the Little America's Cup. Duncan T. MacLane
• The Institute for Marine Dynamics Model Yacht Dynamometer. B. L. Parsons and R. Pallard
• Model Tests of the PACT Base America's Cup Hull in Following Seas, Jesse Falsone
• Appendage Resistance of a Sailing Yacht Hull. J. A. Keuning and B-J. Binkhorst
• Hull - Appendage Interaction of a Sailing Yacht, Investigated with Wave Cut Techniques. Jonathan R. Binns, Kim Klaka and Andrew Dovell
• SPLASH Nonlinear and Unsteady Free-Surface Analysis Code for Grand Prix Yacht Design. Bruce S. Rosen and Joseph P. Laiosa
• The Effect of Pitch Moment of Inertia in Body Axes on the Performance of a Yacht in Waves. C.J. Sutcliffe and A. Millward
• Experimental Determination of Sail Performance and Blockage Corrections. Dr. Robert Ranzenbach and Chris Mairs

The Fourteenth CSYS- January 1999
• Developments in the IMS VPP Formulations. Andrew Claughton
• Experimental Technique for the Determination of Forces Acting on Sailboat Rigging. F. Fossati, G. Moschini, and D. Vitalone

• Fullscale Hydrodynamic Force Measurement on the Berlin Sailing Dynamometer. Karsten Hochkirch and Hartmut Brandt
• USS Constitution Preparations for Sail 200. Howard Chatterton
• In Search of Power, Pace and Windward Performance In Square Rigged Sailing Ships. Philip Goode
• The Windward Performance of Yachts in Rough Water. Jonathan Binns, Bruce McRae, and Giles Thomas
• A 1997-1998 Whitbread Sail Program - Lessons Learned. Robert C. Ranzenbach, Per Andersson and David Flynn
• Sailing Yacht Design Using Advanced Numerical Flow Techniques. Caponnetto et al
• Parametric Design and Optimization of Sailing Yachts. Stefan Harries and Claus Abt
• An Investigation of the Structural Dynamics of a Racing Yacht. Frederic Louarn and Pandeli Temarel
• Use of CFD Techniques in the Preliminary Design of Upwind Sails. Patrick Couser and Norm Deane
• On the Application of RANS Simulation for Downwind Sail Aerodynamics. William Lasher
• Wind Tunnel Testing of Offwind Sails. Robert Ranzenbach and Chris Mairs
• Approximation of the Calm Water Resistance on a Sailing Yacht based on the "Delft Systematic Yacht Hull Series." J.A. Keuning and U.B. Sonnenberg

The Fifteenth CSYS - January 2001
• The Re-Righting of Sailing Yachts in Waves- A Comparison of Different Hull Forms. Martin Renilson, Jonathan R. Binns and Andrew Tuite
• An Improved Upwind Sail Model For VPP's. Peter Jackson
• On-the-water Measurement of Laminar to Turbulent Boundary Layer Transition on Sailboat Appendages. E.A. Lurie
• International America's Cup Class Yacht Design Using Viscous Flow CFD. Paul Jones and Rich Korpus
• Hydrodynamic Modeling of Sailing Yachts. Stefan Harries, Claus Abt and Karsten Hochkirch
• The Effect of Bow Steepness and Flare on the Resistance of Sailing Yachts in Calm Water and Waves. J.A. Keuning, R. Onnink and A. Damman
• A Time-Domain Simulation for Predicting the Downwind Performance of Yachts in Waves. Dougal Harris, Giles Thomas and Martin Renilson

Bibliography of Prior Chesapeake Sailing Yacht Symposia Papers

• An Experimental Investigation of Slamming on Ocean Racing Yachts. Paolo Manganelli and Philip A. Wilson
• Optimising Yacht Routes Under Uncertainty. Andy Philpott and Andrew Mason
• PCSAIL, A Velocity Prediction Program for a Home Computer. David E. Martin and Robert F. Beck
• Schooner *Brilliant* Sail Coefficients and Speed Polars. Howard Grant, Walter Stubner, Walter Alwang, Charles Henry, John Baird and Paul Spens
• Sailing Performance of "Naniwa-maru" - A Full Scale Reconstruction of a Sailing Trader of Japanese Heritage. Kensaku Nomoto, Yutaka Masuyama and Akira Sakurai
• The *Basiliscus* Project - Return of the Cruising Hydrofoil Sailboat. Thomas E. Speer
• Model Tests to Study Capsize and Stability of Sailing Multihulls. Barry Deakin
• Aerodynamic Performance of Offwind Sails Attached to Sprits. Robert Ranzenbach and Jim Teeters

The Sixteenth CSYS – March 2003
• The Yaw Balance of Sailing Yachts Upright and Heeled. J. A. Keuning and K. J. Vermeulen
• Computational Fluid Dynamics for Downwind Sails. Horst J. Richter, Kevin C. Horrigan and J.B. Braun
• Downwind Load Model for Rigs of modern Sailing Yachts for Use in FEA. Guenter Grabe
• Analysis of Hull Shape Effects on Hydrodynamic Drag in Offshore Handicap Racing Rules. Jim Teeters, Rob Pallard and Caroline Muselet
• Changes to Sail Aerodynamics in the IMS Rule Jim Teeters, Robert Ranzenbach and Martyn Prince
• On the Use of CFD to Assist with Sail Design. Andrea Schneider, Andrea Arnone, Marco Savelli, Andrea Ballico and Paolo Scutellaro
• Numerical Simulation using RANS-based Tools for America's Cup Design. Geoff Cowles, Nicola Parolini and Mark L. Sawley
• Sailing Yacht Design for Maximum Speed. Bob Dill
• Composite Sail Batten Design . Audrey Sery and Jean Paul Charles
• Analysis of 2D Coupled Sails: Use of an Optimization Technique Based on Turbulent Viscous Flows. Giovanni Lombardi, Francois Beux and Mattia de. Michieli Vitturi
• Experimental Study of a Directionally Stable Sailing Vehicle with a Free-Raking Rig and a Self-Trimming Sail. Akira Sakurai, Takeshi Nakamura and Yuya Nakamoto
.• Student Research Projects for the New Navy 44 Sail Training Craft. Paul H. Miller and NAOE Naval Architecture and Ocean Engineering Department
• Experimental Force Coefficients for a Parametric Series of Spinnakers. William C. Lasher, James R. Sonnenmeier, David R. Forsman, Cheng Zhang and Kenton White
• The Rise of the Hydrofoil and the Displacement of the Hull: The Design, Construction and Performance Measurement of a 6m Flying Catamaran. Edward Chapman and George Chapman
• Numerical Simulation of Maneuvering of "Naniwamaru," A Full-scale Reconstruction of Sailing Trader of Japanese Heritage. Yutaka Masuyama, Kensaku Nomoto and Akira Sakurai

The Seventeenth CSYS – March 2005
• Toward Numerical VPP with the Full Coupling of Aerodynamic and Hydrodynamic Solvers for ACC Yachts. Erwan Jacquin, Yann Roux, Bertrand Allessandrini
• Time Domain Simulation of a Yacht Sailing Upwind in Waves. D. H. Harris
• Geometry and Resistance of the IACC Systematic Series "Il Moro di Venezia" D. Battistin, D. Peri, E. Campana
• Sailing Yacht Rig Improvements Through Viscous Computational Fluid Dynamics. Vincent G. Chapin, Romaric Neyhousser, Sthephane Jamme, Guillaume Dulliand, Patrick Chassaing
• A New Velocity Prediction Method for Post-Processing of Towing Tank Test Results. Kai Graf, Christoph Bohm
• Hull Form Optimization of Performance Characteristics of Turkish Gulets for Charter. Mark Gammon, Abdi Kukner, Ahmet Alkan
• The Development of an Integrated Ship Design Environment for the Naval Architect on The Linux Operating System. H. James Parker
• Multiobjective Design Optimization of an IACC Sailing Yacht by Means of CFD High- Fidelity Solvers. Daniele Peri, Fabrizio Mandolesi
• Comparison of Tacking and Wearing Performance Between a Japanese Traditional Square Rig and a Chinese Lug Rig. Yutaka Masuyama, Akira Sakurai, Toichi Fukasawa, Kazunori Aoki
• Relative Performance of Conventional Versus Movable-Ballast Racing Yachts. Frank DeBord, Harry Dunning

THE 23rd CHESAPEAKE SAILING YACHT SYMPOSIUM
ANNAPOLIS, MARYLAND, MARCH 15-16th, 2019

Bibliography of Prior Chesapeake Sailing Yacht Symposia Papers

• The Effect of Mast Height and Centre of Gravity on the Re-Righting of Yachts. Jonathan R. Binns, Paul Brandner
• A Generic Mathematical Model for the Maneuvering and Tacking of a Sailing Yacht. J. A. Keuning, K. J. Vermeulen, E. J. de Ridder
• Experimental Methods to Evaluate Underwater Appendages. Robert Ranzenbach, Mathew Zahn
• A Velocity Prediction Program for a Planing Dinghy. Todd Carrico
• Sail Aero-Structures: Studying Primary Load Paths and Distortion. Robert Ranzenbach, Zhenlong Xu Light Weight Sandwich Panels for Yacht Hull Structures. M. C. Rice, C. A. Fleischer, D. D. R. Cartie, Marc Zupan

The Eighteenth CSYS – March 2007
• A Combined Ship Science-Behavioural Science Approach To Create a Winning Yacht-Sailor Combination. Matteo Scarponi, R Ajit Shenoi, Stephen R Turnock, and Paolo Conti
• Database of Sail Shapes vs. Sail Performance and Validation of Numerical Calculation for Upwind Condition. Yutaka Masuyama, Yusuke Tahara, Toichi Fukasawa, and Naotoshi Maeda
• Enhanced Wind Tunnel and Full-Scale Sail Force Comparison. Heikki Hansen, Peter J. Richards, and Peter S. Jackson
• Further Analysis of the Forces on Keel and Rudder of a Sailing Yacht. J. A. Keuning, M. Katgert, and K. J. Vermeulen
• RANSE Calculation of Laminar-to-Turbulent Transition-Flow around Sailing Yacht Appendages. Christoph Böhm and Kai Graf
• Performance Prediction without Empiricism: A RANS-Based VPP and Design Optimization Capability. Richard Korpus
• A Tool for Time Dependent Performance Prediction and Optimization of Sailing Yachts. D. Battistin and M. Ledri
• Hydrodynamic advice of Sailing Yachts through Seakeeping Study. Guilhem Gaillarde, Erik-Jan de Ridder, Frans van Walree and Jos Koning
• Slamming of Composite Yacht Hull Panels. Susan Lake, Michael Eaglen, Brian Jones, and Mark Battley
• ARPRO®: A New Structural Core Material for the Yacht Industry. Corrado Labriola and Vito Tagarielli
• PCLINES, A Parametric Lines Development Program for the Home Computer. David E. Martin

• "That Peculiar Property:" Model Yachting and the Analysis of Balance in Sailing Hulls. Earl Boebert
• SNAME's Stability Letter Improvement Project (SLIP) for Passenger Sailing Vessels. Jan C. Miles, Bruce Johnson, John Womack, and Iver Franzen
• An Aerodynamic Analysis of the U.S. Brig Niagara. William C. Lasher, Terrence D. Musho and Kent C. McKee and Walter Rybka,
• Analysis of Wave Making Resistance And Optimization of Canting Keel Bulbs. Karsten Hochkirch and Claudio Fassardi
• Added Resistance in Seaways and its Impact on Yacht Performance, Kai Graf, Marcus Pelz, Volker Bertram, and H. Söding
• Dynamic Lift Coefficients for Spade Rudders on Yachts. Paul H. Miller

The Nineteenth CSYS – March 2009
• CFD and VPP Challenges in the Design of the New AC90 Americas Cup Yacht. Kai Graf, Christoph Boehm and Hannes Renzsch
• A New Method for the Prediction of the Side Force on Keel and Rudder of a Sailing Yacht Based on the Results of the Delft Systematic Yacht Hull Series. J. A. Keuning and B. Verwerft
• CFD-Based Hydrodynamic Analysis of High Performance Racing Yachts. Len Imas, Bryan Baker, Britton Ward and Gregory Buley
• On the Choice of CFD Codes in the Design Process of Planing Sailing Yachts. Jérémie Raymond, Jean-Marie Finot, Jean-Michel Kobus, Gérard Delhommeau, Patrick Queutey and Aurélien Drouet
• Systematic Series of the IACC yacht "Il Moro di Venezia": Heel and Yaw Analysis. D. Peri, F. Di Ci´o,and M. Roccaldo
• Yacht Design Software 2.0: The Open Source Movement. Mathew Bird, William F. Cook, George S. Hazen and Britton Ward
• Upwind Sail Performance Prediction for a VPP Including "Flying Shape" Analysis. Brian Maskew and Frank DeBord
• Photogrammetric Investigation of the Flying Shape of Spinnakers in a Twisted Flow Wind Tunnel. Kai Graf and Olaf Müller
• Sails Aerodynamic Behavior in Dynamic Conditions. Fabio Fossati and Sara Muggiasca
• Assessing the Wind-Heel Angle Relationship of Traditionally-Rigged Sailing Vessels. William C. Lasher, Diana R. Tinlin, Bruce Johnson , John Womack, Jan C. Miles, Walter Rybka and Wes Heerssen

Bibliography of Prior Chesapeake Sailing Yacht Symposia Papers

Bibliography of Prior Chesapeake Sailing Yacht Symposia Papers

Fossati, Andy Claughton, Will Krzymowski, Tony Anderson
- On The Hydrodynamics Of A Skiff At Different Crew Positions, Ignazio Maria Viola And Joshua Enlander
- A Measurement System for Performance Monitoring on Small Sailing Dinghies, Christoph Boehm, Robert Brehm, Janek Meyer,Lars Duggen, Kai Graf
- A Wind Tunnel Study Of The Interaction Between Two Sailing Yachts, P.J. Richards, D.J. Le Pelley, D. Jowett, J. Little, O. Detlefsen
- The Development of the New Volvo Class, Britton Ward, Chris Cochran and Farr Yacht Design, Ltd.

The Twenty Second CSYS – March 2016

- Prediction and optimization of aerodynamic forces and boat speed of foiling catamaran with a rig of a rigid wing and a jib, H. Renzsch
- A comparison of a RANS based VPP to on the water Sailing Performance, T. Doyle
- Experimental Optimization of Upwind Sail Trim Compared to Computational FSI Results, M. Sacher
- Fully Integrated fluid-structural analysis for the design and performance optimization of fiber reinforced sails, W. Lasher
- Unsteady Sail Dynamics due to Bodyweight motions, R. Schutt
- Large Eddy Simulation of Upwind Sail Aerodynamics, S. Nava
- Memorium for Fabio Fossati, M. Belloli
- Pressure Measurements on Yacht Sails: Development of a new system for wind tunnel and full scale testing, M. Belloli
- Modal Analysis of Pressures on a full scale spinnaker, J. Deparday
- Wind Tunnel investigation of dynamic trimming on upwind sail aerodynamics, B. Augier
- Towards a new mathematical model for investigating course stability and manoeuvring motions of sailing yachts, E. Angelou
- The SYRF Wide Light Project, M. Prince
- Numerical Simulations of a Surface Piercing A-Class Catamaran Hydrofoil and Comparison against model tests, T. Keller
- Advanced CFD Simulations of free-surface flows around modern sailing yachts using a newly developed openFOAM solver, J. Meyer

- Full Scale Load Measurements from the 2014-15 Volvo Ocean Race, G. Vanhollebeke
- Influence of sailor position and motion on the performance prediction of racing dinghies, J. Taylor
- Development of a routing software for Inshore Match Races, F. Tagliaferri
- Teamwork as Joint Activity in Sailing, C. Finnsgard